Cram101 Textbook Outlines to accompany:

Earth`s Dynamic Systems

Hamblin & Christiansen, 10th Edition

An Academic Internet Publishers (AIPI) publication (c) 2007.

You have a discounted membership at www.Cram101.com with this book.

Get all of the practice tests for the chapters of this textbook, and access in-depth reference material for writing essays and papers. Here is an example from a Cram101 Biology text:

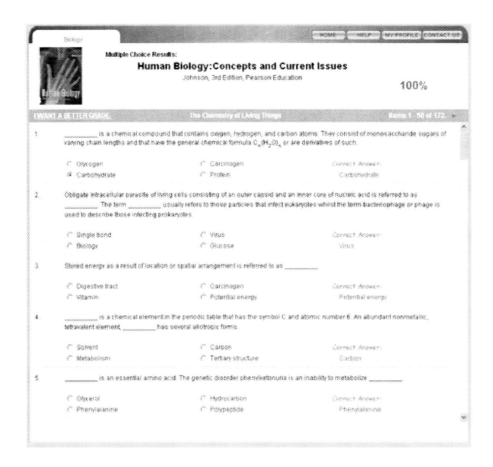

When you need problem solving help with math, stats, and other disciplines, www.Cram101.com will walk through the formulas and solutions step by step.

With Cram101.com online, you also have access to extensive reference material.

You will nail those essays and papers. Here is an example from a Cram101 Biology text:

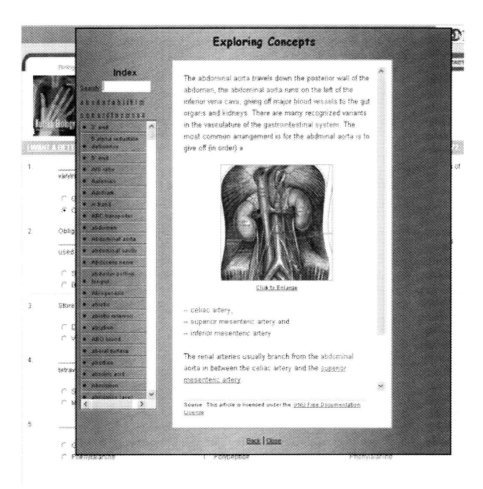

Visit **www.Cram101.com**, click Sign Up at the top of the screen, and enter DK73DW1289 in the promo code box on the registration screen. Access to www.Cram101.com is normally $9.95, but because you have purchased this book, your access fee is only $4.95. Sign up and stop highlighting textbooks forever.

Learning System

Cram101 Textbook Outlines is a learning system. The notes in this book are the highlights of your textbook, you will never have to highlight a book again.

How to use this book. Take this book to class, it is your notebook for the lecture. The notes and highlights on the left hand side of the pages follow the outline and order of the textbook. All you have to do is follow along while your intructor presents the lecture. Circle the items emphasized in class and add other important information on the right side. With Cram101 Textbook Outlines you'll spend less time writing and more time listening. Learning becomes more efficient.

Cram101.com Online

Increase your studying efficiency by using Cram101.com's practice tests and online reference material. It is the perfect complement to Cram101 Textbook Outlines. Use self-teaching matching tests or simulate in-class testing with comprehensive multiple choice tests, or simply use Cram's true and false tests for quick review. Cram101.com even allows you to enter your in-class notes for an integrated studying format combining the textbook notes with your class notes.

Visit **www.Cram101.com**, click Sign Up at the top of the screen, and enter **DK73DW1289** in the promo code box on the registration screen. Access to www.Cram101.com is normally $9.95, but because you have purchased this book, your access fee is only $4.95. Sign up and stop highlighting textbooks forever.

Earth`s Dynamic Systems
Hamblin & Christiansen, 10th

CONTENTS

1. Planet Earth 2
2. Geologic Systems 28
3. Minerals 54
4. Igneous Rocks 66
5. Sedimentary Rocks 84
6. Metamorphic Rocks 116
7. Structure of Rock Bodies 140
8. Geologic Time 156
9. The Atmosphere-Ocean System 182
10. weathering 204
11. Slope Systems 218
12. River systems 234
13. Groundwater Systems 254
14. Glacier systems 272
15. Shoreline Systems 292
16. Eolian systems 314
17. Plate Tectonics 330
18. Seismicity and Earth`s Interior 346
19. Divergent Plate Boundaries 362
20. Transform Plate Boundaries 384
21. Convergent Plate Boundaries 396
22. Hotspots and Mantle Plumes 414
23. Tectonics and Landscapes 432
24. Earth`s Resources 446
25. Other Planets 472

Planet	A planet, as defined by the International Astronomical Union, is a celestial body orbiting a star or stellar remnant that is massive enough to be rounded by its own gravity, not massive enough to cause thermonuclear fusion in its core, and has cleared its neighboring region of planetesimals.
Sedimentary	Sedimentary rock is one of the three main rock groups. Rock formed from these covers 75% of the Earth's land area, and includes common types such as chalk, limestone, dolomite, sandstone, and shale.
Sedimentary rock	Sedimentary rock is one of the three main rock groups. Sedimentary rock covers 75% of the Earth face=symbol>¢s land area. Four basic processes are involved in the formation of a clastic sedimentary rock: weathering caused mainly by friction of waves, transportation where the sediment is carried along by a current, deposition and compaction where the sediment is squashed together to form a rock of this kind.
Mineral	A mineral is a naturally occurring substance formed through geological processes that has a characteristic chemical composition, a highly ordered atomic structure and specific physical properties. A rock, by comparison, is an aggregate of minerals and need not have a specific chemical composition. Minerals range in composition from pure elements and simple salts to very complex silicates with thousands of known forms.
Igneous	Igneous rocks form when molten rock, magma, cools and solidifies, with or without crystallization, either below the surface as intrusive, plutonic rocks or on the surface as extrusive, volcanic, rocks.
Igneous rock	Igneous rock forms when rock cools and solidifies either below the surface as intrusive rocks or on the surface as extrusive rocks. This magma can be derived from partial melts of pre-existing rocks in either the Earth's mantle or crust. Typically, the melting is caused by one or more of the following processes -- an increase in temperature, a decrease in pressure, or a change in composition.
Matter	Matter is the substance of which physical objects are composed. Matter can be solid, liquid, plasma or gas. It constitutes the observable universe.
Space station	A space station is an artificial structure designed for humans to live in outer space. So far only low earth orbit stations are implemented, also known as orbital stations. A space station is distinguished from other manned spacecraft by its lack of major propulsion or landing facilities — instead, other vehicles are used as transport to and from the station.
Atmosphere	An atmosphere is a layer of gases that may surround a material body of sufficient mass. The gases are attracted by the gravity of the body, and are retained for a longer duration if gravity is high and the atmosphere's temperature is low. Some planets consist mainly of various gases, and thus have very deep atmospheres.
Solar system	The Solar System consists of the Sun and the other celestial objects gravitationally bound to it: the eight planets, their 165 known moons, three currently identified dwarf planets and their four known moons, and billions of small bodies.
Hydrosphere	A hydrosphere in physical geography describes the collective mass of water found on, under, and over the surface of a planet.
Ocean basins	Ocean basins are large geologic basins that are below sea level. Geologically, there are other undersea geomorphological features such as the continental shelves, the deep ocean trenches, and the undersea mountain rangeswhich are not considered to be part of the ocean basins.
Continent	A continent is one of several large landmasses on Earth. They are generally identified by convention rather than any strict criteria, but seven areas are commonly reckoned as continents – they are: Asia, Africa, North America, South America, Antarctica, Europe, and Australia.
Mountain	A mountain is a landform that extends above the surrounding terrain in a limited area. A mountain is generally steeper than a hill, but there is no universally accepted standard definition for the height

of a mountain or a hill although a mountain usually has an identifiable summit.

Crust	In geology, a crust is the outermost layer of a planet, part of its lithosphere. They are generally composed of a less dense material than its deeper layers.Earths' is composed mainly of basalt and granite. It is cooler and more rigid than the deeper layers of the mantle and core.
Trench	A trench is a type of excavation or depression in the ground. They are generally defined by being deeper than they are wide, and by being narrow compared to their length.
Ridge	A ridge is a geological feature that is also known as a Rip in the earth causing magma to flow out and forming an undersea volcano, it also has geological features, a continuous elevational crest for some distance. Ridges are usually termed hills or mountains as well, depending on size.
Seamount	A seamount is a mountain rizing from the ocean seafloor that does not reach to the water face=symbol>¢s surface, and thus is not an island. These are typically formed from extinct volcanoes, that rise abruptly and are usually found rizing from a seafloor of 1,000 - 4,000 meters depth. They are defined by oceanographers as independent features that rise to at least 1,000 meters above the seafloor.
Oceanic ridge	A oceanic ridge is an underwater mountain range, formed by plate tectonics. This uplifting of the ocean floor occurs when convection currents rise in the mantle beneath the oceanic crust and create magma where two tectonic plates meet at a divergent boundary.
Abyssal	The abyssal is the pelagic zone that contains the very deep benthic communities near the bottom of oceans.
Mantle	Earth's mantle is a ~2,900 km thick rocky shell comprizing approximately 70% of Earth face=symbol>¢s volume. It is predominantly solid and overlies the Earth face=symbol>¢s iron-rich core, which occupies about 30% of Earth's volume. Past episodes of melting and volcanism at the shallower levels of the mantle have produced a very thin crust of crystallized melt products near the surface, upon which we live.
Inner core	The inner core is a primarily solid sphere about 1220 km in radius situated at Earth face=symbol>¢s center.The existence of an inner core that is different from the liquid outer core was discovered in 1936 by seismologist Inge Lehman using observations of earthquake-generated seismic waves that partly reflect from its boundary and can be detected by sensitive instruments at Earth's surface called seismographs.
Physical property	A physical property is any aspect of an object or substance that can be measured or perceived without changing its identity. Physical properties can be intensive or extensive. An intensive property does not depend on the size or amount of matter in the object, while an extensive property does.
Mesosphere	The mesosphere is the layer of the Earth's atmosphere that is directly above the stratosphere and directly below the thermosphere. The mesosphere is located from about 50 km to 80-90 km altitude above Earth's surface. Within this layer, temperature decreases with increasing altitude. The main dynamical features in this region are atmospheric tides, internal atmospheric gravity waves and planetary waves.
Lithosphere	The lithosphere is the solid outermost shell of a rocky planet. On the Earth, the lithosphere includes the crust and the uppermost mantle which is joined to the crust across the Mohorovièiæ discontinuity. Lithosphere is underlain by asthenosphere, the weaker, hotter, and deeper part of the upper mantle.
Asthenosphere	The asthenosphere is the region of the Earth between 100-200 km below the surface that is the weak or "soft" zone in the upper mantle. It lies just below the lithosphere, which is involved in plate movements and isostatic adjustments. In spite of its heat, pressures keep it plastic, and it has a relatively low density. Seismic waves pass relatively slowly through the asthenosphere.
Geology	Geology is the science and study of the solid matter that constitute the Earth. Encompassing such

Go to **Cram101.com** for the Practice Tests for this Chapter.
And, **NEVER** highlight a book again!

things as rocks, soil, and gemstones, geology studies the composition, structure, physical properties, history, and the processes that shape Earth's components.

Landslide	A landslide is a geological phenomenon which includes a wide range of ground movement, such as rock falls, deep failure of slopes and shallow debris flows. Although gravity's action on an over-steepened slope is the primary reason for a landslide, there are other contributing factors affecting the original slope stability.
Earthquake	An earthquake is the result from the sudden release of stored energy in the Earth face=symbol>¢s crust that creates seismic waves. At the Earth's surface, earthquakes may manifest themselves by a shaking or displacement of the ground. An earthquake is caused by tectonic plates getting stuck and putting a strain on the ground. The strain becomes so great that rocks give way by breaking and sliding along fault planes.
Glacier	A glacier is a large, slow moving river of ice, formed from compacted layers of snow, that slowly deforms and flows in response to gravity. Glacier ice is the largest reservoir of fresh water on Earth, and second only to oceans as the largest reservoir of total water. Glaciers cover vast areas of polar regions but are restricted to the highest mountains in the tropics.
Canals	Canals are artificial channels for water. There are two main types of canals: irrigation canals, which are used for the delivery of water, and waterways, which are transportation canals used for passage of goods and people, often connected to existing lakes, rivers, or oceans.
Dam	A dam is a barrier across flowing water that obstructs, directs or slows down the flow, often creating a reservoir, lake or impoundment.
Beach erosion	Beach erosion is the wearing away of land or the removal of beach or dune sediments by wave action, tidal currents, wave currents, or drainage. Waves, generated by storms or fast moving moter craft, cause beach erosion, which may take the form of long-term losses of sediment and rocks, or merely in the temporary redistribution of coastal sediments; erosion in one location may result in accretion nearby.
Erosion	Erosion is displacement of solids by the agents of ocean currents, wind, water, or ice by downward or down-slope movement in response to gravity or by living organisms.
Natural resources	Natural resources are naturally occurring substances that are considered valuable in their relatively unmodified or natural form. Its value rests in the amount of the material available and the demand for the certain material.
Earth materials	Earth materials is a general term that includes rocks and materials that are not by definition rocks but are commonly regarded as rocks.
Fossil	Fossils are the mineralized or otherwise preserved remains or traces of animals, plants, and other organisms. The totality of fossils, both discovered and undiscovered, and their placement in fossiliferous rock formations and sedimentary layers is known as the fossil record.
Fossil fuel	Fossil fuels are hydrocarbons, primarily coal and petroleum, formed from the fossilized remains of dead plants and animals. In common parlance, the term fossil fuel also includes hydrocarbon-containing natural resources that are not derived from animal or plant sources. Fossil fuels have made large-scale industrial development possible and have largely supplanted water-driven mills, as well as the combustion of wood or peat for heat.
Fossil fuels	Fossil fuels are hydrocarbons, primarily coal and petroleum, formed from the fossilized remains of dead plants and animals by exposure to heat and pressure in the Earth's crust over hundreds of millions of years. The burning of fossil fuels by humans is the largest source of emissions of carbon dioxide, which is one of the greenhouse gases that enhances radiative forcing and contributes to global warming.
Global	The Global telecommunications system is a global network for the transmission of meteorological data

Telecommunic- tions System	from weather stations, satellites and numerical weather prediction centres. The system consists of an integrated network of point-to-point circuits, and multi-point circuits which interconnect meteorological telecommunication centres.
Island	An island is any piece of land that is completely surrounded by water, above high tide. There are two main types of islands: continental islands and oceanic islands. There are also artificial islands. A grouping of geographically and/or geologically related islands is called an archipelago.
Climate	Climate is the average and variations of weather over long periods of time. Climate zones can be defined using parameters such as temperature and rainfall.
Climate change	Climate change refers to the variation in the Earth's global climate or in regional climates over time. It describes changes in the variability or average state of the atmosphere over time scales ranging from decades to millions of years. These changes can be caused by processes internal to the Earth, external forces or, more recently, human activities.
Venus	Venus is the second-closest planet to the Sun, orbiting it every 224.7 Earth days. It is the brightest natural object in the night sky, except for the Moon, reaching an apparent magnitude of −4.6. Because Venus is an inferior planet, from Earth it never appears to venture far from the Sun: its elongation reaches a maximum of 47.8°.
Mars	Mars the fourth planet from the Sun in the Solar System. The planet is named after Mars, the Roman god of war. It is also referred to as the "Red Planet" because of its reddish appearance as seen from Earth.
Mercury	Mercury is a chemical element in the periodic table that has the symbol Hg and atomic number 80. A heavy, silvery transition metal, mercury is one of five elements that are liquid at or near room temperature and pressure.
Moon	The Moon is Earth's only natural satellite. It makes a complete orbit around the Earth every 27.3 days, and the periodic variations in the geometry of the Earth–Moon–Sun system are responsible for the lunar phases that repeat every 29.5 days.
Temperature range	Temperature range is the numerical difference between the minimum and maximum values of temperature observed in a system, such as atmospheric temperature in a given location. A temperature range may refer to a period of time or to an average.
Nuclear fusion	In physics and nuclear chemistry, nuclear fusion is the process by which multiple atomic particles join together to form a heavier nucleus. It is accompanied by the release or absorption of energy.
Star	A star is a massive, luminous ball of plasma. Stars group together to form galaxies, and they dominate the visible universe. The nearest star to Earth is the Sun, which is the source of most of the energy on Earth, including daylight. Other stars are visible in the night sky, when they are not outshone by the Sun. A star shines because nuclear fusion in its core releases energy which traverses the star face=symbol>¢s interior and then radiates into outer space.
Jupiter	Jupiter is the fifth planet from the Sun and the largest planet within the solar system. It is two and a half times as massive as all of the other planets in our solar system combined. Jupiter, along with Saturn, Uranus and Neptune, is classified as a gas giant.
Silt	Silt is soil or rock derived granular material of a specific grain size. Silt may occur as a soil or alternatively as suspended sediment in a water column of any surface water body. It may also exist as deposition soil at the bottom of a water body.
Pluto	Pluto, also designated 134340 Pluto, is the second-largest known dwarf planet in the Solar System and the tenth-largest body observed directly orbiting the Sun. It is primarily composed of rock and ice and is relatively small; approximately a fifth the mass of the Earth's Moon and a third its volume.
Blocks	Blocks in meteorology are large scale patterns in the atmospheric pressure field that are nearly

stationary, effectively "blocking" or redirecting migratory cyclones. These blocks can remain in place for several days or even weeks, causing the areas affected by them to have the same kind of weather for an extended period of time.

Telescope

A telescope is an instrument designed for the observation of remote objects. The term usually refers to optical telescopes, but there are telescopes for most of the spectrum of electromagnetic radiation and for other signal types.

Canyon

A canyon is a deep valley between cliffs often carved from the landscape by a river. Most were formed by a process of long-time erosion from a plateau level. The cliffs form because harder rock strata that are resistant to erosion and weathering remain exposed on the valley walls.

Carbon

Carbon is a chemical element. An abundant nonmetallic, tetravalent element, carbon has several allotropic forms. This element is the basis of the chemistry of all known life.

Carbon dioxide

Carbon dioxide is a chemical compound, normally in a gaseous state, and is composed of one carbon and two oxygen atoms. It is often referred to by its formula CO_2. It is present in the Earth face=symbol>¢s atmosphere at a concentration of approximately .000383 by volume and is an important greenhouse gas due to its ability to absorb many infrared wavelengths of sunlight, and due to the length of time it stays in the atmosphere.

Polar ice cap

A polar ice cap is a high-latitude region of a planet or moon that is covered in ice. There are no requirements with respect to size or composition for a body of ice to be termed a polar ice cap, nor any geological requirement for it to be over land; only that it must be a body of solid phase matter in the polar region.

Ice cap

An ice cap is a dome-shaped ice mass that covers less than 50,000 km² of land area. Masses of ice covering more than 50,000 km² are termed an ice sheet.

Neptune

Neptune is the eighth and farthest known planet from the Sun in the Solar System. It is the fourth largest planet by diameter, and the third largest by mass.

Uranus

Uranus is the seventh planet from the Sun, and the first discovered in modern times. Although, like the five classical planets, Uranus is visible to the naked eye.

Saturn

Saturn is the sixth planet from the Sun. It is a gas giant and the second largest planet in the Solar System after Jupiter. The planet Saturn is primarily composed of hydrogen, with small proportions of helium and trace elements.

Melting

Melting is the process of heating a solid substance to a point where it turns into a liquid. An object that has melted is molten.

Melting point

The melting point of a crystalline solid is the temperature range at which it changes state from solid to liquid. Although the phrase would suggest a specific temperature, most crystalline compounds actually melt over a range of a few degrees or less. At the melting point the solid and liquid phase exist in equilibrium.

Charon

Charon is either the largest moon of Pluto or one member of a double dwarf planet with Pluto being the other member.

Nitrogen

Nitrogen is a chemical element which has the symbol N and atomic number 7. Elemental nitrogen is a colorless, odourless, tasteless and mostly inert diatomic gas at standard conditions, constituting 78.1% by volume of Earth's atmosphere.

Methane

Methane is a chemical compound with the molecular formula CH_4. It is the simplest alkane, and the principal component of natural gas. Burning one molecule of methane in the presence of oxygen releases one molecule. Methane's relative abundance and clean burning process makes it a very attractive fuel.

Methane hydrate

Methane hydrate is a solid form of water that contains a large amount of methane within its crystal

Go to **Cram101.com** for the Practice Tests for this Chapter.

structure. Originally thought to occur only in the outer regions of the solar system where temperatures are low and water ice is common, extremely large deposits of it have been found under sediments on the ocean floors of Earth.

Astronomical unit	The astronomical unit is a unit of length nearly equal to the semi-major axis of Earth face=symbol>¢s orbit around the Sun. The currently accepted value of the AU is 149 597 870 691 ± 30 metres.The symbol "ua" is recommended by the Bureau International des Poids et Mesures but in the United States and other anglophone countries the reverse usage is more common.
Hydrogen	Hydrogen is a chemical element represented by the symbol H and an atomic number of 1. At standard temperature and pressure it is a colorless, odorless, nonmetallic, tasteless, highly flammable diatomic gas . With an atomic mass of 1.00794 g/mol, hydrogen is the lightest element.Hydrogen is the most abundant of the chemical elements, constituting roughly 75% of the universe′s elemental mass.
Vapor	Vapor is the gas phase component of a another state of matter which does not completely fill its container. It is distinguished from the pure gas phase by the presence of the same substance in another state of matter. Hence when a liquid has completely evaporated, it is said that the system has been completely transformed to the gas phase.
Precipitation	Precipitation is any product of the condensation of atmospheric water vapor that is deposited on the earth′s surface. It occurs when the atmosphere becomes saturated with water vapour and the water condenses and falls out of solution. Air becomes saturated via two processes, cooling and adding moisture.
Rainforests	Rainforests are forests characterized by high rainfall, with definitions setting minimum normal annual rainfall.
Tropical rain forest	Tropical rain forest refers to rainforests generally found near the equator. They are common in Asia, Africa, South America, Central America, and on many of the Pacific Islands.
Forest	A forest is an area with a high density of trees, historically, a wooded area set aside for hunting. These plant communities cover large areas of the globe and function as animal habitats, hydrologic flow modulators, and soil conservers, constituting one of the most important aspects of the Earth face=symbol>¢s biosphere.
Seafloor	The seafloor the bottom of the ocean. At the bottom of the continental slope is the continental rise, which is caused by sediment cascading down the continental slope.
Seawater	Seawater is water from a sea or ocean. On average, seawater in the world′s oceans has a salinity of ~3.5%, or 35 parts per thousand. This means that every 1 kg of seawater has approximately 35 grams of dissolved salts.
Limestone	Limestone is a sedimentary rock composed largely of the mineral calcite. Limestone often contains variable amounts of silica in the form of chert or flint, as well as varying amounts of clay, silt and sand as disseminations, nodules, or layers within the rock. The primary source of the calcite in limestone is most commonly marine organisms. These organisms secrete shells that settle out of the water column and are deposited on ocean floors as pelagic ooze or alternatively is conglomerated in a coral reef.
Tropics	The tropics are the geographic region of the Earth centered on the equator and limited in latitude by the Tropic of Cancer in the northern hemisphere, at approximately 23 class="unicode">°30′ N latitude, and the Tropic of Capricorn in the southern hemisphere at 23°30′ S latitude.
Crater	A crater is an approximately circular depression in the surface of a planet, moon or other solid body in the Solar System, formed by the hyper-velocity impact of a smaller body with the surface. Impact craters typically have raised rims, and they range from small, simple, bowl-shaped depressions to large, complex, multi-ringed, impact basins.

Plateau	A plateau is an area of highland, usually consisting of relatively flat rural area.
Meteorite	A meteorite is a natural object originating in outer space that survives an impact with the Earth face=symbol>¢s surface without being destroyed. While in space it is called a meteoroid. When it enters the atmosphere, air resistance causes the body to heat up and emit light, thus forming a fireball.
Desert	In geography, a desert is a landscape form or region that receives very little precipitation. They are defined as areas that receive an average annual precipitation of less than 250 mm. A desert where vegetation cover is exceedingly sparse correspond to the 'hyperarid face=symbol>¢ regions of the earth, where rainfall is exceedingly rare and infrequent.
Lava	Lava is molten rock expelled by a volcano during an eruption. When first extruded from a volcanic vent, it is a liquid at temperatures from 700 °C to 1,200 °C.
Terrain	Terrain is the third or vertical dimension of land surface. When terrain is described underwater, the term bathymetry is used.
Radioactive decay	Radioactive decay is the process in which an unstable atomic nucleus loses energy by emitting radiation in the form of particles or electromagnetic waves.
Uranium	Uranium is approximately 70% more dense than lead and is weakly radioactive. It occurs naturally in low concentrations in soil, rock and water.
Element	An element is a type of atom that is defined by its atomic number; that is, by the number of protons in its nucleus.
Red Sea	The Red Sea is an inlet of the Indian Ocean between Africa and Asia. The connection to the ocean is in the south through the Bab el Mandeb sound and the Gulf of Aden. In the north are the Sinai Peninsula, the Gulf of Aqaba and the Gulf of Suez. The Red Sea is a Global 200 ecoregion.
Rift	In geology, a rift is a place where the Earth's crust and lithosphere are being pulled apart.
East African Rift	The East African Rift is a vast geographical and geological feature, approximately 6,000 kilometres in length, which runs from northern Syria in Southwest Asia to central Mozambique in East Africa. Caused by the geological process of rifting, it is a complex feature where several plates of the earth face=symbol>¢s crust join.
Earths atmosphere	Earths atmosphere is a layer of gases surrounding the planet Earth and retained by the Earth face=symbol>¢s gravity. This mixture of gases is commonly known as air.
Atmospheric pressure	Atmospheric pressure is the pressure at any point in the Earth's atmosphere.
Internal energy	In thermodynamics, the internal energy of a thermodynamic system, or a body with well-defined boundaries, denoted by U, or sometimes E, is the total of the kinetic energy due to the motion of molecules and the potential energy associated with the vibrational and electric energy of atoms within molecules or crystals.
Solar power	Solar power is Solar Radiation emitted from our sun. It has been used in many traditional technologies for centuries, and has come into widespread use where other power supplies are absent, such as in remote locations and in space.
Galaxy	A galaxy is a massive, gravitationally bound system consisting of stars, an interstellar medium of gas and dust, and dark matter. They typically range from dwarfs with as few as ten million stars up to giants with one trillion stars, all orbiting a common center of mass.
Biosphere	The biosphere is the part of the earth, including air, land, surface rocks, and water, within which life occurs, and which biotic processes in turn alter or transform. From the broadest biophysiological point of view, the biosphere is the global ecological system integrating all living beings and their

Go to **Cram101.com** for the Practice Tests for this Chapter.

relationships, including their interaction with the elements of the lithosphere, hydrosphere, and atmosphere. This biosphere is postulated to have evolved, beginning through a process of biogenesis or biopoesis, at least some 3.5 billion years ago.

Organism

In biology and ecology, an organism is a living complex adaptive system of organs that influence each other in such a way that they function in some way as a stable whole.

Weather

The weather is the set of all extant phenomena in a given atmosphere at a given time. The term usually refers to the activity of these phenomena over short periods, as opposed to the term climate, which refers to the average atmospheric conditions over longer periods of time.

Atmospheric circulation

Atmospheric circulation is the large-scale movement of air, and the means by which heat is distributed on the surface of the Earth. The large-scale structure of the atmospheric circulation varies from year to year, but the basic structure remains fairly constant.

Storm

A storm is any disturbed state of an astronomical body's atmosphere, especially affecting its surface, and strongly implying severe weather. It may be marked by strong wind, thunder and lightning, heavy precipitation, such as ice, or wind transporting some substance through the atmosphere.

Latitude

Latitude gives the location of a place on Earth north or south of the equator. Lines of Latitude are the horizontal lines shown running east-to-west on maps. Technically, Latitude is an angular measurement in degrees ranging from 0° at the Equator to 90 class="unicode">° at the poles.

Water vapor

Water vapor is the gas phase of water. Water vapor is one state of the water cycle within the hydrosphere. Water vapor can be produced from the evaporation of liquid water or from the sublimation of ice. Under normal atmospheric conditions, water vapor is continuously evaporating and condensing.

Stream

A stream is a body of water with a current, confined within a bed and banks. Streams are important as conduits in the water cycle, instruments in aquifer recharge, and corridors for fish and wildlife migration.

Groundwater

Groundwater is water located beneath the ground surface in soil pore spaces and in the fractures of geologic formations. Groundwater is recharged from, and eventually flows to, the surface naturally; natural discharge often occurs at springs and seeps, streams and can often form oases or wetlands.

Lake

A lake is a body of water or other liquid of considerable size contained on a body of land. A vast majority are fresh water, and lie in the Northern Hemisphere at higher latitudes. Most have a natural outflow in the form of a river or stream, but some do not, and lose water solely by evaporation and/or underground seepage.

Microorganism

A microorganism is an organism that is microscopic. They can be bacteria, fungi, archaea or protists, but not viruses and prions, which are generally classified as non-living. Micro-organisms are generally single-celled, or unicellular organisms.

Bacteria

Bacteria are unicellular microorganisms. They are typically a few micrometres long and have many shapes including curved rods, spheres, rods, and spirals.

Terrestrial

Terrestrial refers to things having to do with the land or with the planet Earth.

Mountain range

A mountain range is a group of mountains bordered by lowlands or separated from other mountain ranges by passes or rivers. Individual mountains within the same mountain range do not necessarily have the same geology; they may be a mix of different orogeny, for example volcanoes, uplifted mountains or fold mountains and may, therefore, be of different rock.

Snow line

The snow line is the point above which, or poleward of which, snow and ice cover the ground throughout the year.

Sunlight

Sunlight in the broad sense is the total spectrum of the electromagnetic radiation given off by the

Go to **Cram101.com** for the Practice Tests for this Chapter.

Sun. On Earth, it is filtered through the atmosphere, and the solar radiation is obvious as daylight when the Sun is above the horizon.

Sea level	Mean sea level is the average height of the sea, with reference to a suitable reference surface.
Carbonate	In organic chemistry, a carbonate is a salt of carbonic acid.
Natural gas	Natural gas is a gaseous fossil fuel consisting primarily of methane but including significant quantities of ethane, butane, propane, carbon dioxide, nitrogen, helium and hydrogen sulfide.
Coal	Coal is a fossil fuel formed in swamp ecosystems where plant remains were saved by water and mud from oxidization and biodegradation. It is a sedimentary rock, but the harder forms, such as anthracite coal, can be regarded as metamorphic rocks because of later exposure to elevated temperature and pressure. It is composed primarily of carbon along with assorted other elements, including sulfur.
Vegetation	Vegetation is a general term for the plant life of a region; it refers to the ground cover provided by plants, and is, by far, the most abundant biotic element of the biosphere. Primeval redwood forests, coastal mangrove stands, sphagnum bogs, desert soil crusts, roadside weed patches, wheat fields, cultivated gardens and lawns; are all encompassed by the term vegetation.
Phytoplankton	Phytoplankton are the autotrophic component of plankton. Most phytoplankton are too small to be individually seen with the unaided eye. However, when present in high enough numbers, they may appear as a green discoloration of the water due to the presence of chlorophyll within their cells.
Food web	Food web refers to describe the feeding relationships between species in an ecological community. Typically a food web refers to a graph where only connections are recorded, and a food web or ecosystem network refers to a network where the connections are given weights representing the quantity of nutrients or energy being transferred.
Agriculture	Agriculture is the production of food, feed, fiber, fuel and other goods by the systematic raizing of plants and animals.
Chemical reaction	A chemical reaction is a process that results in the interconversion of chemical substances. The substance or substances initially involved in a chemical reaction are called reactants. Chemical reactions are characterized by a chemical change, and they yield one or more products which are, in general, different from the reactants.
Magnetic field	In physics, a magnetic field is a solenoidal vector field in the space surrounding moving electric charges and magnetic dipoles, such as those in electric currents and magnets.
Wave	A wave is a disturbance that propagates through space or spacetime, transferring energy and momentum and sometimes angular momentum.
Magnetism	Magnetism is one of the phenomena by which materials exert attractive or repulsive forces on other materials. Some well known materials that exhibit easily detectable magnetic properties are nickel, iron, some steels, and the mineral magnetite; however, all materials are influenced to greater or lesser degree by the presence of a magnetic field.
Earths magnetic field	Earths magnetic field is approximately a magnetic dipole, with one pole near the north pole and the other near the geographic south pole.
Alcatraz Island	Alcatraz Island is a small island located in the middle of San Francisco Bay in California, United States. It served as a lighthouse, then a military fortification, then a military prison followed by a federal prison until 1963, when it became a national recreation area.
Magnesium	Magnesium has the symbol Mg. It is the ninth most abundant element in the universe by mass. It constitutes about 2% of the Earth's crust by mass, and it is the third most abundant element dissolved in seawater. It is essential to all living cells, and is the 11th most abundant element by mass in the human body.
Iron	Iron is a chemical element metal. It is a lustrous, silvery soft metal. It and nickel are notable for

being the final elements produced by stellar nucleosynthesis, and thus are the heaviest elements which do not require a supernova or similarly cataclysmic event for formation.

Silicate	In geology and astronomy, the term silicate is used to denote types of rock that consist predominantly of silicate minerals. Such rocks include a wide range of igneous, metamorphic and sedimentary types. Most of the Earth's mantle and crust are made up of silicate rocks. The same is true of the Moon and the other rocky planets.
Chemical compound	A chemical compound is a chemical substance of two or more different chemically bonded chemical elements, with a fixed ratio determining the composition. The ratio of each element is usually expressed by chemical formula.
Rotation	A rotation is a movement of an object in a circular motion. A two-dimensional object rotates around a center of rotation. A three-dimensional object rotates around a line called an axis. A circular motion about an external point, e.g. the Earth about the Sun, is called an orbit or more properly an orbital revolution.
Oceanic crust	Oceanic crust is the part of Earth's lithosphere that surfaces in the ocean basins. Oceanic crust is primarily composed of mafic rocks, or sima. It is thinner than continental crust, or sial, generally less than 10 kilometers thick, however it is more dense, having a mean density of about 3.3 grams per cubic centimeter.
Continental crust	The continental crust is the layer of granitic, sedimentary, and metamorphic rocks which form the continents and the areas of shallow seabed close to their shores, known as continental shelves. It is less dense than the material of the Earth's mantle and thus "floats" on top of it. Continental crust is also less dense than oceanic crust, though it is considerably thicker. About 40% of the Earth's surface is now underlain by continental crust.
Topography	Topography is the study of Earth's surface features or those of other planets, moons, and asteroids
Shoreline	A shoreline is the fringe of land at the edge of a large body of water, such as an ocean, sea, or lake. A strict definition is the strip of land along a water body that is alternately exposed and covered by waves and tides.
Crystal	A crystal is a solid in which the constituent atoms, molecules, or ions are packed in a regularly ordered, repeating pattern extending in all three spatial dimensions. Most metals encountered in everyday life are polycrystals. Crystals are often symmetrically intergrown to form crystal twins.
Metamorphic rock	Metamorphic rock is the result of the transformation of a pre-existing rock type, the protolith, in a process called metamorphism, which means "change in form". The protolith is subjected to heat and extreme pressure causing profound physical and/or chemical change. The protolith may be sedimentary rock, igneous rock or another older rock.
Metamorphism	Metamorphism can be defined as the solid state recrystallisation of pre-existing rocks due to changes in heat and/or pressure and/or introduction of fluids. There will be mineralogical, chemical and crystallographic changes. Metamorphism produced with increasing pressure and temperature conditions is known as prograde metamorphism. Conversely, decreasing temperatures and pressure characterize retrograde metamorphism.
Rocky Mountains	The Rocky Mountains are a broad mountain range in western North America. The Rocky Mountains stretch more than 4,800 kilometers from northernmost British Columbia, in Canada, to New Mexico, in the United States.
Lake Superior	Lake Superior, bounded by Ontario, Canada and Minnesota, USA, to the north and Wisconsin and Michigan, USA, to the south, is the largest of North America's Great Lakes. It is the largest freshwater lake in the world by surface area and is the world's third-largest freshwater lake by volume.

Appalachian Mountains	The Appalachian Mountains are a vast system of mountains in eastern North America.
Craton	A craton is an old and stable part of the continental crust that has survived the merging and splitting of continents and supercontinents. Cratons are generally found in the interiors of continents and are characteristically composed of ancient crystalline basement crust of lightweight felsic igneous rock such as granite. They have a thick crust and deep roots that extend into the mantle beneath to depths of 200 km.
Dome	In geology, a dome is a deformational feature consisting of symmetrically-dipping anticlines; their general outline on a geologic map is circular or oval.
Landform	A landform comprises a geomorphological unit, and is largely defined by its surface form and location in the landscape, as part of the terrain, and as such, is typically an element of topography. They are categorised by features such as elevation, slope, orientation, stratification, rock exposure, and soil type. They include berms, mounds, hills, cliffs, valleys, rivers and numerous other elements.
Basalt	Basalt is a common gray to black extrusive volcanic rock. It is usually fine-grained due to rapid cooling of lava on the Earth's surface. It may be porphyritic containing larger crystals in a fine matrix, or vesicular, or frothy scoria.
Canadian Shield	The Canadian Shield is a large shield covered by a thin layer of soil that forms the nucleus of the North American craton. It has a deep, common, joined bedrock region in eastern and central Canada and stretches North from the Great Lakes to the Arctic Ocean, covering half the country.
Topsoil	Topsoil is the uppermost layer of soil, usually the top 2 to 6 inches. It has the highest concentration of organic matter and microorganisms, and is where most of the Earth's biological soil activity occurs. Plants generally concentrate their roots in, and obtain most of their nutrients from this layer. The actual depth of the topsoil layer can be measured as the depth from the surface to the first densely packed soil layer known as hardpan.
Bogs	Bogs are wetland types that accumulate acidic peat, a deposit of dead plant material.
Fold	The term fold is used in geology when one or a stack of originally flat and planar surfaces, such as sedimentary strata, are bent or curved as a result of plastic, i.e. permanent, deformation.
Sandstone	Sandstone is a sedimentary rock composed mainly of sand-size mineral or rock grains. Most sandstone is composed of quartz and/or feldspar because these are the most common minerals in the Earth face=symbol>¢s crust. Like sand, sandstone may be any color, but the most common colors are tan, brown, yellow, red, gray and white.
Amazon Rainforest	The Amazon Rainforest is a moist broadleaf forest in that Basin of South America. The area encompasses seven million square kilometers, though the forest itself occupies some 5.5 million square kilometers. It represents over half of the planet's remaining rainforests and comprises the largest and most species-rich tract of tropical rainforest in the world.
Andes Mountains	The Andes Mountains are South America's longest mountain range, forming a continuous chain of highland along the western coast of South America.
Landfill	A landfill, is a site for the disposal of waste materials by burial and is the oldest form of waste treatment.
Amazon River	The Amazon River of South America is the largest river in the world by volume, with greater total river flow than the next eight largest rivers combined, and with the largest drainage basin in the world. Because of its vast dimensions it is sometimes called The River Sea.
Amazon River system	The Amazon River System is made up of the tributaries of The Amazon River, the largest river in the world by volume, with greater total river flow than the next eight largest rivers combined, and with the largest drainage basin in the world.

Go to **Cram101.com** for the Practice Tests for this Chapter.

Atlantic Ocean	The Atlantic Ocean is the second-largest of the world's oceanic divisions; with a total area of about 106.4 million square kilometres , it covers approximately one-fifth of the Earth's surface. The Atlantic Ocean occupies an elongated, S-shaped basin extending longitudinally between the Americas to the west, and Eurasia and Africa to the east.
Tectonics	Tectonics is a field of study within geology concerned generally with the structures within the crust of the Earth, or other planets, and particularly with the forces and movements that have operated in a region to create these structures.
Reservoir	Most often, a reservoir refers to an artificial lake, used to store water for various uses. Reservoirs are created first by building a sturdy dam, usually out of cement, earth, rock, or a mixture. Once the dam is completed, a stream is allowed to flow behind it and eventually fill it to capacity.
Sediment	Sediment is any particulate matter that can be transported by fluid flow and which eventually is deposited as a layer of solid particles on the bed or bottom of a body of water or other liquid.
Abyssal plain	An abyssal plain is a flat or very gently sloping area of the deep ocean basin floor. They are among the Earth's flattest and smoothest regions and the least explored. They cover approximately 40% of the ocean floor and generally lie between the foot of a continental rise and a mid-oceanic ridge.
Arctic	The Arctic is the region around the Earth's North Pole, opposite the Antarctic region around the South Pole. In the northern hemisphere, the Arctic includes the Arctic Ocean and parts of Canada, Greenland, Russia, the United States, Iceland, Norway, Sweden and Finland. The word Arctic comes from the Greek word arktos, which means bear. This is due to the location of the constellation Ursa Major, the "Great Bear", above the Arctic region.
Indian Ocean	The Indian Ocean is the third largest of the world's oceanic divisions, covering about 20% of the Earth's water surface. It is bounded on the north by Asia on the west by Africa; on the east by the Malay Peninsula, the Sunda Islands, and Australia; and on the south by the Southern Ocean.
Rift valley	A rift valley in geology is a valley created by the formation of a rift.
Valley	In geology, a valley is a depression with predominant extent in one direction. The terms U-shaped and V-shaped are descriptive terms of geography to characterize the form of valleys. Most valleys belong to one of these two main types or a mixture of them, at least with respect of the cross section of the slopes or hillsides.
Submarine volcanoes	Submarine volcanoes are underwater fissures in the earth's surface from which magma can erupt. They estimated to account for 75% of annual magma output. The vast majority are located near areas of tectonic plate movement, known as mid-ocean ridges.
Alvin	Alvin is a 16-ton, manned deep-ocean research submersible owned by the United States Navy and operated by the Woods Hole Oceanographic Institution in Woods Hole, Massachusetts. The three-person vessel allows for two scientists and one pilot to dive for up to nine hours at 4500 metersor 15,000 feet.
Hawaiian Islands	The Hawaiian Islands form an archipelago of nineteen islands and atolls, numerous smaller islets, and undersea seamounts trending northwest by southeast in the North Pacific Ocean between latitudes 19 class="unicode">° N and 29° N. The archipelago takes its name from the largest island in the group and extends some 1500 miles from the Island of Hawai i in the south to northernmost Kure Atoll.
Continental shelf	The continental shelf is the extended perimeter of each continent and associated coastal plain, which is covered during interglacial periods such as the current epoch by relatively shallow seas and gulfs. The shelf usually ends at a point of increasing slope.
Echo sounding	Echo sounding is the technique of using sound pulses directed from the surface or from a submarine vertically down to measure the distance to the bottom by means of sound waves.

Hydrophone   A hydrophone is a sound-to-electricity transducer for use in water or other liquids, analogous to a ear for listening to underwater sound.

27

Planet	A planet, as defined by the International Astronomical Union, is a celestial body orbiting a star or stellar remnant that is massive enough to be rounded by its own gravity, not massive enough to cause thermonuclear fusion in its core, and has cleared its neighboring region of planetesimals.
Tectonics	Tectonics is a field of study within geology concerned generally with the structures within the crust of the Earth, or other planets, and particularly with the forces and movements that have operated in a region to create these structures.
Atmosphere	An atmosphere is a layer of gases that may surround a material body of sufficient mass. The gases are attracted by the gravity of the body, and are retained for a longer duration if gravity is high and the atmosphere's temperature is low. Some planets consist mainly of various gases, and thus have very deep atmospheres.
Gravitation	Gravitation, in everyday life, is most familiar as the agency that endows objects with weight. Gravitation is responsible for keeping the Earth and the other planets in their orbits around the Sun; for the formation of tides; and for various other phenomena that we observe. Gravitation is also the reason for the very existence of the Earth, the Sun, and most macroscopic objects in the universe; without it, matter would not have coalesced into these large masses, and life, as we know it, would not exist.
Water vapor	Water vapor is the gas phase of water. Water vapor is one state of the water cycle within the hydrosphere. Water vapor can be produced from the evaporation of liquid water or from the sublimation of ice. Under normal atmospheric conditions, water vapor is continuously evaporating and condensing.
Vapor	Vapor is the gas phase component of a another state of matter which does not completely fill its container. It is distinguished from the pure gas phase by the presence of the same substance in another state of matter. Hence when a liquid has completely evaporated, it is said that the system has been completely transformed to the gas phase.
Groundwater	Groundwater is water located beneath the ground surface in soil pore spaces and in the fractures of geologic formations. Groundwater is recharged from, and eventually flows to, the surface naturally; natural discharge often occurs at springs and seeps, streams and can often form oases or wetlands.
Glacier	A glacier is a large, slow moving river of ice, formed from compacted layers of snow, that slowly deforms and flows in response to gravity. Glacier ice is the largest reservoir of fresh water on Earth, and second only to oceans as the largest reservoir of total water. Glaciers cover vast areas of polar regions but are restricted to the highest mountains in the tropics.
Lithosphere	The lithosphere is the solid outermost shell of a rocky planet. On the Earth, the lithosphere includes the crust and the uppermost mantle which is joined to the crust across the Mohorovièiæ discontinuity. Lithosphere is underlain by asthenosphere, the weaker, hotter, and deeper part of the upper mantle.
Continent	A continent is one of several large landmasses on Earth. They are generally identified by convention rather than any strict criteria, but seven areas are commonly reckoned as continents – they are: Asia, Africa, North America, South America, Antarctica, Europe, and Australia.
Canadian Shield	The Canadian Shield is a large shield covered by a thin layer of soil that forms the nucleus of the North American craton. It has a deep, common, joined bedrock region in eastern and central Canada and stretches North from the Great Lakes to the Arctic Ocean, covering half the country.
Mountain	A mountain is a landform that extends above the surrounding terrain in a limited area. A

	mountain is generally steeper than a hill, but there is no universally accepted standard definition for the height of a mountain or a hill although a mountain usually has an identifiable summit.
Ridge	A ridge is a geological feature that is also known as a Rip in the earth causing magma to flow out and forming an undersea volcano, it also has geological features, a continuous elevational crest for some distance. Ridges are usually termed hills or mountains as well, depending on size.
Lake	A lake is a body of water or other liquid of considerable size contained on a body of land. A vast majority are fresh water, and lie in the Northern Hemisphere at higher latitudes. Most have a natural outflow in the form of a river or stream, but some do not, and lose water solely by evaporation and/or underground seepage.
Bahamas	The Commonwealth of The Bahamas is an English-speaking nation consisting of two thousand cays and seven hundred islands that form an archipelago. It is located in the Atlantic Ocean, east of Florida and the United States, north of Cuba and the Caribbean, and northwest of the British overseas territory of the Turks and Caicos Islands.
Stream	A stream is a body of water with a current, confined within a bed and banks. Streams are important as conduits in the water cycle, instruments in aquifer recharge, and corridors for fish and wildlife migration.
Erosion	Erosion is displacement of solids by the agents of ocean currents, wind, water, or ice by downward or down-slope movement in response to gravity or by living organisms.
Geology	Geology is the science and study of the solid matter that constitute the Earth. Encompassing such things as rocks, soil, and gemstones, geology studies the composition, structure, physical properties, history, and the processes that shape Earth's components.
Desert	In geography, a desert is a landscape form or region that receives very little precipitation. They are defined as areas that receive an average annual precipitation of less than 250 mm. A desert where vegetation cover is exceedingly sparse correspond to the face=symbol>¢hyperarid' regions of the earth, where rainfall is exceedingly rare and infrequent.
Mediterranean Sea	The Mediterranean Sea is a sea of the Atlantic Ocean almost completely enclosed by land: on the north by Europe, on the south by Africa, and on the east by Asia. It covers an approximate area of 2.5 million km², but its connection to the Atlantic is only 14 km wide.
Drainage	Drainage is the natural or artificial removal of surface and sub-surface water from a given area. Many agricultural soils need drainage to improve production or to manage water supplies.
Delta	A delta is a landform where the mouth of a river flows into an ocean, sea, desert, estuary or lake. It builds up sediment outwards into the flat area which the river face=symbol>¢s flow encounters transported by the water and set down as the currents slow.
Nile River	The Nile River is a major north-flowing river in Africa, generally regarded as the longest river in the world. The Nile River has two major tributaries, the White Nile and Blue Nile, the latter being the source of most of the Nile's water and fertile soil, but the former being the longer of the two. It ends in a large delta that empties into the Mediterranean Sea.
Vegetation	Vegetation is a general term for the plant life of a region; it refers to the ground cover provided by plants, and is, by far, the most abundant biotic element of the biosphere. Primeval redwood forests, coastal mangrove stands, sphagnum bogs, desert soil crusts,

roadside weed patches, wheat fields, cultivated gardens and lawns; are all encompassed by the term vegetation.

Sediment	Sediment is any particulate matter that can be transported by fluid flow and which eventually is deposited as a layer of solid particles on the bed or bottom of a body of water or other liquid.
Dead Sea	The Dead Sea is a salt lake between the West Bank and Israel to the west, and Jordan to the east. It is said to be the lowest point on Earth, at 420 m below sea level; its shores are actually the lowest point on dry land, as there are deeper points on Earth under water or ice. At 330m deep , the Dead Sea is the deepest hypersaline lake in the world.
Red Sea	The Red Sea is an inlet of the Indian Ocean between Africa and Asia. The connection to the ocean is in the south through the Bab el Mandeb sound and the Gulf of Aden. In the north are the Sinai Peninsula, the Gulf of Aqaba and the Gulf of Suez. The Red Sea is a Global 200 ecoregion.
Rift	In geology, a rift is a place where the Earth's crust and lithosphere are being pulled apart.
Valley	In geology, a valley is a depression with predominant extent in one direction. The terms U-shaped and V-shaped are descriptive terms of geography to characterize the form of valleys. Most valleys belong to one of these two main types or a mixture of them, at least with respect of the cross section of the slopes or hillsides.
Reservoir	Most often, a reservoir refers to an artificial lake, used to store water for various uses. Reservoirs are created first by building a sturdy dam, usually out of cement, earth, rock, or a mixture. Once the dam is completed, a stream is allowed to flow behind it and eventually fill it to capacity.
Earths atmosphere	Earths atmosphere is a layer of gases surrounding the planet Earth and retained by the Earth's gravity. This mixture of gases is commonly known as air.
Atmospheric circulation	Atmospheric circulation is the large-scale movement of air, and the means by which heat is distributed on the surface of the Earth. The large-scale structure of the atmospheric circulation varies from year to year, but the basic structure remains fairly constant.
Sleet	Sleet is a term used in a variety of ways to describe precipitation intermediate between rain and snow but distinct from hail.
Hail	Hail is a form of precipitation which consists of balls or irregular lumps of ice, 5 mm–50 mm in diameter on average, with much larger hailstones from severe thunderstorms. Hail is always produced by thunderstorms, and is composed of transparent ice or alternating layers of transparent and translucent ice at least 1 mm thick.
Surface runoff	Surface runoff is a term used to describe the flow of water, from rain, snowmelt, or other sources, over the land surface, and is a major component of the water cycle.
Evaporation	Evaporation is the process by which molecules in a liquid state become a gas.
Seep	A seep is a wet place where a liquid, usually groundwater, has oozed from the ground to the surface. They are usually not flowing, with the liquid sourced only from underground. The term may also refer to the movement of liquid hydrocarbons to the surface through fractures and fissures in the rock and between geological layers. It may be a significant source of pollution.
Melting	Melting is the process of heating a solid substance to a point where it turns into a liquid. An object that has melted is molten.
Blue ice	Blue ice occurs when snow falls on a glacier, is compressed, and becomes part of a glacier

Go to **Cram101.com** for the Practice Tests for this Chapter.

that winds its way toward a body of water.

Transpiration	Transpiration is the evaporation of water from aerial parts and of plants, especially leaves but also stems, flowers and fruits. Transpiration is a side effect of the plant needing to open its stomata in order to obtain carbon dioxide gas from the air for photosynthesis. Transpiration also cools plants and enables mass flow of mineral nutrients from roots to shoots.
Mercury	Mercury is a chemical element in the periodic table that has the symbol Hg and atomic number 80. A heavy, silvery transition metal, mercury is one of five elements that are liquid at or near room temperature and pressure.
Crater	A crater is an approximately circular depression in the surface of a planet, moon or other solid body in the Solar System, formed by the hyper-velocity impact of a smaller body with the surface. Impact craters typically have raised rims, and they range from small, simple, bowl-shaped depressions to large, complex, multi-ringed, impact basins.
Moon	The Moon is Earth's only natural satellite. It makes a complete orbit around the Earth every 27.3 days, and the periodic variations in the geometry of the Earth–Moon–Sun system are responsible for the lunar phases that repeat every 29.5 days.
Sea level	Mean sea level is the average height of the sea, with reference to a suitable reference surface.
Precipitation	Precipitation is any product of the condensation of atmospheric water vapor that is deposited on the earth's surface. It occurs when the atmosphere becomes saturated with water vapour and the water condenses and falls out of solution. Air becomes saturated via two processes, cooling and adding moisture.
Ocean basins	Ocean basins are large geologic basins that are below sea level. Geologically, there are other undersea geomorphological features such as the continental shelves, the deep ocean trenches, and the undersea mountain rangeswhich are not considered to be part of the ocean basins.
Ice Age	An ice age is a period of long-term reduction in the temperature of Earth face=symbol>¢s climate, resulting in an expansion of the continental ice sheets, polar ice sheets and mountain glaciers .
Canyon	A canyon is a deep valley between cliffs often carved from the landscape by a river. Most were formed by a process of long-time erosion from a plateau level. The cliffs form because harder rock strata that are resistant to erosion and weathering remain exposed on the valley walls.
Grand Canyon	The Grand Canyon is a very colorful, steep-sided gorge, carved by the Colorado River in the U.S. state of Arizona. It is one of the first national parks in the United States.
Ocean current	An ocean current is any more or less continuous, directed movement of ocean water that flows in one of the Earth's oceans.Ocean Currents are rivers of hot or cold water within the ocean. The currents are generated from the forces acting upon the water like the earth's rotation, the wind, the temperature and salinity differences and the gravitation of the moon. The depth contours, the shoreline and other currents influence the current's direction and strength.
Currents	Ocean currents are any more or less continuous, directed movement of ocean water that flows in one of the Earth's oceans.They are rivers of hot or cold water within the ocean. They are generated from the forces acting upon the water like the earth's rotation, the wind, the temperature and salinity differences and the gravitation of the moon.

Convection	Convection in the most general terms refers to the movement of currents within fluids. Convection is one of the major modes of Heat and mass transfer. In fluids, convective heat and mass transfer take place through both diffusion and by advection, in which matter or heat is transported by the larger-scale motion of currents in the fluid.
Drainage system	A drainage system is the pattern formed by the streams, rivers, and laked in a particular watershed. They are governed by the topography of the land, whether a particular region is dominated by hard or soft rocks, and the gradient of the land.
Landform	A landform comprises a geomorphological unit, and is largely defined by its surface form and location in the landscape, as part of the terrain, and as such, is typically an element of topography. They are categorised by features such as elevation, slope, orientation, stratification, rock exposure, and soil type. They include berms, mounds, hills, cliffs, valleys, rivers and numerous other elements.
Silt	Silt is soil or rock derived granular material of a specific grain size. Silt may occur as a soil or alternatively as suspended sediment in a water column of any surface water body. It may also exist as deposition soil at the bottom of a water body.
Wave	A wave is a disturbance that propagates through space or spacetime, transferring energy and momentum and sometimes angular momentum.
Mars	Mars the fourth planet from the Sun in the Solar System. The planet is named after Mars, the Roman god of war. It is also referred to as the "Red Planet" because of its reddish appearance as seen from Earth.
Mountain range	A mountain range is a group of mountains bordered by lowlands or separated from other mountain ranges by passes or rivers. Individual mountains within the same mountain range do not necessarily have the same geology; they may be a mix of different orogeny, for example volcanoes, uplifted mountains or fold mountains and may, therefore, be of different rock.
Agriculture	Agriculture is the production of food, feed, fiber, fuel and other goods by the systematic raizing of plants and animals.
Climate	Climate is the average and variations of weather over long periods of time. Climate zones can be defined using parameters such as temperature and rainfall.
Ice sheet	An ice sheet is a mass of glacier ice that covers surrounding terrain and is greater than 19,305 mile². The only current ice sheets are in Antarctica and Greenland. Ice sheets are bigger than ice shelves or glaciers. Masses of ice covering less than 50,000 km² are termed an ice cap. An ice cap will typically feed a series of glaciers around its periphery. Although the surface is cold, the base of an ice sheet is generally warmer. This process produces fast-flowing channels in the ice sheet.
Great Lakes	The Laurentian Great Lakes are a group of five large lakes in North America on or near the Canada-United States border. They are the largest group of fresh water lakes on Earth.
Porosity	Porosity is a measure of the void spaces in a material, and is measured as a fraction, between 0–1, or as a percentage between 0–100%.
Sinkhole	A sinkhole is a natural depression or hole in the surface topography caused by the removal of soil or bedrock, often both, by water. They may vary in size from less than a meter to several hundred meters both in diameter and depth, and vary in form from soil-lined bowls to bedrock-edged chasms.
Cave	A cave is a natural underground void large enough for a human to enter. Some people suggest that the term 'cave' should only apply to cavities that have some part which does not receive daylight; however, in popular usage, the term includes smaller spaces like a sea cave, rock shelters, and grottos.

Limestone	Limestone is a sedimentary rock composed largely of the mineral calcite. Limestone often contains variable amounts of silica in the form of chert or flint, as well as varying amounts of clay, silt and sand as disseminations, nodules, or layers within the rock. The primary source of the calcite in limestone is most commonly marine organisms. These organisms secrete shells that settle out of the water column and are deposited on ocean floors as pelagic ooze or alternatively is conglomerated in a coral reef.
Tide	Tide refers to the cyclic rizing and falling of Earth's ocean surface caused by the tidal forces of the Moon and the sun acting on the oceans. They cause changes in the depth of the marine and estuarine water bodies and produce oscillating currents known as tidal streams, making prediction of tides important for coastal navigation.
Tides	Tides are the cyclic rizing and falling of Earth's ocean surface caused by the tidal forces of the Moon and the sun acting on the oceans. Tides cause changes in the depth of the marine and estuarine water bodies and produce oscillating currents known as tidal streams, making prediction of tides important for coastal navigation.
Coast	The coast is defined as the part of the land adjoining or near the ocean. A coastline is properly a line on a map indicating the disposition of a coast, but the word is often used to refer to the coast itself. The adjective coastal describes something as being on, near to, or associated with a coast.
Shoreline	A shoreline is the fringe of land at the edge of a large body of water, such as an ocean, sea, or lake. A strict definition is the strip of land along a water body that is alternately exposed and covered by waves and tides.
Terrace	In agriculture, a terrace is a leveled section of a hilly cultivated area, designed as a method of soil conservation to slow or prevent the rapid surface runoff of irrigation water
Lagoon	A lagoon is a body of comparatively shallow salt or brackish water separated from the deeper sea by a shallow or exposed sandbank, coral reef, or similar feature. Thus, the enclosed body of water behind a barrier reef or barrier islands or enclosed by an atoll reef is called a lagoon.
Cliff	A cliff is a significant vertical, or near vertical, rock exposure. Cliffs are categorized as erosion landforms due to the processes of erosion and weathering that produce them. Cliffs are common on coasts, in mountainous areas, escarpments and along rivers. Cliffs are usually formed by rock that is resistant to erosion and weathering.
Eolian	Eolian processes pertain to the activity of the winds and more specifically, to the winds' ability to shape the surface of the Earth and other planets.
Dune	A dune is a hill of sand built by eolian processes. Dunes are subject to different forms and sizes based on their interaction with the wind. Most kinds of dune are longer on the windward side where the sand is pushed up the dune, and a shorter in the lee of the wind. The trough between dunes is called a slack. A "dune field" is an area covered by extensive sand dunes. Large dune fields are known as ergs.
Volcano	A volcano is an opening, or rupture, in the Earth's surface or crust, which allows hot, molten rock, ash and gases to escape from deep below the surface.
Seafloor	The seafloor the bottom of the ocean. At the bottom of the continental slope is the continental rise, which is caused by sediment cascading down the continental slope.
Seafloor spreading	Seafloor spreading occurs at mid-ocean ridges, where new oceanic crust is formed through volcanic activity and then gradually moves away from the ridge. Seafloor spreading helps explain continental drift in the theory of plate tectonics.
Orogeny	Orogeny is the process of building mountains, and may be studied as a tectonic structural

event, as a geographical event and a chronological event, in that orogenic events cause distinctive structural phenomena and related tectonic activity, affect certain regions of rocks and crust and happen within a time frame.

Earthquake

An earthquake is the result from the sudden release of stored energy in the Earth face=symbol>¢s crust that creates seismic waves. At the Earth face=symbol>¢s surface, earthquakes may manifest themselves by a shaking or displacement of the ground. An earthquake is caused by tectonic plates getting stuck and putting a strain on the ground. The strain becomes so great that rocks give way by breaking and sliding along fault planes.

Internal energy

In thermodynamics, the internal energy of a thermodynamic system, or a body with well-defined boundaries, denoted by U, or sometimes E, is the total of the kinetic energy due to the motion of molecules and the potential energy associated with the vibrational and electric energy of atoms within molecules or crystals.

Crust

In geology, a crust is the outermost layer of a planet, part of its lithosphere. They are generally composed of a less dense material than its deeper layers.Earths face=symbol>¢ is composed mainly of basalt and granite. It is cooler and more rigid than the deeper layers of the mantle and core.

Mesosphere

The mesosphere is the layer of the Earth's atmosphere that is directly above the stratosphere and directly below the thermosphere. The mesosphere is located from about 50 km to 80-90 km altitude above Earth's surface. Within this layer, temperature decreases with increasing altitude. The main dynamical features in this region are atmospheric tides, internal atmospheric gravity waves and planetary waves.

Mantle

Earth's mantle is a ~2,900 km thick rocky shell comprizing approximately 70% of Earth's volume. It is predominantly solid and overlies the Earth's iron-rich core, which occupies about 30% of Earth face=symbol>¢s volume. Past episodes of melting and volcanism at the shallower levels of the mantle have produced a very thin crust of crystallized melt products near the surface, upon which we live.

Asthenosphere

The asthenosphere is the region of the Earth between 100-200 km below the surface that is the weak or "soft" zone in the upper mantle. It lies just below the lithosphere, which is involved in plate movements and isostatic adjustments. In spite of its heat, pressures keep it plastic, and it has a relatively low density. Seismic waves pass relatively slowly through the asthenosphere.

Element

An element is a type of atom that is defined by its atomic number; that is, by the number of protons in its nucleus.

Plate tectonics

Plate tectonics is a theory of geology that has been developed to explain the observed evidence for large scale motions of the Earth's lithosphere. The theory encompassed and superseded the older theory of continental drift.

Trench

A trench is a type of excavation or depression in the ground. They are generally defined by being deeper than they are wide, and by being narrow compared to their length.

Island

An island is any piece of land that is completely surrounded by water, above high tide. There are two main types of islands: continental islands and oceanic islands. There are also artificial islands. A grouping of geographically and/or geologically related islands is called an archipelago.

Subduction

In geology, a subduction zone is an area on Earth where two tectonic plates meet and move towards one another, with one sliding underneath the other and moving down into the mantle,

Go to **Cram101.com** for the Practice Tests for this Chapter.

at rates typically measured in centimeters per year. An oceanic plate ordinarily slides underneath a continental plate; this often creates an orogenic zone with many volcanoes and earthquakes.

Subduction zone	A subduction zone is an area on Earth where two tectonic plates meet and move towards one another, with one sliding underneath the other and moving down into the mantle, at rates typically measured in centimeters per year. In a sense, subduction zones are the opposite of divergent boundaries, areas where material rises up from the mantle and plates are moving apart.
Continental crust	The continental crust is the layer of granitic, sedimentary, and metamorphic rocks which form the continents and the areas of shallow seabed close to their shores, known as continental shelves. It is less dense than the material of the Earth's mantle and thus "floats" on top of it. Continental crust is also less dense than oceanic crust, though it is considerably thicker. About 40% of the Earth's surface is now underlain by continental crust.
Blocks	Blocks in meteorology are large scale patterns in the atmospheric pressure field that are nearly stationary, effectively "blocking" or redirecting migratory cyclones. These blocks can remain in place for several days or even weeks, causing the areas affected by them to have the same kind of weather for an extended period of time.
Mantle plume	A mantle plume is an upwelling of abnormally hot rock within the Earth face=symbol>¢s mantle. As the heads of mantle plumes can partly melt when they reach shallow depths, they are thought to be the cause of volcanic centers known as hotspots and probably also to have caused flood basalts.
Divergent plate boundary	In plate tectonics, a divergent plate boundary a linear feature that exists between two tectonic plates that are moving away from each other. These areas can form in the middle of continents but eventually form ocean basins.
Lava	Lava is molten rock expelled by a volcano during an eruption. When first extruded from a volcanic vent, it is a liquid at temperatures from 700 °C to 1,200 °C.
Oceanic crust	Oceanic crust is the part of Earth's lithosphere that surfaces in the ocean basins. Oceanic crust is primarily composed of mafic rocks, or sima. It is thinner than continental crust, or sial, generally less than 10 kilometers thick, however it is more dense, having a mean density of about 3.3 grams per cubic centimeter.
Indian Ocean	The Indian Ocean is the third largest of the world's oceanic divisions, covering about 20% of the Earth's water surface. It is bounded on the north by Asia on the west by Africa; on the east by the Malay Peninsula, the Sunda Islands, and Australia; and on the south by the Southern Ocean.
Gulf of Suez	The northern end of the Red Sea is bifurcated by the Sinai Peninsula, creating the Gulf of Suez It is a relatively young rift basin, dating back 40 million years.
Fault	Faults are planar rock fractures, which show evidence of relative movement. Large faults within the Earth's crust are the result of shear motion and active fault zones are the causal locations of most earthquakes. Earthquakes are caused by energy release during rapid slippage along faults. The largest examples are at tectonic plate boundaries but many faults occur far from active plate boundaries. Since faults do not usually consist of a single, clean fracture, the term fault zone is used when referring to the zone of complex deformation that is associated with the fault plane.
Oceanic ridge	A oceanic ridge is an underwater mountain range, formed by plate tectonics. This uplifting of the ocean floor occurs when convection currents rise in the mantle beneath the oceanic crust and create magma where two tectonic plates meet at a divergent boundary.

Transform boundary	In plate tectonics, a transform boundary is said to occur when tectonic plates slide and grind against each other along a transform fault. The relative motion of such plates is horizontal in either sinistral or dextral direction. Many transform boundaries are locked in tension before suddenly releasing, and causing earthquakes.
San Andreas fault	The San Andreas Fault is a geological fault that runs a length of roughly 800 miles through western and southern California in the United States. The fault, a right-lateral strike-slip fault, marks a transform boundary between the Pacific Plate and the North American Plate.
Pacific plate	The Pacific Plate is an oceanic tectonic plate beneath the Pacific Ocean.
North American Plate	The North American Plate is a tectonic plate covering most of North America, extending eastward to the Mid-Atlantic Ridge and westward to the Cherskiy Range in East Siberia.
Alcatraz Island	Alcatraz Island is a small island located in the middle of San Francisco Bay in California, United States. It served as a lighthouse, then a military fortification, then a military prison followed by a federal prison until 1963, when it became a national recreation area.
Alvin	Alvin is a 16-ton, manned deep-ocean research submersible owned by the United States Navy and operated by the Woods Hole Oceanographic Institution in Woods Hole, Massachusetts. The three-person vessel allows for two scientists and one pilot to dive for up to nine hours at 4500 metersor 15,000 feet.
Compression	In geology the term compression refers to the system of forces that tend to decrease the volume of or shorten rocks. Compressive strength refers to the maximum compressive stress that can be applied to a material before failure occurs.
Transform fault	A transform fault is a geological fault that is a special case of strike-slip faulting which terminates abruptly, at both ends, at a major transverse geological feature. Also known as a conservative plate boundary.
Butte	A butte is an isolated hill with steep sides and a small flat top, smaller than mesas and plateaus. Buttes are prevalent in the western United States and on the Hawaiian Islands, especially around Honolulu.
Chalk	Chalk is a soft, white, porous sedimentary rock, a form of limestone composed of the mineral calcite. It forms under relatively deep marine conditions from the gradual accumulation of minute calcite plates shed from micro-organisms called coccolithophores. It is common to find flint nodules embedded in it.
Volcanic arc	A volcanic arc is a chain of volcanic islands or mountains formed by plate tectonics as an oceanic tectonic plate subducts under another tectonic plate and produces magma.
Continental shelf	The continental shelf is the extended perimeter of each continent and associated coastal plain, which is covered during interglacial periods such as the current epoch by relatively shallow seas and gulfs. The shelf usually ends at a point of increasing slope.
Pacific Ocean	The Pacific Ocean is the largest of the Earth's oceanic divisions. It extends from the Arctic in the north to the Antarctic in the south, bounded by Asia and Australia on the west and the Americas on the east. At 169.2 million square kilometres in area, this largest division of the World Ocean – and, in turn, the hydrosphere – covers about 46% of the Earth's water surface and about 32% of its total surface area, making it larger than all of the Earth's land area combined.
Andes Mountains	The Andes Mountains are South America's longest mountain range, forming a continuous chain of highland along the western coast of South America.
Sedimentary	Sedimentary rock is one of the three main rock groups. Rock formed from these covers 75% of the Earth's land area, and includes common types such as chalk, limestone, dolomite, sandstone, and shale.

Go to **Cram101.com** for the Practice Tests for this Chapter.

Sedimentary rock	Sedimentary rock is one of the three main rock groups. Sedimentary rock covers 75% of the Earth's land area. Four basic processes are involved in the formation of a clastic sedimentary rock: weathering caused mainly by friction of waves, transportation where the sediment is carried along by a current, deposition and compaction where the sediment is squashed together to form a rock of this kind.
Nazca plate	The Nazca Plate, named after the Nazca region of southern Peru, is an oceanic tectonic plate in the eastern Pacific Ocean basin off the west coast of South America.
Rocky Mountains	The Rocky Mountains are a broad mountain range in western North America. The Rocky Mountains stretch more than 4,800 kilometers from northernmost British Columbia, in Canada, to New Mexico, in the United States.
Yellowstone National Park	The Yellowstone National Park is the centerpiece of the Greater Yellowstone Ecosystem, the largest intact ecosystem in the Earth's northern temperate zone. Located mostly in the U.S. state of Wyoming, the park extends into Montana and Idaho. The park is known for its wildlife and geothermal features; Old Faithful Geyser is one of the most popular features in the park.
Geyser	A geyser is a type of hot spring that erupts periodically, ejecting a column of hot water and steam into the air.
Juan de Fuca	Juan de Fuca was a Greek captain employed by Spain to sail northward from Mexico and look for a northern passage from the Pacific Ocean to the Atlantic Ocean. In 1592, his exploration took him into the body of water, the Strait of Juan de Fuca.
Juan de Fuca Ridge	The Juan de Fuca Ridge is a tectonic spreading center located off the coasts of the state of Washington in the United States and the province of British Columbia in Canada. It runs northward from a transform boundary, the Blanco Fracture Zone, to a triple junction with the Nootka Fault and the Sovanco Fracture Zone.
Sea of Cortez	The Sea of Cortez is a body of water that separates the Baja California Peninsula from the Mexican mainland. It is bordered by the states of Baja California, Baja California Sur, Sonora, and Sinaloa.
Thermal	A thermal column is a column of rizing air in the lower altitudes of the Earth face=symbol>¢s atmosphere. Thermals are created by the uneven heating of the Earth's surface from solar radiation, and are an example of convection. The Sun warms the ground, which in turn warms the air directly above it.
Thermal energy	In thermal physics, thermal energy is the energy portion of a system that increases with its temperature. In thermodynamics, thermal energy is the internal energy present in a system in a state of thermodynamic equilibrium by virtue of its temperature.
Geothermal	In geology, geothermal refers to heat sources within the planet. The planet face=symbol>¢s internal heat was originally generated during its accretion, due to gravitational binding energy, and since then additional heat has continued to be generated by the radioactive decay of elements such as uranium, thorium, and potassium.
Geothermal gradient	The geothermal gradient is the rate of increase in temperature per unit depth in the Earth. It varies with location and is typically measured by determining the bottom open-hole temperature after the drilling of a borehole.
Uranium	Uranium is approximately 70% more dense than lead and is weakly radioactive. It occurs naturally in low concentrations in soil, rock and water.
Radioactive decay	Radioactive decay is the process in which an unstable atomic nucleus loses energy by emitting radiation in the form of particles or electromagnetic waves.
Matter	Matter is the substance of which physical objects are composed. Matter can be solid, liquid,

	plasma or gas. It constitutes the observable universe.
Igneous	Igneous rocks form when molten rock, magma, cools and solidifies, with or without crystallization, either below the surface as intrusive, plutonic rocks or on the surface as extrusive, volcanic, rocks.
Anomalous	An anomalous phenomenon is an observed event which deviates from what is expected according to existing rules or scientific theory.
Fold	The term fold is used in geology when one or a stack of originally flat and planar surfaces, such as sedimentary strata, are bent or curved as a result of plastic, i.e. permanent, deformation.
Volcanic cone	A volcanic cone is among the simplest volcano formations in the world. They are built by fragments thrown up from a volcanic vent, piling up around the vent in the shape of a cone with a central crater. A volcanic cone is of different types, depending upon the nature and size of the fragments ejected during the eruption. Types typically differentiated are spatter cone, cinder cone, ash cone, and tuff cone.
Arctic	The Arctic is the region around the Earth's North Pole, opposite the Antarctic region around the South Pole. In the northern hemisphere, the Arctic includes the Arctic Ocean and parts of Canada, Greenland, Russia, the United States, Iceland, Norway, Sweden and Finland. The word Arctic comes from the Greek word arktos, which means bear. This is due to the location of the constellation Ursa Major, the "Great Bear", above the Arctic region.
South American plate	The South American Plate is a tectonic plate covering the continent of South America and extending eastward to the Mid-Atlantic Ridge. The easterly side is a divergent boundary with the African Plate forming the southern part of the Mid-Atlantic Ridge. The southerly side is a complex boundary with the Antarctic Plate and the Scotia Plate.
African plate	The African Plate is a tectonic plate covering the continent of Africa and extending westward to the Mid-Atlantic Ridge.
Southern Ocean	The Southern Ocean is the oceanic division completely in Earth's southern hemisphere encircling Antarctica, comprizing the southernmost waters of the World Ocean south of 60° S latitude. However, the Southern Ocean face=symbol>¢s northern boundary is not precise. Instead, the Antarctic Convergence separates the Southern Ocean from other oceans.This dynamic, natural boundary is formed by the convergence of two circumpolar currents.
Hot spots	Hot spots is a location on the Earth's surface that has experienced active volcanism for a long period of time. J. Tuzo Wilson came up with the idea in 1963 that volcanic chains like the Hawaiian Islands result from the slow movement of a tectonic plate across a "fixed" hot spot deep beneath the surface of the planet.
Isostasy	Isostasy is a term used in Geology to refer to the state of gravitational equilibrium between the Earth's lithosphere and asthenosphere such that the tectonic plates "float" at an elevation which depends on their thickness and density. It is invoked to explain how different topographic heights can exist at the Earth's surface.
Universe	The Universe is defined as the summation of all particles and energy that exist and the space-time in which all events occur.
Solar system	The Solar System consists of the Sun and the other celestial objects gravitationally bound to it: the eight planets, their 165 known moons, three currently identified dwarf planets and their four known moons, and billions of small bodies.

Go to **Cram101.com** for the Practice Tests for this Chapter.

Meteorite	A meteorite is a natural object originating in outer space that survives an impact with the Earth's surface without being destroyed. While in space it is called a meteoroid. When it enters the atmosphere, air resistance causes the body to heat up and emit light, thus forming a fireball.
Post glacial rebound	Post glacial rebound is the rise of land masses that were depressed by the huge weight of ice sheets during the last ice age, through a process known as isostatic depression. It affects northern Europe, especially Scotland and Scandinavia, Siberia, Canada, and the Great Lakes of Canada and the United States.
Post-glacial rebound	Post-glacial rebound is the rise of land masses that were depressed by the huge weight of ice sheets during the last ice age, through a process known as isostatic depression.
Subsidence	In geology, engineering, and surveying, subsidence is the motion of a surface as it shifts downward relative to a datum such as sea-level. The opposite of subsidence is uplift, which results in an increase in elevation. In meteorology, subsidence refers to the downward movement of air.
Lake Mead	Lake Mead is the largest man-made lake and reservoir in the United States. It is located on the Colorado River about 30 miles southeast of Las Vegas, Nevada, in the states of Nevada and Arizona. Formed by water impounded by Hoover Dam, it extends 110 mi behind the dam, holding approximately 28.5 million acre feet of water. The water held in Lake Mead is released to communities in southern California, via aqueducts, and Nevada.
Dam	A dam is a barrier across flowing water that obstructs, directs or slows down the flow, often creating a reservoir, lake or impoundment.
Greenland	Greenland is a self-governed Danish territory lying between the Arctic and Atlantic Oceans. Though geographically and ethnically an Arctic island nation associated with the continent of North America, politically and historically Greenland is closely tied to Europe. It is the largest island in the world that is not also considered a continent.
Baltic Sea	The Baltic Sea is located in Northern Europe. It is bounded by the Scandinavian Peninsula, the mainland of Europe, and the Danish islands. It drains into the Kattegat by way of the Øresund, the Great Belt and the Little Belt. The Kattegat continues through the Skagerrak into the North Sea and the Atlantic Ocean.
Great Salt Lake	Great Salt Lake, located in the northern part of the U.S. state of Utah, is the largest salt lake in the Western Hemisphere, the fourth-largest terminal lake in the world, and the 33rd largest lake on Earth.
Tropics	The tropics are the geographic region of the Earth centered on the equator and limited in latitude by the Tropic of Cancer in the northern hemisphere, at approximately 23 class="unicode">°30′ N latitude, and the Tropic of Capricorn in the southern hemisphere at 23°30′ S latitude.
Biosphere	The biosphere is the part of the earth, including air, land, surface rocks, and water, within which life occurs, and which biotic processes in turn alter or transform. From the broadest biophysiological point of view, the biosphere is the global ecological system integrating all living beings and their relationships, including their interaction with the elements of the lithosphere, hydrosphere, and atmosphere. This biosphere is postulated to have evolved, beginning through a process of biogenesis or biopoesis, at least some 3.5 billion years ago.
Convection cell	A convection cell is a phenomenon of fluid dynamics which occurs in situations where there are temperature differences within a body of liquid or gas.
South Pole	The South Pole is the southernmost point on the surface of the Earth, on the opposite side of the Earth from the North Pole. It is the site of the US Amundsen-Scott South Pole Station,

which was established in 1956 and has been permanently staffed since that date

Storm	A storm is any disturbed state of an astronomical body's atmosphere, especially affecting its surface, and strongly implying severe weather. It may be marked by strong wind, thunder and lightning, heavy precipitation, such as ice, or wind transporting some substance through the atmosphere.
Solar radiation	Solar radiation is radiant energy emitted by the sun from a nuclear fusion reaction that creates electromagnetic energy. The spectrum of solar radiation is close to that of a black body with a temperature of about 5800 K. About half of the radiation is in the visible short-wave part of the electromagnetic spectrum. The other half is mostly in the near-infrared part, with some in the ultraviolet part of the spectrum.
Rotation	A rotation is a movement of an object in a circular motion. A two-dimensional object rotates around a center of rotation. A three-dimensional object rotates around a line called an axis. A circular motion about an external point, e.g. the Earth about the Sun, is called an orbit or more properly an orbital revolution.
Radiation	Radiation as used in physics, is energy in the form of waves or moving subatomic particles.
Polar ice cap	A polar ice cap is a high-latitude region of a planet or moon that is covered in ice. There are no requirements with respect to size or composition for a body of ice to be termed a polar ice cap, nor any geological requirement for it to be over land; only that it must be a body of solid phase matter in the polar region.
Ice cap	An ice cap is a dome-shaped ice mass that covers less than 50,000 km² of land area. Masses of ice covering more than 50,000 km² are termed an ice sheet.
Continental drift	Continental drift refers to the movement of the Earth's continents relative to each other. Continental drift is a concept that said the shapes of continents on either side of the Atlantic Ocean seem to fit together and the similarity of southern continent fossil faunae could mean that all the continents had once been joined into a supercontinent. It was suggested that the continents had been pulled apart by the centrifugal pseudoforce of the Earth's rotation.
Hydrologic cycle	The Earth's water is always in movement, and the hydrologic cycle, describes the continuous movement of water on, above, and below the surface of the Earth. Since the hydrologic cycle is truly a "cycle," there is no beginning or end. Water can change states among liquid, vapor, and ice at various places in the hydrologic cycle, with these processes happening in the blink of an eye and over millions of years. Although the balance of water on Earth remains fairly constant over time, individual water molecules can come and go in a hurry.
Global change	Global change is the term used to encompass a multitude of environmental and ecological changes that have been noticed, measured and studied on Earth. It encompasses the study of climate change, species extinction, land use change, changes in the carbon cycle and hydrologic cycle.

Mineral	A mineral is a naturally occurring substance formed through geological processes that has a characteristic chemical composition, a highly ordered atomic structure and specific physical properties. A rock, by comparison, is an aggregate of minerals and need not have a specific chemical composition. Minerals range in composition from pure elements and simple salts to very complex silicates with thousands of known forms.
Crystal	A crystal is a solid in which the constituent atoms, molecules, or ions are packed in a regularly ordered, repeating pattern extending in all three spatial dimensions. Most metals encountered in everyday life are polycrystals. Crystals are often symmetrically intergrown to form crystal twins.
Ion	An ion is an atom or group of atoms which have lost or gained one or more electrons, making them negatively or positively charged.
Molecule	In chemistry, a molecule is defined as a sufficiently stable electrically neutral group of at least two atoms in a definite arrangement held together by strong chemical bonds.
Atoms	Atoms are the fundamental building blocks of chemistry, and are conserved in chemical reactions.
Matter	Matter is the substance of which physical objects are composed. Matter can be solid, liquid, plasma or gas. It constitutes the observable universe.
Weather	The weather is the set of all extant phenomena in a given atmosphere at a given time. The term usually refers to the activity of these phenomena over short periods, as opposed to the term climate, which refers to the average atmospheric conditions over longer periods of time.
Sediment	Sediment is any particulate matter that can be transported by fluid flow and which eventually is deposited as a layer of solid particles on the bed or bottom of a body of water or other liquid.
Crystal structure	A crystal structure is a unique arrangement of atoms in a crystal. It is composed of a unit cell, a set of atoms arranged in a particular way, which is periodically repeated in three dimensions on a lattice. The spacing between unit cells in various directions is called its lattice parameters. The symmetry properties of the crystal are embodied in its space group.
Tectonics	Tectonics is a field of study within geology concerned generally with the structures within the crust of the Earth, or other planets, and particularly with the forces and movements that have operated in a region to create these structures.
Continent	A continent is one of several large landmasses on Earth. They are generally identified by convention rather than any strict criteria, but seven areas are commonly reckoned as continents – they are: Asia, Africa, North America, South America, Antarctica, Europe, and Australia.
Element	An element is a type of atom that is defined by its atomic number; that is, by the number of protons in its nucleus.
Neutron	In physics, the neutron is a subatomic particle with no net electric charge.
Proton	In physics, the proton is a subatomic particle with an electric charge of one positive fundamental unit a diameter of about 1.5×10 class="unicode">-15 m, and a mass of 938.27231(28) MeV/c2 (1.6726 class="unicode">$\times 10^{-27}$ kg), 1.007 276 466 88(13) u or about 1836 times the mass of an electron.
Electron	The electron is a fundamental subatomic particle that carries a negative electric charge.
Isotopes	Isotopes are any of the several different forms of an element each having different atomic mass. Isotopes of an element have nuclei with the same number of protons but different

Go to **Cram101.com** for the Practice Tests for this Chapter.

numbers of neutrons.

Silicate

In geology and astronomy, the term silicate is used to denote types of rock that consist predominantly of silicate minerals. Such rocks include a wide range of igneous, metamorphic and sedimentary types. Most of the Earth's mantle and crust are made up of silicate rocks. The same is true of the Moon and the other rocky planets.

Hardness

Hardness is the characteristic of a solid material expressing its resistance to permanent deformation.

Crust

In geology, a crust is the outermost layer of a planet, part of its lithosphere. They are generally composed of a less dense material than its deeper layers.Earths face=symbol>¢ is composed mainly of basalt and granite. It is cooler and more rigid than the deeper layers of the mantle and core.

Cleavage

Cleavage, in mineralogy, is the tendency of crystalline materials to split along definite planes, creating smooth surfaces.

Mica

The mica group of sheet silicate minerals includes several closely related materials having highly perfect basal cleavage. All are monoclinic with a tendency towards pseudo-hexagonal crystals and are similar in chemical composition. The highly perfect cleavage, which is the most prominent characteristic of mica, is explained by the hexagonal sheet-like arrangement of its atoms.

Olivine

The mineral olivine is a magnesium iron silicate. It is one of the most common minerals on Earth, and has also been identified on the Moon, Mars, and comet Wild 2.

Quartz

Quartz is the second most common mineral in the Earth's continental crust. It is made up of a lattice of silica tetrahedra. Quartz belongs to the rhombohedral crystal system. In nature quartz crystals are often twinned, distorted, or so intergrown with adjacent crystals of quartz or other minerals as to only show part of this shape, or to lack obvious crystal faces altogether and appear massive.

Feldspar

Feldspar is the name of a group of rock-forming minerals which make up as much as sixty percent of the Earth's crust. Feldspars crystallize from magma in both intrusive and extrusive rocks, and they can also occur as compact minerals, as veins, and are also present in many types of metamorphic rock.

Amphibole

Amphibole defines an important group of generally dark-colored rock-forming inosilicate minerals linked at the vertices and generally containing ions of iron and/or magnesium in their structures. Amphiboles crystallize into two crystal systems, monoclinic and orthorhombic.

Clay

Clay is a term used to describe a group of hydrous aluminium phyllosilicate minerals, that are typically less than 2 micrometres in diameter. Clay consists of a variety of phyllosilicate minerals rich in silicon and aluminium oxides and hydroxides which include variable amounts of structural water. Clays are generally formed by the chemical weathering of silicate-bearing rocks by carbonic acid but some are formed by hydrothermal activity.

Clay minerals

Clay minerals are hydrous aluminium phyllosilicates, sometimes with variable amounts of iron, magnesium, alkali metals, alkaline earths and other cations. Clays have structures similar to the micas and therefore form flat hexagonal sheets. Clay minerals are common weathering products and low temperature hydrothermal alteration products.

Calcite

The carbonate mineral Calcite is a chemical or biochemical calcium carbonate and is one of the most widely distributed minerals on the Earth's surface. It is a common constituent of sedimentary rocks, limestone in particular. It is also the primary mineral in metamorphic marble

Go to **Cram101.com** for the Practice Tests for this Chapter.

Gypsum	Gypsum is a very soft mineral composed of calcium sulfate dihydrate, with the chemical formula CaSO4·2H2O. Gypsum occurs in nature as flattened and often twinned crystals and transparent cleavable masses. It may also occur silky and fibrous. Finally it may also be granular or quite compact.
Halite	Halite is the mineral form of sodium chloride. Halite forms isometric crystals. It commonly occurs with other evaporite deposit minerals such as several of the sulfates, halides and borates. Halite occurs in vast lakes of sedimentary evaporite minerals that result from the drying up of enclosed beds, playas, and seas.
Dolomite	Dolomite is the name of a sedimentary carbonate rock and a mineral, both composed of calcium magnesium carbonate found in crystals. Dolomite rock is composed predominantly of the mineral dolomite. Limestone that is partially replaced by dolomite is referred to as dolomitic limestone.
Physical property	A physical property is any aspect of an object or substance that can be measured or perceived without changing its identity. Physical properties can be intensive or extensive. An intensive property does not depend on the size or amount of matter in the object, while an extensive property does.
Chemical formula	A chemical formula is a concise way of expressing information about the atoms that constitute a particular chemical compound. A chemical formula is also a short way of showing how a chemical reaction occurs. For molecular compounds, it identifies each constituent element by its chemical symbol and indicates the number of atoms of each element found in each discrete molecule of that compound.
Atomic mass	The atomic mass is the mass of an atom at rest, most often expressed in unified atomic mass units.[
Atomic number	In chemistry and physics, the atomic number is the number of protons found in the nucleus of an atom. It is traditionally represented by the symbol Z.
Hydrogen	Hydrogen is a chemical element represented by the symbol H and an atomic number of 1. At standard temperature and pressure it is a colorless, odorless, nonmetallic, tasteless, highly flammable diatomic gas . With an atomic mass of 1.00794 g/mol, hydrogen is the lightest element.Hydrogen is the most abundant of the chemical elements, constituting roughly 75% of the universe's elemental mass.
Iron	Iron is a chemical element metal. It is a lustrous, silvery soft metal. It and nickel are notable for being the final elements produced by stellar nucleosynthesis, and thus are the heaviest elements which do not require a supernova or similarly cataclysmic event for formation.
Radioactive decay	Radioactive decay is the process in which an unstable atomic nucleus loses energy by emitting radiation in the form of particles or electromagnetic waves.
Noble gas	The noble gas is the element in group 18 of the periodic table. It is also called helium family or neon family. Chemically, they are very stable due to having the maximum number of valence electrons their outer shell can hold. A thorough explanation requires an understanding of electronic configuration, with references to quantum mechanics.
Ionic bond	An ionic bond is a type of chemical bond based on electrostatic forces between two oppositely-charged ions. In this formation, a metal donates an electron, due to a low electronegativity to form a positive ion or cation.
Gold	Gold is a highly sought-after precious metal which, for many centuries, has been used as money, a store of value and in jewelery. The metal occurs as nuggets or grains in rocks, underground "veins" and in alluvial deposits. It is one of the coinage metals. Itis dense, soft, shiny and the most malleable and ductile of the known metals.

Metallic bond	Metallic bond is the bonding between atoms within metals. It involves the delocalized sharing of free electrons among a lattice of metal atoms. Thus, metallic bonds may be compared to molten salts. Metallic bonding is the electrostatic attraction between the metal atoms or ions and the delocalized electrons, also called conduction electrons.
Metallic bonding	Metallic bonding is the bonding between atoms within metals. It involves the delocalized sharing of free electrons among a lattice of metal atoms. Thus, they may be compared to molten salts.
Crystalline solids	Crystalline solids are a class of solids that have regular or nearly regular crystalline structures.
Amorphous	An amorphous solid is a solid in which there is no long-range order of the positions of the atoms.These materials are often prepared by rapidly cooling molten material, such as glass. The cooling reduces the mobility of the material's molecules before they can pack into a more thermodynamically favorable crystalline state.
Amorphous solid	An amorphous solid is a solid in which there is no long-range order of the positions of the atoms. Most classes of solid materials can be found or prepared in an amorphous form.
Melting	Melting is the process of heating a solid substance to a point where it turns into a liquid. An object that has melted is molten.
Rapid	A rapid is a section of a river of relatively steep gradient causing an increase in water flow and turbulence. A rapid is a hydrological feature between a run and a cascade. It is characterized by the river becoming shallower and having some rocks exposed above the flow surface.
Diffusion	Diffusion is the net action of matter, particles or molecules, heat, momentum, or light whose end is to minimize a concentration gradient.
Temperature range	Temperature range is the numerical difference between the minimum and maximum values of temperature observed in a system, such as atmospheric temperature in a given location. A temperature range may refer to a period of time or to an average.
State of matter	In the physical sciences, a state of matter is one of the many ways that matter can interact with itself to form a macroscopic, homogenous phase. The most familiar examples of states of matter are solids, liquids, and gases; the most common state of matter in the visible universe is plasma.
Weathering	Weathering is the process of breaking down rocks, soils and their minerals through direct contact with the atmosphere. Weathering occurs without movement. Two main classifications of weathering processes exist. Mechanical or physical weathering involves the breakdown of rocks and soils through direct contact with atmospheric conditions. The second classification, chemical weathering, involves the direct effect of atmospheric chemicals in the breakdown of rocks, soils and minerals.
Climate	Climate is the average and variations of weather over long periods of time. Climate zones can be defined using parameters such as temperature and rainfall.
Ice Age	An ice age is a period of long-term reduction in the temperature of Earth face=symbol>¢s climate, resulting in an expansion of the continental ice sheets, polar ice sheets and mountain glaciers .
Equilibrium	Equilibrium is the condition of a system in which competing influences are balanced.
Luster	Luster is a description of the way light interacts with the surface of a crystal, rock, or mineral. For example, a diamond is said to have an adamantine luster and pyrite is said to have a metallic luster.

Diamonds	Diamonds are the hardest natural material known to man and the third-hardest known material. Its hardness and high dispersion of light make it useful for industrial applications and jewelry.
Chemical compound	A chemical compound is a chemical substance of two or more different chemically bonded chemical elements, with a fixed ratio determining the composition. The ratio of each element is usually expressed by chemical formula.
Petroleum	Petroleum is a naturally occurring liquid found in formations in the Earth consisting of a complex mixture of hydrocarbons of various lengths.
Coal	Coal is a fossil fuel formed in swamp ecosystems where plant remains were saved by water and mud from oxidization and biodegradation. It is a sedimentary rock, but the harder forms, such as anthracite coal, can be regarded as metamorphic rocks because of later exposure to elevated temperature and pressure. It is composed primarily of carbon along with assorted other elements, including sulfur.
Organic compound	An organic compound is any member of a large class of chemical compounds whose molecules contain carbon.
Organism	In biology and ecology, an organism is a living complex adaptive system of organs that influence each other in such a way that they function in some way as a stable whole.
Gemstone	A gemstone is a highly attractive and valuable piece of mineral, which, when cut and polished, is used in jewelry or other adornments.
Island	An island is any piece of land that is completely surrounded by water, above high tide. There are two main types of islands: continental islands and oceanic islands. There are also artificial islands. A grouping of geographically and/or geologically related islands is called an archipelago.
Silver	Silver is a soft white lustrous transition metal, it has the highest electrical and thermal conductivity for a metal.
Sulfur	Sulfur or sulphur is the chemical element that has the symbol S and atomic number 16. It is an abundant, tasteless, multivalent non-metal. Sulfur, in its native form, is a yellow crystalline solid. In nature, it can be found as the pure element or as sulfide and sulfate minerals. It is an essential element for life and is found in two amino acids, cysteine and methionine.
Copper	Copper is a ductile metal with excellent electrical conductivity, and finds extensive use as an electrical conductor, heat conductor, as a building material, and as a component of various alloys.
Interference	Interference is the addition of two or more waves that results in a new wave pattern. As most commonly used, the term interference usually refers to the interaction of waves which are correlated or coherent with each other, either because they come from the same source or because they have the same or nearly the same frequency.
Carbon	Carbon is a chemical element. An abundant nonmetallic, tetravalent element, carbon has several allotropic forms. This element is the basis of the chemistry of all known life.
Graphite	Graphite is one of the allotropes of carbon. It holds the distinction of being the most stable form of solid carbon ever discovered. It may be considered to be the highest grade of coal, just above anthracite, although it is not normally used as fuel because it is hard to ignite.
Nicolaus Steno	Nicolaus Steno was a pioneer in both anatomy and geology. He first studied anatomy, beginning with a focus on the muscular system and the nature of muscle contraction. He used geometry to show that a contracting muscle changes its shape but not its volume.

Pyroxenes	The pyroxenes are a group of important rock-forming silicate minerals found in many igneous and metamorphic rocks. They share a common structure comprised of single chains of silica tetrahedra and they crystalise in the monoclinic and orthorhombic system.
Magnesium	Magnesium has the symbol Mg. It is the ninth most abundant element in the universe by mass. It constitutes about 2% of the Earth's crust by mass, and it is the third most abundant element dissolved in seawater. It is essential to all living cells, and is the 11th most abundant element by mass in the human body.
Nickel	Nickel is a silvery white metal that takes on a high polish. It belongs to the transition metals, and is hard and ductile. It occurs most usually in combination with sulfur and iron in pentlandite, with sulfur in millerite, with arsenic in the mineral niccolite, and with arsenic and sulfur.
Aluminum	Aluminum is a silvery and ductile member of the poor metal group of chemical elements. It has the symbol Al and atomic number 13.
Pyrite	The mineral pyrite is iron disulfide, FeS_2. It has isometric crystals that usually appear as cubes. Its metallic luster and pale-to-normal, brass-yellow hue have earned it a nickname due to many miners mistaking it for the real thing.
Conchoidal fracture	Conchoidal fracture describes the way that brittle materials break when they do not follow any natural planes of separation. Materials that break in this way include flint and other fine-grained minerals, as well as most amorphous solids, such as obsidian and other types of glass.
Facets	Facets are flat faces on geometric shapes. They reflect the underlying symmetry of the crystal structure. Gemstones commonly have them cut into them in order to improve their appearance.
Mohs scale of mineral hardness	The Mohs scale of mineral hardness characterizes the scratch resistance of various minerals through the ability of a harder material to scratch a softer material.
Abrasion	Abrasion is mechanical scraping of a rock surface by friction between rocks and moving particles during their transport in wind, glacier, waves, gravity or running water.
Impurities	Impurities are substances inside a confined amount of liquid, gas, or solid, which differ from the chemical composition of the material or compound.
Spectrum	A spectrum is a condition or value that is not limited to a specific set of values but can vary infinitely within a continuum. The word saw its first scientific use within the field of optics to describe the rainbow of colors in visible light when separated using a prism; it has since been applied by analogy to many fields.

Thermal	A thermal column is a column of rizing air in the lower altitudes of the Earth face=symbol>¢s atmosphere. Thermals are created by the uneven heating of the Earth's surface from solar radiation, and are an example of convection. The Sun warms the ground, which in turn warms the air directly above it.
Igneous	Igneous rocks form when molten rock, magma, cools and solidifies, with or without crystallization, either below the surface as intrusive, plutonic rocks or on the surface as extrusive, volcanic, rocks.
Igneous rock	Igneous rock forms when rock cools and solidifies either below the surface as intrusive rocks or on the surface as extrusive rocks. This magma can be derived from partial melts of pre-existing rocks in either the Earth's mantle or crust. Typically, the melting is caused by one or more of the following processes -- an increase in temperature, a decrease in pressure, or a change in composition.
Crust	In geology, a crust is the outermost layer of a planet, part of its lithosphere. They are generally composed of a less dense material than its deeper layers.Earths face=symbol>¢ is composed mainly of basalt and granite. It is cooler and more rigid than the deeper layers of the mantle and core.
Granite	Granite is a common and widely occurring type of intrusive, felsic, igneous rock. Granites are usually medium to coarsely crystalline, occasionally with some individual crystals larger than the groundmass forming a rock known as porphyry. Granites can be pink to dark gray or even black, depending on their chemistry and mineralogy.
Mountain	A mountain is a landform that extends above the surrounding terrain in a limited area. A mountain is generally steeper than a hill, but there is no universally accepted standard definition for the height of a mountain or a hill although a mountain usually has an identifiable summit.
Erosion	Erosion is displacement of solids by the agents of ocean currents, wind, water, or ice by downward or down-slope movement in response to gravity or by living organisms.
Volcano	A volcano is an opening, or rupture, in the Earth's surface or crust, which allows hot, molten rock, ash and gases to escape from deep below the surface.
Kilauea	Kilauea is an active volcano in the Hawaiian Islands, one of five shield volcanoes that together form the Island of Hawai i. In Hawaiian, the word Kilauea means "spewing" or "much spreading", in reference to the mountain's frequent outpouring of lava. It is presently the most active volcano and one of the most visited active volcanoes on the planet.
Lava	Lava is molten rock expelled by a volcano during an eruption. When first extruded from a volcanic vent, it is a liquid at temperatures from 700 °C to 1,200 °C.
Forest	A forest is an area with a high density of trees, historically, a wooded area set aside for hunting. These plant communities cover large areas of the globe and function as animal habitats, hydrologic flow modulators, and soil conservers, constituting one of the most important aspects of the Earth's biosphere.
Island	An island is any piece of land that is completely surrounded by water, above high tide. There are two main types of islands: continental islands and oceanic islands. There are also artificial islands. A grouping of geographically and/or geologically related islands is called an archipelago.
Magma	Magma is molten rock located beneath the surface of the Earth, and which often collects in a magma chamber. Magma is a complex high-temperature fluid substance. Most are silicate solutions. It is capable of intrusion into adjacent rocks or of extrusion onto the surface as lava or ejected explosively as tephra to form pyroclastic rock. Environments of magma

formation include subduction zones, continental rift zones, mid-oceanic ridges, and hotspots, some of which are interpreted as mantle plumes.

Stream
A stream is a body of water with a current, confined within a bed and banks. Streams are important as conduits in the water cycle, instruments in aquifer recharge, and corridors for fish and wildlife migration.

Mantle
Earth's mantle is a ~2,900 km thick rocky shell comprizing approximately 70% of Earth's volume. It is predominantly solid and overlies the Earth's iron-rich core, which occupies about 30% of Earth face=symbol>¢s volume. Past episodes of melting and volcanism at the shallower levels of the mantle have produced a very thin crust of crystallized melt products near the surface, upon which we live.

Melting
Melting is the process of heating a solid substance to a point where it turns into a liquid. An object that has melted is molten.

Pyroclastics
Pyroclastics are clastic rocks composed solely or primarily of volcanic materials.

Mafic
In geology, mafic minerals and rocks are silicate minerals, magmas, and volcanic and intrusive igneous rocks that have relatively high concentrations of the heavier elements. The term is a combination of "magnesium" and ferrum.

Quartz
Quartz is the second most common mineral in the Earth's continental crust. It is made up of a lattice of silica tetrahedra. Quartz belongs to the rhombohedral crystal system. In nature quartz crystals are often twinned, distorted, or so intergrown with adjacent crystals of quartz or other minerals as to only show part of this shape, or to lack obvious crystal faces altogether and appear massive.

Biotite
Biotite is a common phyllosilicate mineral within the mica group. Primarily a solid-solution series between the iron-endmember annite, and the magnesium-endmember phlogopite; more aluminous endmembers include siderophyllite.

Amphibole
Amphibole defines an important group of generally dark-colored rock-forming inosilicate minerals linked at the vertices and generally containing ions of iron and/or magnesium in their structures. Amphiboles crystallize into two crystal systems, monoclinic and orthorhombic.

Olivine
The mineral olivine is a magnesium iron silicate. It is one of the most common minerals on Earth, and has also been identified on the Moon, Mars, and comet Wild 2.

Pyroxenes
The pyroxenes are a group of important rock-forming silicate minerals found in many igneous and metamorphic rocks. They share a common structure comprised of single chains of silica tetrahedra and they crystalise in the monoclinic and orthorhombic system.

Gabbro
Gabbro is a dark, coarse-grained, intrusive igneous rock chemically equivalent to basalt. It is a plutonic rock, formed when molten magma is trapped beneath the Earth face=symbol>¢s surface and cools into a crystalline mass.

Diorite
Diorite is a grey to dark grey intermediate intrusive igneous rock composed principally of plagioclase feldspar, biotite, hornblende, and/or pyroxene. It may contain small amounts of quartz, microcline and olivine.

Shield volcano
A shield volcano is a large volcano with shallowly-sloping sides. A shield volcano is formed by lava flows of low viscosity — lava that flows easily. Consequently, a volcanic mountain having a broad profile is built up over time by flow after flow of relatively fluid basaltic lava issuing from vents or fissures on the surface of the volcano.

Extrusive
Extrusive refers to the mode of igneous volcanic rock formation in which hot magma from inside the Earth flows out onto the surface as lava or explodes violently into the atmosphere

to fall back as pyroclastics or tuff.

Basalt	Basalt is a common gray to black extrusive volcanic rock. It is usually fine-grained due to rapid cooling of lava on the Earth's surface. It may be porphyritic containing larger crystals in a fine matrix, or vesicular, or frothy scoria.
Cinder	A cinder is a fragment of cooled pyroclastic material, lava or magma.
Cinder cones	Cinder cones are steep, conical hills of volcanic fragments that accumulate around and downwind from a volcanic vent. The rock fragments, often called cinders are glassy and contain numerous gas bubbles "frozen" into place as magma exploded into the air and then cooled quickly.
Calderas	Calderas are volcanic features formed by the collapse of land following a volcanic eruption. They are often confused with volcanic craters.
Tuff	Tuff is a type of rock consisting of consolidated volcanic ash ejected from vents during a volcanic eruption.
Stratovolcano	A stratovolcano, is a tall, conical volcano composed of many layers of hardened lava, tephra, and volcanic ash. These volcanoes are characterized by a steep profile and periodic, explosive eruptions. The lava that flows from them is viscous, and cools and hardens before spreading very far.
Abundance	Abundance is an ecological concept referring to the relative representation of a species in a particular ecosystem. It is usually measured as the mean number of individuals found per sample.
Intrusion	An intrusion is a body of igneous rock that has crystallized from a molten magma below the surface of the Earth.
Pluton	A pluton in geology is an intrusive igneous rock body that crystallized from a magma below the surface of the Earth. Plutons include batholiths, dikes, sills, laccoliths, lopoliths, and other igneous bodies. In practice, "pluton" usually refers to a distinctive mass of igneous rock, typically kilometers in dimension, without a tabular shape like those of dikes and sills.
Batholiths	Batholiths are large emplacements of igneous intrusive rock that forms from cooled magma deep in the Earth's crust. They are almost always made mostly of felsic or intermediate rock-types, such as granite, quartz monzonite, or diorite.
Sill	In geology, a sill is a tabular pluton that has intruded between older layers of sedimentary rock, beds of volcanic lava or tuff, or even along the direction of foliation in metamorphic rock. The term sill is synonymous with concordant intrusive sheet. This means that the sill does not cut across preexisting rocks. Contrast this with dikes.
Dike	A dike is an intrusion into a cross-cutting fissure, meaning a dike cuts across other pre-existing layers or bodies of rock, this means that a dike is always younger than the rocks that contain it. The thickness is usually much smaller than the other two dimensions. Thickness can vary from sub-centimeter scale to many meters in thickness and the lateral dimensions can extend over many kilometers.
Crystallization	Crystallization is the process of formation of solid crystals from a uniform solution. It is also a chemical solid-liquid separation technique, in which mass transfer of a solute from the liquid solution to a pure solid crystalline phase occurs.
Mantle plume	A mantle plume is an upwelling of abnormally hot rock within the Earth face=symbol>¢s mantle. As the heads of mantle plumes can partly melt when they reach shallow depths, they are thought to be the cause of volcanic centers known as hotspots and probably also to have caused flood basalts.

Go to **Cram101.com** for the Practice Tests for this Chapter.

Rift	In geology, a rift is a place where the Earth's crust and lithosphere are being pulled apart.
Continental crust	The continental crust is the layer of granitic, sedimentary, and metamorphic rocks which form the continents and the areas of shallow seabed close to their shores, known as continental shelves. It is less dense than the material of the Earth's mantle and thus "floats" on top of it. Continental crust is also less dense than oceanic crust, though it is considerably thicker. About 40% of the Earth's surface is now underlain by continental crust.
Crystal	A crystal is a solid in which the constituent atoms, molecules, or ions are packed in a regularly ordered, repeating pattern extending in all three spatial dimensions. Most metals encountered in everyday life are polycrystals. Crystals are often symmetrically intergrown to form crystal twins.
Mineral	A mineral is a naturally occurring substance formed through geological processes that has a characteristic chemical composition, a highly ordered atomic structure and specific physical properties. A rock, by comparison, is an aggregate of minerals and need not have a specific chemical composition. Minerals range in composition from pure elements and simple salts to very complex silicates with thousands of known forms.
Buoyancy	In physics, buoyancy is the upward force on an object produced by the surrounding fluid in which it is fully or partially immersed, due to the pressure difference of the fluid between the top and bottom of the object. The net upward buoyancy force is equal to the magnitude of the weight of fluid displaced by the body.
Dome	In geology, a dome is a deformational feature consisting of symmetrically-dipping anticlines; their general outline on a geologic map is circular or oval.
Alcatraz Island	Alcatraz Island is a small island located in the middle of San Francisco Bay in California, United States. It served as a lighthouse, then a military fortification, then a military prison followed by a federal prison until 1963, when it became a national recreation area.
Equilibrium	Equilibrium is the condition of a system in which competing influences are balanced.
Silicate	In geology and astronomy, the term silicate is used to denote types of rock that consist predominantly of silicate minerals. Such rocks include a wide range of igneous, metamorphic and sedimentary types. Most of the Earth's mantle and crust are made up of silicate rocks. The same is true of the Moon and the other rocky planets.
Terrestrial	Terrestrial refers to things having to do with the land or with the planet Earth.
Magnesium	Magnesium has the symbol Mg. It is the ninth most abundant element in the universe by mass. It constitutes about 2% of the Earth's crust by mass, and it is the third most abundant element dissolved in seawater. It is essential to all living cells, and is the 11th most abundant element by mass in the human body.
Iron	Iron is a chemical element metal. It is a lustrous, silvery soft metal. It and nickel are notable for being the final elements produced by stellar nucleosynthesis, and thus are the heaviest elements which do not require a supernova or similarly cataclysmic event for formation.
Physical property	A physical property is any aspect of an object or substance that can be measured or perceived without changing its identity. Physical properties can be intensive or extensive. An intensive property does not depend on the size or amount of matter in the object, while an extensive property does.
Viscosity	Viscosity is a measure of the resistance of a fluid to deform under shear stress. It is commonly perceived as "thickness", or resistance to flow. Viscosity describes a fluid

Go to **Cram101.com** for the Practice Tests for this Chapter.

	face=symbol>¢s internal resistance to flow and may be thought of as a measure of fluid friction.
Potassium	Potassium is a chemical element. It is a soft silvery-white metallic alkali metal that occurs naturally bound to other elements in seawater and many minerals. It oxidizes rapidly in air and is very reactive, especially towards water. In many respects, it and sodium are chemically similar, although organisms in general, and animal cells in particular, treat them very differently.
Aluminum	Aluminum is a silvery and ductile member of the poor metal group of chemical elements. It has the symbol Al and atomic number 13.
Element	An element is a type of atom that is defined by its atomic number; that is, by the number of protons in its nucleus.
Ash fall	Ash fall consists of very fine rock and mineral particles less than 2 mm in diameter that are ejected from a volcanic vent. The very fine particles may be carried for many miles, settling out as a dust-like layer across the landscape
Pumice	Pumice is a highly vesicular pyroclastic extrusive igneous rock of intermediate to siliceous magmas including rhyolite, trachyte and phonolite. Pumice is usually light in color ranging from white, yellowish, gray, gray brown, and a dull red. Most of the time, it is white. As an extrusive rock it was made from a volcanic eruption.
Ion	An ion is an atom or group of atoms which have lost or gained one or more electrons, making them negatively or positively charged.
Feldspar	Feldspar is the name of a group of rock-forming minerals which make up as much as sixty percent of the Earth's crust. Feldspars crystallize from magma in both intrusive and extrusive rocks, and they can also occur as compact minerals, as veins, and are also present in many types of metamorphic rock.
Vesicle	In cell biology, a vesicle is a relatively small and enclosed compartment, separated from the cytosol by at least one lipid bilayer.
Planet	A planet, as defined by the International Astronomical Union, is a celestial body orbiting a star or stellar remnant that is massive enough to be rounded by its own gravity, not massive enough to cause thermonuclear fusion in its core, and has cleared its neighboring region of planetesimals.
Vesicular texture	Vesicular texture is a volcanic rock texture characterized by, or containing, many vesicles. The texture is often found in extrusive aphanitic, or glassy, igneous rock. The vesicles are small cavities formed by the expansion of bubbles of gas or steam during the solidification of the rock.
Electron	The electron is a fundamental subatomic particle that carries a negative electric charge.
Oxide	An oxide is a chemical compound containing an oxygen atom and other elements. Most of the earth's crust consists of them. They result when elements are oxidized by air.
Pyrite	The mineral pyrite is iron disulfide, FeS_2. It has isometric crystals that usually appear as cubes. Its metallic luster and pale-to-normal, brass-yellow hue have earned it a nickname due to many miners mistaking it for the real thing.
Wavelength	In physics, wavelength is the distance between repeating units of a propagating wave of a given frequency. It is commonly designated by the Greek letter lambda. Examples of wave-like phenonomena are light, water waves, and sound waves. Wavelength of a sine wave.In a wave, a property varies with the position.

Metamorphism	Metamorphism can be defined as the solid state recrystallisation of pre-existing rocks due to changes in heat and/or pressure and/or introduction of fluids. There will be mineralogical, chemical and crystallographic changes. Metamorphism produced with increasing pressure and temperature conditions is known as prograde metamorphism. Conversely, decreasing temperatures and pressure characterize retrograde metamorphism.
Sedimentary	Sedimentary rock is one of the three main rock groups. Rock formed from these covers 75% of the Earth's land area, and includes common types such as chalk, limestone, dolomite, sandstone, and shale.
Magnetism	Magnetism is one of the phenomena by which materials exert attractive or repulsive forces on other materials. Some well known materials that exhibit easily detectable magnetic properties are nickel, iron, some steels, and the mineral magnetite; however, all materials are influenced to greater or lesser degree by the presence of a magnetic field.
Plagioclase	Plagioclase is a very important series of tectosilicate minerals within the feldspar family. Rather than referring to a particular mineral with a specific chemical composition, it is a solid solution series.
Groundmass	Groundmass rock is the fine-grained mass of material in which larger grains or crystals are embedded. The groundmass of an igneous rock consists of fine-grained, often microscopic, crystals in which larger crystals are embedded. This porphyritic texture is indicative of multi-stage cooling of magma.
Andesite	Andesite is an igneous, volcanic rock, of intermediate composition, with aphanitic to porphyritic texture.
Peridotite	Peridotite is a dense, coarse-grained igneous rock, consisting mostly of the minerals olivine and pyroxene. Peridotite is ultramafic and ultrabasic, as the rock contains less than 45% silica. This type of rock is derived from the Earth's mantle, either as solid blocks and fragments, or as crystals accumulated from magmas that formed in the mantle.
Oceanic crust	Oceanic crust is the part of Earth's lithosphere that surfaces in the ocean basins. Oceanic crust is primarily composed of mafic rocks, or sima. It is thinner than continental crust, or sial, generally less than 10 kilometers thick, however it is more dense, having a mean density of about 3.3 grams per cubic centimeter.
Continent	A continent is one of several large landmasses on Earth. They are generally identified by convention rather than any strict criteria, but seven areas are commonly reckoned as continents – they are: Asia, Africa, North America, South America, Antarctica, Europe, and Australia.
Alps	The Alps is the name for one of the great mountain range systems of Europe, stretching from Austria and Slovenia in the east, through Italy, Switzerland, Liechtenstein and Germany to France in the west.
Atlantic Ocean	The Atlantic Ocean is the second-largest of the world's oceanic divisions; with a total area of about 106.4 million square kilometres , it covers approximately one-fifth of the Earth's surface. The Atlantic Ocean occupies an elongated, S-shaped basin extending longitudinally between the Americas to the west, and Eurasia and Africa to the east.
Rhyolite	Rhyolite is an igneous, volcanic rock, of felsic composition. It may have any texture from aphanitic to porphyritic. The mineral assemblage is usually quartz, alkali feldspar and plagioclase. Biotite and pyroxene are common accessory minerals.
Phenocryst	A phenocryst is a relatively large and usually conspicuous crystal distinctly larger than the grains of the rock groundmass of a porphyritic igneous rock. They often have euhedral forms

	either due to early growth within a magma or by post-emplacement recrystallization.
Silica	Silica is the oxide of silicon, chemical formula SiO_2, and is known for its hardness as early as the 16th century. It is a principle component in most types of glass and substances such as concrete.
Ridge	A ridge is a geological feature that is also known as a Rip in the earth causing magma to flow out and forming an undersea volcano, it also has geological features, a continuous elevational crest for some distance. Ridges are usually termed hills or mountains as well, depending on size.
Andes Mountains	The Andes Mountains are South America's longest mountain range, forming a continuous chain of highland along the western coast of South America.
Oceanic ridge	A oceanic ridge is an underwater mountain range, formed by plate tectonics. This uplifting of the ocean floor occurs when convection currents rise in the mantle beneath the oceanic crust and create magma where two tectonic plates meet at a divergent boundary.
Volcanic rock	Volcanic rock is an igneous rock of volcanic origin. They often have a vesicular texture, which is the result voids left by volatiles escaping from the molten lava. Pumice is a rock, which is an example of explosive volcanic eruption. It is so vesicular that it floats in water.
Seafloor	The seafloor the bottom of the ocean. At the bottom of the continental slope is the continental rise, which is caused by sediment cascading down the continental slope.
Pāhoehoe	A pāhoehoe flow typically advances as a series of small lobes and toes that continually break out from a cooled crust. Also forms lava tubes where the minimal heat loss maintains low viscosity.
Fold	The term fold is used in geology when one or a stack of originally flat and planar surfaces, such as sedimentary strata, are bent or curved as a result of plastic, i.e. permanent, deformation.
Spatter cone	A spatter cone is formed of molten lava ejected from a vent somewhat like taffy. Expanding gases in the lava fountains tear the liquid rock into irregular gobs that fall back to earth, forming a heap around the vent.
Tephra	Tephra is air-fall material produced by a volcanic eruption regardless of composition or fragment size. It is typically rhyolitic in composition as most explosive volcanoes are the product of the more viscous felsic or high silica magmas.
Volcanic bomb	A volcanic bomb is a globe of molten rock larger than 65 mm in diameter, formed when a volcano ejects viscous fragments of lava during an eruption. They cool into solid fragments before they reach the ground. Lava bombs can be thrown many kilometres from an erupting vent, and often acquire aerodynamic shapes during their flight.
Volcanic Bombs	Volcanic bombs are globes of molten rock larger than 65 mm in diameter, formed when a volcano ejects viscous fragments of lava during an eruption. They cool into solid fragments before they reach the ground.
Lake	A lake is a body of water or other liquid of considerable size contained on a body of land. A vast majority are fresh water, and lie in the Northern Hemisphere at higher latitudes. Most have a natural outflow in the form of a river or stream, but some do not, and lose water solely by evaporation and/or underground seepage.
Crater	A crater is an approximately circular depression in the surface of a planet, moon or other solid body in the Solar System, formed by the hyper-velocity impact of a smaller body with the surface. Impact craters typically have raised rims, and they range from small, simple, bowl-shaped depressions to large, complex, multi-ringed, impact basins.

Crater Lake	Crater Lake is a caldera lake in the U.S. state of Oregon. It is the main feature of Crater Lake National Park and famous for its deep blue color and water clarity. The lake partly fills a nearly 4,000 feet deep caldera that was formed around 5,677 by the collapse of the volcano Mount Mazama.
Wizard Island	Wizard Island is a volcanic cinder cone which forms an island at the west end of Crater Lake in Crater Lake National Park, Oregon. The Wizard Island was created after Mount Mazama, a large stratovolcano, erupted violently approximately 7700 years ago, forming the caldera which now contains Crater Lake.
Atmosphere	An atmosphere is a layer of gases that may surround a material body of sufficient mass. The gases are attracted by the gravity of the body, and are retained for a longer duration if gravity is high and the atmosphere's temperature is low. Some planets consist mainly of various gases, and thus have very deep atmospheres.
Velocity	In physics, velocity is defined as the rate of change of displacement or the rate of displacement. Simply put, it is distance per units of time.
Columnar jointing	Columnar jointing is when long fractures form vertically in rock as it cools and contracts.
Catastrophism	Catastrophism is the theory that Earth has been affected by sudden, short-lived, violent events that were sometimes worldwide in scope. The dominant paradigm of geology has been uniformitarianism, but recently a more inclusive and integrated view of geologic events has developed resulting in a gradual change in the scientific consensus, reflecting acceptance of some catastrophic events.
Landslide	A landslide is a geological phenomenon which includes a wide range of ground movement, such as rock falls, deep failure of slopes and shallow debris flows. Although gravity face=symbol>¢s action on an over-steepened slope is the primary reason for a landslide, there are other contributing factors affecting the original slope stability.
Earthquake	An earthquake is the result from the sudden release of stored energy in the Earth face=symbol>¢s crust that creates seismic waves. At the Earth face=symbol>¢s surface, earthquakes may manifest themselves by a shaking or displacement of the ground. An earthquake is caused by tectonic plates getting stuck and putting a strain on the ground. The strain becomes so great that rocks give way by breaking and sliding along fault planes.
Seismograph	A seismograph is used by seismologists to measure and record the size and force of seismic waves.
Water vapor	Water vapor is the gas phase of water. Water vapor is one state of the water cycle within the hydrosphere. Water vapor can be produced from the evaporation of liquid water or from the sublimation of ice. Under normal atmospheric conditions, water vapor is continuously evaporating and condensing.
Vapor	Vapor is the gas phase component of a another state of matter which does not completely fill its container. It is distinguished from the pure gas phase by the presence of the same substance in another state of matter. Hence when a liquid has completely evaporated, it is said that the system has been completely transformed to the gas phase.
Gravitation	Gravitation, in everyday life, is most familiar as the agency that endows objects with weight. Gravitation is responsible for keeping the Earth and the other planets in their orbits around the Sun; for the formation of tides; and for various other phenomena that we observe. Gravitation is also the reason for the very existence of the Earth, the Sun, and most macroscopic objects in the universe; without it, matter would not have coalesced into these large masses, and life, as we know it, would not exist.

Mono Lake	Mono Lake is an alkaline and hypersaline lake in California, United States that is a critical nesting habitat for several bird species and is an unusually productive ecosystem.
Mount Mazama	Mount Mazama is a destroyed stratovolcano in the Oregon part of the Cascade Volcanic Belt and the Cascade Range. The volcano's collapsed caldera holds Crater Lake, and the entire mountain is located in Crater Lake National Park.
Magma chamber	A magma chamber is a large underground pool of molten rock lying under the surface of the earth's crust. The molten rock in such a chamber is under great pressure, and given enough time and pressure can gradually fracture the rock around it creating outlets for the magma.
Sedimentary rock	Sedimentary rock is one of the three main rock groups. Sedimentary rock covers 75% of the Earth's land area. Four basic processes are involved in the formation of a clastic sedimentary rock: weathering caused mainly by friction of waves, transportation where the sediment is carried along by a current, deposition and compaction where the sediment is squashed together to form a rock of this kind.
Volcanic neck	A volcanic neck is a volcanic landform created when lava hardens within a vent on an active volcano. When forming, a plug can cause an extreme build-up of pressure if volatile-charged magma is trapped beneath it, and this can sometimes lead to an explosive eruption.
Batholith	A batholith is a large emplacement of igneous intrusive rock that forms from cooled magma deep in the Earth's crust. They are almost always made mostly of felsic or intermediate rock-types, such as granite, quartz monzonite, or diorite.
Discordant	A discordant coastline occurs where bands of differing rock type run perpendicular to the coast.
Anticline	An anticline is a fold that is convex up or to the youngest beds. Anticlines are usually recognized by a sequence of rock layers that are progressively older toward the center of the fold because the uplifted core of the fold is preferentially eroded to a deeper stratigraphic level relative to the topographically lower flanks. If an anticline plunges, the surface strata will form Vs that point in the direction of the plunge.
Reef	A reef is a rock, sandbar, or other feature lying beneath the surface of the water yet shallow enough to be a hazard to ships. They result from abiotic processes—deposition of sand, wave erosion planning down rock outcrops, and other natural processes.

Sedimentary	Sedimentary rock is one of the three main rock groups. Rock formed from these covers 75% of the Earth's land area, and includes common types such as chalk, limestone, dolomite, sandstone, and shale.
Sedimentary rock	Sedimentary rock is one of the three main rock groups. Sedimentary rock covers 75% of the Earth's land area. Four basic processes are involved in the formation of a clastic sedimentary rock: weathering caused mainly by friction of waves, transportation where the sediment is carried along by a current, deposition and compaction where the sediment is squashed together to form a rock of this kind.
Mountain	A mountain is a landform that extends above the surrounding terrain in a limited area. A mountain is generally steeper than a hill, but there is no universally accepted standard definition for the height of a mountain or a hill although a mountain usually has an identifiable summit.
Mountain range	A mountain range is a group of mountains bordered by lowlands or separated from other mountain ranges by passes or rivers. Individual mountains within the same mountain range do not necessarily have the same geology; they may be a mix of different orogeny, for example volcanoes, uplifted mountains or fold mountains and may, therefore, be of different rock.
Erosion	Erosion is displacement of solids by the agents of ocean currents, wind, water, or ice by downward or down-slope movement in response to gravity or by living organisms.
Mudstone	Mudstone is a fine-grained sedimentary rock whose original constituents were clays or muds. Grain size is up to 0.0625 mm with individual grains too small to be distinguished without a microscope.
Siltstone	Siltstone is a sedimentary rock which has a composition intermediate in grain size between the coarser sandstones and the finer mudstones and shales.
Shale	Shale is a fine-grained sedimentary rock whose original constituents were clays or muds. It is characterized by thin laminae breaking with an irregular curving fracture, often splintery and usually parallel to the often-indistinguishable bedding plane.
Ripple marks	In geology, ripple marks are sedimentary structures that indicate agitation by or wind.
Alcatraz Island	Alcatraz Island is a small island located in the middle of San Francisco Bay in California, United States. It served as a lighthouse, then a military fortification, then a military prison followed by a federal prison until 1963, when it became a national recreation area.
Cliff	A cliff is a significant vertical, or near vertical, rock exposure. Cliffs are categorized as erosion landforms due to the processes of erosion and weathering that produce them. Cliffs are common on coasts, in mountainous areas, escarpments and along rivers. Cliffs are usually formed by rock that is resistant to erosion and weathering.
Deposition	Deposition is the geological process whereby material is added to a landform. This is the process by which wind and water create a sediment deposit, through the laying down of granular material that has been eroded and transported from another geographical location.
Global change	Global change is the term used to encompass a multitude of environmental and ecological changes that have been noticed, measured and studied on Earth. It encompasses the study of climate change, species extinction, land use change, changes in the carbon cycle and hydrologic cycle.
Fossil	Fossils are the mineralized or otherwise preserved remains or traces of animals, plants, and other organisms. The totality of fossils, both discovered and undiscovered, and their placement in fossiliferous rock formations and sedimentary layers is known as the fossil record.
Delta	A delta is a landform where the mouth of a river flows into an ocean, sea, desert, estuary or

Go to **Cram101.com** for the Practice Tests for this Chapter.

	lake. It builds up sediment outwards into the flat area which the river face=symbol>¢s flow encounters transported by the water and set down as the currents slow.
Sediment	Sediment is any particulate matter that can be transported by fluid flow and which eventually is deposited as a layer of solid particles on the bed or bottom of a body of water or other liquid.
Flint	Flint is a hard, sedimentary cryptocrystalline silicate form of the mineral quartz, categorized as a variety of chalcedony and broadly part of the mineral group known as silicas. When struck against steel, it will produce sparks, which when directed onto tinder can be used to start a fire.
Chert	Chert is a fine-grained silica-rich cryptocrystalline sedimentary rock that may contain small fossils. It varies greatly in color from white to black, but most often manifests as gray, brown, grayish brown and light green to rusty red; its color is an expression of trace elements present in the rock, and both red and green are most often related to traces of iron.
Limestone	Limestone is a sedimentary rock composed largely of the mineral calcite. Limestone often contains variable amounts of silica in the form of chert or flint, as well as varying amounts of clay, silt and sand as disseminations, nodules, or layers within the rock. The primary source of the calcite in limestone is most commonly marine organisms. These organisms secrete shells that settle out of the water column and are deposited on ocean floors as pelagic ooze or alternatively is conglomerated in a coral reef.
Mineral	A mineral is a naturally occurring substance formed through geological processes that has a characteristic chemical composition, a highly ordered atomic structure and specific physical properties. A rock, by comparison, is an aggregate of minerals and need not have a specific chemical composition. Minerals range in composition from pure elements and simple salts to very complex silicates with thousands of known forms.
Natural gas	Natural gas is a gaseous fossil fuel consisting primarily of methane but including significant quantities of ethane, butane, propane, carbon dioxide, nitrogen, helium and hydrogen sulfide.
Petroleum	Petroleum is a naturally occurring liquid found in formations in the Earth consisting of a complex mixture of hydrocarbons of various lengths.
Coal	Coal is a fossil fuel formed in swamp ecosystems where plant remains were saved by water and mud from oxidization and biodegradation. It is a sedimentary rock, but the harder forms, such as anthracite coal, can be regarded as metamorphic rocks because of later exposure to elevated temperature and pressure. It is composed primarily of carbon along with assorted other elements, including sulfur.
Gravel	Gravel is rock that is of a certain particle size range. In geology, gravel is any loose rock that is at least two millimeters in its largest dimension and no more than 75 millimeters.
Gold	Gold is a highly sought-after precious metal which, for many centuries, has been used as money, a store of value and in jewelery. The metal occurs as nuggets or grains in rocks, underground "veins" and in alluvial deposits. It is one of the coinage metals. Itis dense, soft, shiny and the most malleable and ductile of the known metals.
Reservoir	Most often, a reservoir refers to an artificial lake, used to store water for various uses. Reservoirs are created first by building a sturdy dam, usually out of cement, earth, rock, or a mixture. Once the dam is completed, a stream is allowed to flow behind it and eventually fill it to capacity.
Zinc	Zinc is a chemical element in the periodic table that has the symbol Zn and atomic number 30.

	In some historical and sculptural contexts, it is known as spelter.
Uranium	Uranium is approximately 70% more dense than lead and is weakly radioactive. It occurs naturally in low concentrations in soil, rock and water.
Diamonds	Diamonds are the hardest natural material known to man and the third-hardest known material. Its hardness and high dispersion of light make it useful for industrial applications and jewelry.
Groundwater	Groundwater is water located beneath the ground surface in soil pore spaces and in the fractures of geologic formations. Groundwater is recharged from, and eventually flows to, the surface naturally; natural discharge often occurs at springs and seeps, streams and can often form oases or wetlands.
Copper	Copper is a ductile metal with excellent electrical conductivity, and finds extensive use as an electrical conductor, heat conductor, as a building material, and as a component of various alloys.
Stratification	In horticulture, stratification is the process of pretreating seeds to simulate natural conditions that a seed must endure before germination.
Shoreline	A shoreline is the fringe of land at the edge of a large body of water, such as an ocean, sea, or lake. A strict definition is the strip of land along a water body that is alternately exposed and covered by waves and tides.
Eolian	Eolian processes pertain to the activity of the winds and more specifically, to the winds' ability to shape the surface of the Earth and other planets.
Organism	In biology and ecology, an organism is a living complex adaptive system of organs that influence each other in such a way that they function in some way as a stable whole.
Alluvial	An alluvial plain is a relatively flat and gently sloping landform found at the base of a range of hills or mountains, formed by the deposition of alluvial soil over a long period of time by one or more rivers coming from the mountains.
Alvin	Alvin is a 16-ton, manned deep-ocean research submersible owned by the United States Navy and operated by the Woods Hole Oceanographic Institution in Woods Hole, Massachusetts. The three-person vessel allows for two scientists and one pilot to dive for up to nine hours at 4500 metersor 15,000 feet.
Glacier	A glacier is a large, slow moving river of ice, formed from compacted layers of snow, that slowly deforms and flows in response to gravity. Glacier ice is the largest reservoir of fresh water on Earth, and second only to oceans as the largest reservoir of total water. Glaciers cover vast areas of polar regions but are restricted to the highest mountains in the tropics.
Tectonics	Tectonics is a field of study within geology concerned generally with the structures within the crust of the Earth, or other planets, and particularly with the forces and movements that have operated in a region to create these structures.
Plate tectonics	Plate tectonics is a theory of geology that has been developed to explain the observed evidence for large scale motions of the Earth's lithosphere. The theory encompassed and superseded the older theory of continental drift.
Continent	A continent is one of several large landmasses on Earth. They are generally identified by convention rather than any strict criteria, but seven areas are commonly reckoned as continents – they are: Asia, Africa, North America, South America, Antarctica, Europe, and Australia.
Climate	Climate is the average and variations of weather over long periods of time. Climate zones can

be defined using parameters such as temperature and rainfall.

Precipitation	Precipitation is any product of the condensation of atmospheric water vapor that is deposited on the earth's surface. It occurs when the atmosphere becomes saturated with water vapour and the water condenses and falls out of solution. Air becomes saturated via two processes, cooling and adding moisture.
Carbonate	In organic chemistry, a carbonate is a salt of carbonic acid.
Reef	A reef is a rock, sandbar, or other feature lying beneath the surface of the water yet shallow enough to be a hazard to ships. They result from abiotic processes—deposition of sand, wave erosion planning down rock outcrops, and other natural processes.
Swamp	A swamp is a wetland that features temporary or permanent inundation of large areas of land by shallow bodies of water, generally with a substantial number of hummocks, or dry-land protrusions, and covered by aquatic vegetation, or vegetation that tolerates periodical inundation.
Vegetation	Vegetation is a general term for the plant life of a region; it refers to the ground cover provided by plants, and is, by far, the most abundant biotic element of the biosphere. Primeval redwood forests, coastal mangrove stands, sphagnum bogs, desert soil crusts, roadside weed patches, wheat fields, cultivated gardens and lawns; are all encompassed by the term vegetation.
Gypsum	Gypsum is a very soft mineral composed of calcium sulfate dihydrate, with the chemical formula $CaSO_4 \cdot 2H_2O$. Gypsum occurs in nature as flattened and often twinned crystals and transparent cleavable masses. It may also occur silky and fibrous. Finally it may also be granular or quite compact.
Lake	A lake is a body of water or other liquid of considerable size contained on a body of land. A vast majority are fresh water, and lie in the Northern Hemisphere at higher latitudes. Most have a natural outflow in the form of a river or stream, but some do not, and lose water solely by evaporation and/or underground seepage.
Coral	Coral refer to marine animals from the class Anthozoa and exist as small sea anemone-like polyps, typically in colonies of many identical individuals. The group includes the important reef builders that are found in tropical oceans, which secrete calcium carbonate to form a hard skeleton.
Coral reef	A coral reef is an aragonite structure produced by living organisms, found in shallow, tropical marine waters with little to no nutrients in the water.
Abundance	Abundance is an ecological concept referring to the relative representation of a species in a particular ecosystem. It is usually measured as the mean number of individuals found per sample.
Canyon	A canyon is a deep valley between cliffs often carved from the landscape by a river. Most were formed by a process of long-time erosion from a plateau level. The cliffs form because harder rock strata that are resistant to erosion and weathering remain exposed on the valley walls.
Geological time scale	The geological time scale is used by geologists and other scientists to describe the timing and relationships between events that have occurred during the history of Earth.
Weathering	Weathering is the process of breaking down rocks, soils and their minerals through direct contact with the atmosphere. Weathering occurs without movement. Two main classifications of weathering processes exist. Mechanical or physical weathering involves the breakdown of rocks and soils through direct contact with atmospheric conditions. The second classification, chemical weathering, involves the direct effect of atmospheric chemicals in the breakdown of

rocks, soils and minerals.

Grand Canyon	The Grand Canyon is a very colorful, steep-sided gorge, carved by the Colorado River in the U.S. state of Arizona. It is one of the first national parks in the United States.
Weather	The weather is the set of all extant phenomena in a given atmosphere at a given time. The term usually refers to the activity of these phenomena over short periods, as opposed to the term climate, which refers to the average atmospheric conditions over longer periods of time.
Terrace	In agriculture, a terrace is a leveled section of a hilly cultivated area, designed as a method of soil conservation to slow or prevent the rapid surface runoff of irrigation water
Metamorphic rock	Metamorphic rock is the result of the transformation of a pre-existing rock type, the protolith, in a process called metamorphism, which means "change in form". The protolith is subjected to heat and extreme pressure causing profound physical and/or chemical change. The protolith may be sedimentary rock, igneous rock or another older rock.
Metamorphic rocks	Metamorphic rock is the result of the transformation of a pre-existing rock type, the protolith, in a process called metamorphism. The protolith is subjected to heat and extreme pressure causing profound physical and/or chemical change. Metamorphic rocks make up a large part of the Earth's crust. They are formed deep beneath the Earth's surface by great stresses from rocks above and high pressures and temperatures.
Metamorphism	Metamorphism can be defined as the solid state recrystallisation of pre-existing rocks due to changes in heat and/or pressure and/or introduction of fluids. There will be mineralogical, chemical and crystallographic changes. Metamorphism produced with increasing pressure and temperature conditions is known as prograde metamorphism. Conversely, decreasing temperatures and pressure characterize retrograde metamorphism.
Igneous	Igneous rocks form when molten rock, magma, cools and solidifies, with or without crystallization, either below the surface as intrusive, plutonic rocks or on the surface as extrusive, volcanic, rocks.
Bedding planes	Bedding planes are where one sedimetary deposit ends and another one begins. The rock is prone to breakage at these points because of the weakness between the layers.
Particle size	Particle size refers to the diameter of individual grains of sediment, or the lithified particles in clastic rocks. The term may also be applied to other granular materials. This is different from the crystallite size, which is the size of a single crystal inside the particles or grains.
Sandstone	Sandstone is a sedimentary rock composed mainly of sand-size mineral or rock grains. Most sandstone is composed of quartz and/or feldspar because these are the most common minerals in the Earth's crust. Like sand, sandstone may be any color, but the most common colors are tan, brown, yellow, red, gray and white.
Calcite	The carbonate mineral Calcite is a chemical or biochemical calcium carbonate and is one of the most widely distributed minerals on the Earth's surface. It is a common constituent of sedimentary rocks, limestone in particular. It is also the primary mineral in metamorphic marble
Dolomite	Dolomite is the name of a sedimentary carbonate rock and a mineral, both composed of calcium magnesium carbonate found in crystals. Dolomite rock is composed predominantly of the mineral dolomite. Limestone that is partially replaced by dolomite is referred to as dolomitic limestone.
Abrasion	Abrasion is mechanical scraping of a rock surface by friction between rocks and moving particles during their transport in wind, glacier, waves, gravity or running water.

Mudflats	Mudflats are coastal wetlands that form when mud is deposited by the tides or rivers, sea and oceans. They are found in sheltered areas such as bays, bayous, lagoons, and estuaries. Mudflats may be viewed geologically as exposed layers of bay mud, resulting from deposition of estuarine silts, clays and marine animal detritus.
Stream	A stream is a body of water with a current, confined within a bed and banks. Streams are important as conduits in the water cycle, instruments in aquifer recharge, and corridors for fish and wildlife migration.
Canadian Shield	The Canadian Shield is a large shield covered by a thin layer of soil that forms the nucleus of the North American craton. It has a deep, common, joined bedrock region in eastern and central Canada and stretches North from the Great Lakes to the Arctic Ocean, covering half the country.
Currents	Ocean currents are any more or less continuous, directed movement of ocean water that flows in one of the Earth's oceans.They are rivers of hot or cold water within the ocean. They are generated from the forces acting upon the water like the earth's rotation, the wind, the temperature and salinity differences and the gravitation of the moon.
Clastic	Clastic rocks are rocks formed from fragments of pre-existing rock.
Clastic rock	Clastic rock is rock formed from fragments of pre-existing rock. The most common usage is for clastic or detrital sedimentary rocks. However, the usage is not restricted to sediments.
Glaciers	A glacier is a large, slow moving river of ice, formed from compacted layers of snow, that slowly deforms and flows in response to gravity. Glacier ice is the largest reservoir of fresh water on Earth, and second only to oceans as the largest reservoir of total water. Glaciers cover vast areas of polar regions but are restricted to the highest mountains in the tropics.
Conglomerate	A conglomerate is a rock consisting of individual stones that have become cemented together. Conglomerates are sedimentary rocks consisting of rounded fragements and are thus differentiated from breccias, which consist of angular clasts. Both conglomerates and breccias are characterized by clasts larger than sand.
Carrying capacity	Carrying capacity is the supportable population of an organism, given the food, habitat, water and other necessities available. For the human population other variables such as sanitation and medical care are sometimes considered as infrastructure.
Igneous rock	Igneous rock forms when rock cools and solidifies either below the surface as intrusive rocks or on the surface as extrusive rocks. This magma can be derived from partial melts of pre-existing rocks in either the Earth's mantle or crust. Typically, the melting is caused by one or more of the following processes -- an increase in temperature, a decrease in pressure, or a change in composition.
Clasts	Clastic sedimentary rocks are rocks composed predominantly of broken pieces or clasts of older weathered and eroded rocks.
Velocity	In physics, velocity is defined as the rate of change of displacement or the rate of displacement. Simply put, it is distance per units of time.
Quartz	Quartz is the second most common mineral in the Earth's continental crust. It is made up of a lattice of silica tetrahedra. Quartz belongs to the rhombohedral crystal system. In nature quartz crystals are often twinned, distorted, or so intergrown with adjacent crystals of quartz or other minerals as to only show part of this shape, or to lack obvious crystal faces altogether and appear massive.
Oxide	An oxide is a chemical compound containing an oxygen atom and other elements. Most of the

earth's crust consists of them. They result when elements are oxidized by air.

Iron	Iron is a chemical element metal. It is a lustrous, silvery soft metal. It and nickel are notable for being the final elements produced by stellar nucleosynthesis, and thus are the heaviest elements which do not require a supernova or similarly cataclysmic event for formation.
Mica	The mica group of sheet silicate minerals includes several closely related materials having highly perfect basal cleavage. All are monoclinic with a tendency towards pseudo-hexagonal crystals and are similar in chemical composition. The highly perfect cleavage, which is the most prominent characteristic of mica, is explained by the hexagonal sheet-like arrangement of its atoms.
Olivine	The mineral olivine is a magnesium iron silicate. It is one of the most common minerals on Earth, and has also been identified on the Moon, Mars, and comet Wild 2.
Feldspar	Feldspar is the name of a group of rock-forming minerals which make up as much as sixty percent of the Earth's crust. Feldspars crystallize from magma in both intrusive and extrusive rocks, and they can also occur as compact minerals, as veins, and are also present in many types of metamorphic rock.
Claystone	Claystone is a geological term used to describe a sedimentary rock that is composed primarily of clay-sized particles.
Clay	Clay is a term used to describe a group of hydrous aluminium phyllosilicate minerals, that are typically less than 2 micrometres in diameter. Clay consists of a variety of phyllosilicate minerals rich in silicon and aluminium oxides and hydroxides which include variable amounts of structural water. Clays are generally formed by the chemical weathering of silicate-bearing rocks by carbonic acid but some are formed by hydrothermal activity.
Lagoon	A lagoon is a body of comparatively shallow salt or brackish water separated from the deeper sea by a shallow or exposed sandbank, coral reef, or similar feature. Thus, the enclosed body of water behind a barrier reef or barrier islands or enclosed by an atoll reef is called a lagoon.
Organic matter	Organic matter is matter that has come from a recently living organism; is capable of decay, or the product of decay; or is composed of organic compounds. The definition of organic matter varies upon the subject it is being used for.
Decomposition	Decomposition refers to the reduction of the body of a formerly living organism into simpler forms of matter.
Jasper	Jasper is an opaque, impure variety of quartz, usually red, yellow or brown in color. This mineral breaks with a smooth surface, and is used for ornamentation or as a gemstone.
Nodule	A nodule in petrology or mineralogy is an irregular rounded to spherical concretion. They are typically solid replacement bodies of chert or iron oxides formed during diagenesis of a sedimentary rock.
Carbonate rocks	Carbonate rocks are a class of sedimentary rocks composed primarily of carbonate minerals. The two major types are limestone and dolomite, composed of calcite and the mineral dolomite respectively. Chalk and tufa are also minor sedimentary carbonates.
Diatoms	Diatoms are a major group of eukaryotic algae, and are one of the most common types of phytoplankton. Most diatoms are unicellular, although some form chains or simple colonies. A characteristic feature of diatom cells is that they are encased within a unique cell wall made of silica called a frustule.
Radiolarian	Radiolarian refers to an amoeboid protozoa that produce intricate mineral skeletons,

typically with a central capsule dividing the cell into inner and outer portions, called endoplasm and ectoplasm. They are found as zooplankton throughout the ocean, and because of their rapid turn-over of species, their tests are important diagnostic fossils found from the Cambrian onwards.

Seafloor	The seafloor the bottom of the ocean. At the bottom of the continental slope is the continental rise, which is caused by sediment cascading down the continental slope.
Pelagic sediment	Pelagic sediment is an accumulate in the abyssal plain of the deep ocean, far away from terrestrial sources that provide terrigenous sediments; the latter are primarily limited to the continental shelf, and deposited by rivers.
Carbonate minerals	Carbonate minerals are those minerals containing the carbonate ion: $CO_{3}2-$.
Seawater	Seawater is water from a sea or ocean. On average, seawater in the world face=symbol>¢s oceans has a salinity of ~3.5%, or 35 parts per thousand. This means that every 1 kg of seawater has approximately 35 grams of dissolved salts.
Organic compound	An organic compound is any member of a large class of chemical compounds whose molecules contain carbon.
Hydrocarbon	In organic chemistry, a hydrocarbon is an organic compound consisting entirely of hydrogen and carbon. With relation to chemical terminology, aromatic hydrocarbons or arenes, alkanes, alkenes and alkyne-based compounds composed entirely of carbon or hydrogen are referred to as "Pure" hydrocarbons, whereas other hydrocarbons with bonded compounds or impurities of sulphur or nitrogen, are referred to as "impure", and remain somewhat erroneously referred to as hydrocarbons.
Chemical compound	A chemical compound is a chemical substance of two or more different chemically bonded chemical elements, with a fixed ratio determining the composition. The ratio of each element is usually expressed by chemical formula.
Ion	An ion is an atom or group of atoms which have lost or gained one or more electrons, making them negatively or positively charged.
Evaporation	Evaporation is the process by which molecules in a liquid state become a gas.
Halite	Halite is the mineral form of sodium chloride. Halite forms isometric crystals. It commonly occurs with other evaporite deposit minerals such as several of the sulfates, halides and borates. Halite occurs in vast lakes of sedimentary evaporite minerals that result from the drying up of enclosed beds, playas, and seas.
Brine	Brine is water saturated or nearly saturated with salt and is a common fluid used in the transport of heat from place to place. It is used because the addition of salt to water lowers the freezing temperature of the solution and a relatively great efficiency in the transport can be obtained for the low cost of the material.
Dead Sea	The Dead Sea is a salt lake between the West Bank and Israel to the west, and Jordan to the east. It is said to be the lowest point on Earth, at 420 m below sea level; its shores are actually the lowest point on dry land, as there are deeper points on Earth under water or ice. At 330m deep , the Dead Sea is the deepest hypersaline lake in the world.
Desert	In geography, a desert is a landscape form or region that receives very little precipitation. They are defined as areas that receive an average annual precipitation of less than 250 mm. A desert where vegetation cover is exceedingly sparse correspond to the face=symbol>¢hyperarid' regions of the earth, where rainfall is exceedingly rare and infrequent.
Great Salt Lake	Great Salt Lake, located in the northern part of the U.S. state of Utah, is the largest salt

lake in the Western Hemisphere, the fourth-largest terminal lake in the world, and the 33rd largest lake on Earth.

Sulfate	In inorganic chemistry, a sulfate is a salt of sulfuric acid
Wave	A wave is a disturbance that propagates through space or spacetime, transferring energy and momentum and sometimes angular momentum.
Migration	Migration refers to directed, regular, or systematic movement of a group of objects, organisms, or people.
Dune	A dune is a hill of sand built by eolian processes. Dunes are subject to different forms and sizes based on their interaction with the wind. Most kinds of dune are longer on the windward side where the sand is pushed up the dune, and a shorter in the lee of the wind. The trough between dunes is called a slack. A "dune field" is an area covered by extensive sand dunes. Large dune fields are known as ergs.
Island	An island is any piece of land that is completely surrounded by water, above high tide. There are two main types of islands: continental islands and oceanic islands. There are also artificial islands. A grouping of geographically and/or geologically related islands is called an archipelago.
Season	A season is one of the major divisions of the year, generally based on yearly periodic changes in weather. They are recognized as: spring, summer, autumn, and winter.
Sedimentary basin	The term sedimentary basin is used to refer to any geographical feature exhibiting subsidence and consequent infilling by sedimentation. As the sediments are buried, they are subjected to increasing pressure and begin the process of lithification.
Sahara Desert	The Sahara Desert is the world's largest hot desert, and second largest desert after Antarctica. At over 9,000,000 square kilometres, it is almost as large as the United States, and is larger than Australia. It's name derives from the Arabic word "çahra"; to refer to the Sahara as the face=symbol>¢Sahara Desert' is a pleonasm.
Zion National Park	Zion National Park is a United States National Park located in the Southwestern United States, near Springdale, Utah. Located at the junction of the Colorado Plateau, Great Basin, and Mojave Desert regions, this unique geography and variety of life zones allow for unusual plant and animal diversity.
Silt	Silt is soil or rock derived granular material of a specific grain size. Silt may occur as a soil or alternatively as suspended sediment in a water column of any surface water body. It may also exist as deposition soil at the bottom of a water body.
Valley	In geology, a valley is a depression with predominant extent in one direction. The terms U-shaped and V-shaped are descriptive terms of geography to characterize the form of valleys. Most valleys belong to one of these two main types or a mixture of them, at least with respect of the cross section of the slopes or hillsides.
Boulders	In geology, boulders are rock s with a grain size of usually no less than 256 mm diameter.
Meltwater	Meltwater is the water released by the melting of snow or ice, including glacial ice. Meltwater provides drinking water for a large proportion of the world face=symbol>¢s population, as well as providing water for irrigation and hydroelectric plants.
Nile River	The Nile River is a major north-flowing river in Africa, generally regarded as the longest river in the world. The Nile River has two major tributaries, the White Nile and Blue Nile, the latter being the source of most of the Nile's water and fertile soil, but the former being the longer of the two. It ends in a large delta that empties into

the Mediterranean Sea.	
Mediterranean Sea	The Mediterranean Sea is a sea of the Atlantic Ocean almost completely enclosed by land: on the north by Europe, on the south by Africa, and on the east by Asia. It covers an approximate area of 2.5 million km², but its connection to the Atlantic is only 14 km wide.
Colorado Plateau	Colorado Plateau is a physiographic region of the Intermontane Plateaus, roughly centered on the Four Corners region of the southwestern United States. Development of the province has in large part been influenced by structural features in its oldest rocks. Part of the Wasatch Line and its various faults form the western edge of the province.
Tertiary	The Tertiary covers roughly the time span between the demise of the non-avian dinosaurs and beginning of the most recent Ice Age. Each epoch of the Tertiary was marked by striking developments in mammalian life. The earliest recognizable hominoid relatives of humans appeared. Tectonic activity continued as Gondwana finally split completely apart.
Plateau	A plateau is an area of highland, usually consisting of relatively flat rural area.
Coast	The coast is defined as the part of the land adjoining or near the ocean. A coastline is properly a line on a map indicating the disposition of a coast, but the word is often used to refer to the coast itself. The adjective coastal describes something as being on, near to, or associated with a coast.
Spit	A spit is a deposition landform found off coasts. A spit is a type of bar or beach that develops where a re-entrant occurs, such as at a cove, bay, ria, or river mouth. A spit is formed by the movement of sediment along a shore by a process known as longshore drift. Where the direction of the shore turns inland the longshore current spreads out or dissipates. No longer able to carry the full load, much of the sediment is dropped. This causes a bar to build out from the shore, eventually becoming a spit.
Wave power	Wave power refers to the energy of ocean surface waves and the capture of that energy to do useful work - including electricity generation, desalination, and the pumping of water. Wave power is a form of renewable energy. Though often co-mingled, wave power is distinct from the diurnal flux of tidal power and the steady gyre of ocean currents.
Matter	Matter is the substance of which physical objects are composed. Matter can be solid, liquid, plasma or gas. It constitutes the observable universe.
Sea level	Mean sea level is the average height of the sea, with reference to a suitable reference surface.
Tide	Tide refers to the cyclic rizing and falling of Earth's ocean surface caused by the tidal forces of the Moon and the sun acting on the oceans. They cause changes in the depth of the marine and estuarine water bodies and produce oscillating currents known as tidal streams, making prediction of tides important for coastal navigation.
Evaporite	Evaporite refers to water-soluble, mineral sediments that result from the evaporation of bodies of surficial water.
Great Barrier Reef	The Great Barrier Reef in Australia is the world's largest coral reef system, composed of roughly 3,000 individual reefs and 900 islands. The Great Barrier Reef can be seen from outer space and is the world's biggest single structure made by living organisms. This reef structure is composed of and built by billions of tiny organisms, known as coral polyps.
Paleozoic	The Paleozoic is the earliest of three geologic eras of the Phanerozoic eon. The Paleozoic is subdivided into six geologic periods; from oldest to youngest they are: the Cambrian, Ordovician, Silurian, Devonian, Carboniferous, and Permian.
Era	An era is a long period of time with different technical and colloquial meanings, and usages

Go to **Cram101.com** for the Practice Tests for this Chapter.

	in language. It begins with some beginning event known as an epoch, epochal date, epochal event or epochal moment.
Guadalupe Mountains	The Guadalupe Mountains are a mountain range located in western Texas and southeastern New Mexico. The range includes the highest summit in Texas, Guadalupe Peak, 8,749 ft, and the "signature peak" of West Texas, El Capitan.
Blocks	Blocks in meteorology are large scale patterns in the atmospheric pressure field that are nearly stationary, effectively "blocking" or redirecting migratory cyclones. These blocks can remain in place for several days or even weeks, causing the areas affected by them to have the same kind of weather for an extended period of time.
Atoll	An atoll is an oceanic reef formation, often having a characteristic ring-like shape surrounding a lagoon. Atolls are formed when coral reef grows around a volcanic island that later subsides into the ocean.
Subsidence	In geology, engineering, and surveying, subsidence is the motion of a surface as it shifts downward relative to a datum such as sea-level. The opposite of subsidence is uplift, which results in an increase in elevation. In meteorology, subsidence refers to the downward movement of air.
Bahamas	The Commonwealth of The Bahamas is an English-speaking nation consisting of two thousand cays and seven hundred islands that form an archipelago. It is located in the Atlantic Ocean, east of Florida and the United States, north of Cuba and the Caribbean, and northwest of the British overseas territory of the Turks and Caicos Islands.
Ridge	A ridge is a geological feature that is also known as a Rip in the earth causing magma to flow out and forming an undersea volcano, it also has geological features, a continuous elevational crest for some distance. Ridges are usually termed hills or mountains as well, depending on size.
Baltic Sea	The Baltic Sea is located in Northern Europe. It is bounded by the Scandinavian Peninsula, the mainland of Europe, and the Danish islands. It drains into the Kattegat by way of the Øresund, the Great Belt and the Little Belt. The Kattegat continues through the Skagerrak into the North Sea and the Atlantic Ocean.
Distributary channel	A distributary channel, is a stream that branches off and flows away from a main stream channel. They are a common feature of river deltas. The phenomenon is known as river bifurcation.
Continental slope	The sea floor below the break is the continental slope. Below the slope is the continental rise, which finally merges into the deep ocean floor, the abyssal plain. As the continental shelf and the slope are part of the continental margin, both are covered in this article.
Turbidite	Turbidite geological formations have their origins in turbidity current deposits, deposits from a form of underwater avalanche that are responsible for distributing vast amounts of clastic sediment into the deep ocean.
Graded bed	In geology, a graded bed is one characterized by coarse sediments at its base, which grade upward into progressively finer ones. They are perhaps best represented in turbidite strata, where they indicate a sudden strong current that deposits heavy, coarse sediments first, with finer ones following as the current weakens.
Submarine canyon	A Submarine canyon is a steep-sided valley on the sea floor of the continental slope. They are formed by powerful turbidity currents, volcanic and earthquake activity. Many continue as submarine channels across continental rise areas and may extend for hundreds of kilometers.
Ocean basins	Ocean basins are large geologic basins that are below sea level. Geologically, there are other undersea geomorphological features such as the continental shelves, the deep ocean

Go to **Cram101.com** for the Practice Tests for this Chapter.

trenches, and the undersea mountain rangeswhich are not considered to be part of the ocean basins.

Stratigraphy	Stratigraphy, a branch of geology, studies rock layers and layering. It is primarily used in the study of sedimentary and layered volcanic rocks. Stratigraphy includes two related subfields: lithologic or lithostratigraphy and biologic stratigraphy or biostratigraphy.
Transgression	A transgression is a geologic event during which sea level rises relative to the land and the shoreline moves toward higher ground, resulting in flooding. Transgressions can be caused either by the land sinking or the ocean basins filling with water.
Continental shelf	The continental shelf is the extended perimeter of each continent and associated coastal plain, which is covered during interglacial periods such as the current epoch by relatively shallow seas and gulfs. The shelf usually ends at a point of increasing slope.
Unconformity	An unconformity is a buried erosion surface separating two rock masses or strata of different ages, indicating that sediment deposition was not continuous. In general, the older layer was exposed to erosion for an interval of time before deposition of the younger, but the term is used to describe any break in the sedimentary geologic record.
Floodplain	A floodplain is flat or nearly flat land adjacent to a stream or river that experiences occasional or periodic flooding. It includes the floodway, which consists of the stream channel and adjacent areas that carry flood flows, and the flood fringe, which are areas covered by the flood, but which do not experience a strong current.
Sea level rise	Sea level rise is an increase in sea level. Multiple complex factors may influence such changes.
Rapid	A rapid is a section of a river of relatively steep gradient causing an increase in water flow and turbulence. A rapid is a hydrological feature between a run and a cascade. It is characterized by the river becoming shallower and having some rocks exposed above the flow surface.
Seafloor spreading	Seafloor spreading occurs at mid-ocean ridges, where new oceanic crust is formed through volcanic activity and then gradually moves away from the ridge. Seafloor spreading helps explain continental drift in the theory of plate tectonics.
Ocean current	An ocean current is any more or less continuous, directed movement of ocean water that flows in one of the Earth's oceans.Ocean Currents are rivers of hot or cold water within the ocean. The currents are generated from the forces acting upon the water like the earth's rotation, the wind, the temperature and salinity differences and the gravitation of the moon. The depth contours, the shoreline and other currents influence the current's direction and strength.
Topography	Topography is the study of Earth's surface features or those of other planets, moons, and asteroids
Rift	In geology, a rift is a place where the Earth's crust and lithosphere are being pulled apart.
Divergent plate boundary	In plate tectonics, a divergent plate boundary a linear feature that exists between two tectonic plates that are moving away from each other. These areas can form in the middle of continents but eventually form ocean basins.
Turbidity	Turbidity is a cloudiness or haziness of water caused by individual particles that are generally invisible to the naked eye, thus being much like smoke in air. Turbidity is generally caused by phytoplankton. Measurement of turbidity is a key test of water quality.
Turbidity current	A turbidity current is a current of rapidly moving, sediment-laden water moving down a slope through air, water, or another fluid. The current moves because it has a higher density and

Go to **Cram101.com** for the Practice Tests for this Chapter.
And, **NEVER** highlight a book again!

turbidity than the fluid through which it flows.

Abyssal	The abyssal is the pelagic zone that contains the very deep benthic communities near the bottom of oceans.
Abyssal plain	An abyssal plain is a flat or very gently sloping area of the deep ocean basin floor. They are among the Earth's flattest and smoothest regions and the least explored. They cover approximately 40% of the ocean floor and generally lie between the foot of a continental rise and a mid-oceanic ridge.
Cambrian	The Cambrian is a major division of the geologic timescale. The Cambrian is the earliest period in whose rocks are found numerous large, distinctly fossilizable multicellular organisms that are more complex than sponges or medusoids. During this time, roughly fifty separate major groups of organisms or "phyla" emerged suddenly, in most cases without evident precursors. This radiation of animal phyla is referred to as the Cambrian explosion.
Silurian	The Silurian is a major division of the geologic timescale that extends from the end of the Ordovician period to the beginning of the Devonian period. The base of the Silurian is set at a major extinction event when 60% of marine species were wiped out.
Pennsylvanian	The Pennsylvanian is an epoch of the Carboniferous period lasting from roughly 325 Ma to 299 Ma. As with most other geologic periods, the rock beds that define the period are well identified, but the exact date of the start and end are uncertain by a few million years.
Permian	The Permian is the last period of the Palaeozoic Era. As the Permian opened, the Earth was still in the grip of an ice age, so the polar regions were covered with deep layers of ice. During the Permian, all the Earth's major land masses except portions of East Asia were collected into a single supercontinent known as Pangaea. The Permian ended with the most extensive extinction event recorded in paleontology: the Permian-Triassic extinction event.
Devonian	The Devonian is a geologic period of the Paleozoic era. During the Devonian the first fish evolved legs and started to walk on land as tetrapods and the first insects and spiders also started to colonize terrestrial habitats. The first seed-bearing plants spread across dry land, forming huge forests. In the oceans, Primitive sharks became more numerous. The first ammonite mollusks appeared, and trilobites as well as great coral reefs were still common.
Mississippian	The Mississippian was an epoch of the Carboniferous period lasting from roughly 360 to 325 Ma. As with most other geologic periods, the rock beds that define the period are well identified, but the exact start and end dates are uncertain by a few million years.
Ordovician	The Ordovician is the second of the six periods of the Paleozoic era. It follows the Cambrian period and is followed by the Silurian period. The Ordovician started at a major extinction called the Cambrian-Ordovician extinction and lasted for about 44.6 million years. It ended with another major extinction event that wiped out 60% of marine genera.
Stem	A stem is one of two main structural axes of a vascular plant. It is normally divided into nodes and internodes, the nodes hold buds which grow into one or more leaves, inflorescence, cones etc.
Ocean Drilling Program	The Ocean Drilling Program was an international cooperative effort to explore and study the composition and structure of the earth's ocean basins. Ocean Drilling Program, which began in 1985, was the direct successor to the highly successful Deep Sea Drilling Project initiated in 1968 by the United States.
JOIDES Resolution	JOIDES Resolution is a scientific drilling ship once used by the Ocean Drilling Program, then by its successor, the Integrated Ocean Drilling Program. It is the successor of the Glomar Challenger.

Arctic	The Arctic is the region around the Earth's North Pole, opposite the Antarctic region around the South Pole. In the northern hemisphere, the Arctic includes the Arctic Ocean and parts of Canada, Greenland, Russia, the United States, Iceland, Norway, Sweden and Finland. The word Arctic comes from the Greek word arktos, which means bear. This is due to the location of the constellation Ursa Major, the "Great Bear", above the Arctic region.
Arctic Ocean	The Arctic Ocean, located in the northern hemisphere and mostly in the Arctic north polar region, is the smallest of the world's five major oceanic divisions and the shallowest.
Deep sea	The deep sea is the lowest layer in the ocean, existing below the thermocline. Little or no light penetrates this area of the ocean, and most of its organisms rely on falling organic matter produced in the photic zone for subsistence. For this reason life is much more sparse, becoming rarer still with increasing depth. The other essential ingredient in for life is oxygen, which is brought to the ocean's depths via the thermohaline circulation.
Oceanic crust	Oceanic crust is the part of Earth's lithosphere that surfaces in the ocean basins. Oceanic crust is primarily composed of mafic rocks, or sima. It is thinner than continental crust, or sial, generally less than 10 kilometers thick, however it is more dense, having a mean density of about 3.3 grams per cubic centimeter.
Crust	In geology, a crust is the outermost layer of a planet, part of its lithosphere. They are generally composed of a less dense material than its deeper layers.Earths face=symbol>¢ is composed mainly of basalt and granite. It is cooler and more rigid than the deeper layers of the mantle and core.
Meteorite	A meteorite is a natural object originating in outer space that survives an impact with the Earth's surface without being destroyed. While in space it is called a meteoroid. When it enters the atmosphere, air resistance causes the body to heat up and emit light, thus forming a fireball.
Milankovitch	Milankovitch cycles are the collective effect of changes in the Earth face=symbol>¢s movements upon its climate.
Milankovitch cycles	Milankovitch cycles are the collective effect of changes in the Earth face=symbol>¢s movements upon its climate, named after Serbian civil engineer and mathematician Milutin Milankoviæ.
Climate change	Climate change refers to the variation in the Earth's global climate or in regional climates over time. It describes changes in the variability or average state of the atmosphere over time scales ranging from decades to millions of years. These changes can be caused by processes internal to the Earth, external forces or, more recently, human activities.
Mesozoic	The Mesozoic is one of three geologic eras of the Phanerozoic eon. The Mesozoic was a time of tectonic, climatic and evolutionary activity, shifting from a state of connectedness into their present configuration. The climate was exceptionally warm throughout the period, also playing an important role in the evolution and diversification of new animal species. By the end of the era, the basis of modern life was in place.
Lava	Lava is molten rock expelled by a volcano during an eruption. When first extruded from a volcanic vent, it is a liquid at temperatures from 700 °C to 1,200 °C.
Lithosphere	The lithosphere is the solid outermost shell of a rocky planet. On the Earth, the lithosphere includes the crust and the uppermost mantle which is joined to the crust across the Mohorovièiæ discontinuity. Lithosphere is underlain by asthenosphere, the weaker, hotter, and

deeper part of the upper mantle.

Alluvial fans	An alluvial fan is a fan-shaped deposit formed where a fast flowing stream flattens, slows, and spreads typically at the exit of a canyon onto a flatter plain. A convergence of neighboring alluvial fans into a single apron of deposits against a slope is called a bajada, or compound alluvial fan.
Drainage	Drainage is the natural or artificial removal of surface and sub-surface water from a given area. Many agricultural soils need drainage to improve production or to manage water supplies.
Drainage basin	A drainage basin is a region of land where water from rain or snow melt drains downhill into a body of water, such as a river, lake, dam, estuary, wetland, sea or ocean. The drainage basin includes both the streams and rivers that convey the water as well as the land surfaces from which water drains into those channels. The drainage basin acts like a funnel - collecting all the water within the area covered by the basin and channeling it into a waterway.
Drainage basins	Drainage basins are regions of land where water from rain or snow melt drains downhill into a body of water, such as a river, lake, dam, estuary, wetland, sea or ocean. It includes both the streams and rivers that convey the water as well as the land surfaces from which water drains into those channels. It acts like a funnel - collecting all the water within the area covered by it and channeling it into a waterway.
Landform	A landform comprises a geomorphological unit, and is largely defined by its surface form and location in the landscape, as part of the terrain, and as such, is typically an element of topography. They are categorised by features such as elevation, slope, orientation, stratification, rock exposure, and soil type. They include berms, mounds, hills, cliffs, valleys, rivers and numerous other elements.
Compaction	Compaction is the process of a material being more closely packed together.
Cementation	Cementation is the process of deposition of dissolved mineral components in the interstices of sediments. It is an important factor in the consolidation of coarse-grained clastic sedimentary rocks such as sandstones, conglomerates, or breccias during diagenesis or lithification. Cementing materials may include silica, carbonates, iron oxides, or clay minerals.
Oolite	An oolite is a sedimentary rock formed from ooids, spherical grains composed of concentric layers.
Dolostone	Dolostone is a sedimentary carbonate rock that contains a high percentage of the mineral dolomite. It is usually referred to as dolomite rock. In old U.S.G.S. publications it was referred to as magnesian limestone.
Facies	A facies should ideally be a distinctive rock that forms under certain conditions of sedimentation, reflecting a particular process or environment.
Chalk	Chalk is a soft, white, porous sedimentary rock, a form of limestone composed of the mineral calcite. It forms under relatively deep marine conditions from the gradual accumulation of minute calcite plates shed from micro-organisms called coccolithophores. It is common to find flint nodules embedded in it.
Petrology	Petrology is a field of geology which focuses on the study of rocks and the conditions by which they form. There are three branches of petrology, corresponding to the three types of rocks: igneous, metamorphic, and sedimentary. Petrology utilizes the classical fields of mineralogy, petrography, optical mineralogy, and chemical analyses to describe the composition and texture of rocks.

Sedimentology	Sedimentology encompasses the study of modern sediments and understanding the processes that deposit them. It also compares these observations to studies of ancient sedimentary rocks. Sedimentologists apply their understanding of modern processes to historically formed sedimentary rocks, allowing them to understand how they formed.

Mountain	A mountain is a landform that extends above the surrounding terrain in a limited area. A mountain is generally steeper than a hill, but there is no universally accepted standard definition for the height of a mountain or a hill although a mountain usually has an identifiable summit.
Metamorphic rock	Metamorphic rock is the result of the transformation of a pre-existing rock type, the protolith, in a process called metamorphism, which means "change in form". The protolith is subjected to heat and extreme pressure causing profound physical and/or chemical change. The protolith may be sedimentary rock, igneous rock or another older rock.
Metamorphic rocks	Metamorphic rock is the result of the transformation of a pre-existing rock type, the protolith, in a process called metamorphism. The protolith is subjected to heat and extreme pressure causing profound physical and/or chemical change. Metamorphic rocks make up a large part of the Earth's crust. They are formed deep beneath the Earth's surface by great stresses from rocks above and high pressures and temperatures.
Metamorphism	Metamorphism can be defined as the solid state recrystallisation of pre-existing rocks due to changes in heat and/or pressure and/or introduction of fluids. There will be mineralogical, chemical and crystallographic changes. Metamorphism produced with increasing pressure and temperature conditions is known as prograde metamorphism. Conversely, decreasing temperatures and pressure characterize retrograde metamorphism.
Sedimentary	Sedimentary rock is one of the three main rock groups. Rock formed from these covers 75% of the Earth's land area, and includes common types such as chalk, limestone, dolomite, sandstone, and shale.
Igneous	Igneous rocks form when molten rock, magma, cools and solidifies, with or without crystallization, either below the surface as intrusive, plutonic rocks or on the surface as extrusive, volcanic, rocks.
Mineral	A mineral is a naturally occurring substance formed through geological processes that has a characteristic chemical composition, a highly ordered atomic structure and specific physical properties. A rock, by comparison, is an aggregate of minerals and need not have a specific chemical composition. Minerals range in composition from pure elements and simple salts to very complex silicates with thousands of known forms.
Canyon	A canyon is a deep valley between cliffs often carved from the landscape by a river. Most were formed by a process of long-time erosion from a plateau level. The cliffs form because harder rock strata that are resistant to erosion and weathering remain exposed on the valley walls.
Grand Canyon	The Grand Canyon is a very colorful, steep-sided gorge, carved by the Colorado River in the U.S. state of Arizona. It is one of the first national parks in the United States.
Cliff	A cliff is a significant vertical, or near vertical, rock exposure. Cliffs are categorized as erosion landforms due to the processes of erosion and weathering that produce them. Cliffs are common on coasts, in mountainous areas, escarpments and along rivers. Cliffs are usually formed by rock that is resistant to erosion and weathering.
Magma	Magma is molten rock located beneath the surface of the Earth, and which often collects in a magma chamber. Magma is a complex high-temperature fluid substance. Most are silicate solutions. It is capable of intrusion into adjacent rocks or of extrusion onto the surface as lava or ejected explosively as tephra to form pyroclastic rock. Environments of magma formation include subduction zones, continental rift zones, mid-oceanic ridges, and hotspots, some of which are interpreted as mantle plumes.
Crust	In geology, a crust is the outermost layer of a planet, part of its lithosphere. They are

	generally composed of a less dense material than its deeper layers.Earths face=symbol>¢ is composed mainly of basalt and granite. It is cooler and more rigid than the deeper layers of the mantle and core.
Sill	In geology, a sill is a tabular pluton that has intruded between older layers of sedimentary rock, beds of volcanic lava or tuff, or even along the direction of foliation in metamorphic rock. The term sill is synonymous with concordant intrusive sheet. This means that the sill does not cut across preexisting rocks. Contrast this with dikes.
Dike	A dike is an intrusion into a cross-cutting fissure, meaning a dike cuts across other pre-existing layers or bodies of rock, this means that a dike is always younger than the rocks that contain it. The thickness is usually much smaller than the other two dimensions. Thickness can vary from sub-centimeter scale to many meters in thickness and the lateral dimensions can extend over many kilometers.
Igneous rock	Igneous rock forms when rock cools and solidifies either below the surface as intrusive rocks or on the surface as extrusive rocks. This magma can be derived from partial melts of pre-existing rocks in either the Earth's mantle or crust. Typically, the melting is caused by one or more of the following processes -- an increase in temperature, a decrease in pressure, or a change in composition.
Fold	The term fold is used in geology when one or a stack of originally flat and planar surfaces, such as sedimentary strata, are bent or curved as a result of plastic, i.e. permanent, deformation.
Alcatraz Island	Alcatraz Island is a small island located in the middle of San Francisco Bay in California, United States. It served as a lighthouse, then a military fortification, then a military prison followed by a federal prison until 1963, when it became a national recreation area.
Tectonics	Tectonics is a field of study within geology concerned generally with the structures within the crust of the Earth, or other planets, and particularly with the forces and movements that have operated in a region to create these structures.
Melting	Melting is the process of heating a solid substance to a point where it turns into a liquid. An object that has melted is molten.
Butterfly	A butterfly is an insect of the order Lepidoptera. They are notable for their unusual life cycle with a larval caterpillar stage, an inactive pupal stage and a spectacular metamorphosis into a familiar and colorful winged adult form, and most species being day-flying, they regularly attract attention.
Metamorphosis	Metamorphosis is a biological process by which an animal physically develops after birth or hatching, involving a conspicuous and relatively abrupt change in the animal face=symbol>¢s form or structure through cell growth and differentiation.
Erosion	Erosion is displacement of solids by the agents of ocean currents, wind, water, or ice by downward or down-slope movement in response to gravity or by living organisms.
Continent	A continent is one of several large landmasses on Earth. They are generally identified by convention rather than any strict criteria, but seven areas are commonly reckoned as continents – they are: Asia, Africa, North America, South America, Antarctica, Europe, and Australia.
Recrystalliz-tion	Recrystallization is an essentially physical process that has meanings in chemistry, metallurgy and geology. In geology, solid-state recrystallization is a metamorphic process that occurs under situations of intense temperature and pressure where grains, atoms or molecules of a rock or mineral are packed closer together, creating a new crystal structure.
Thermal	A thermal column is a column of rizing air in the lower altitudes of the Earth

face=symbol>¢s atmosphere. Thermals are created by the uneven heating of the Earth's surface from solar radiation, and are an example of convection. The Sun warms the ground, which in turn warms the air directly above it.

Limestone	Limestone is a sedimentary rock composed largely of the mineral calcite. Limestone often contains variable amounts of silica in the form of chert or flint, as well as varying amounts of clay, silt and sand as disseminations, nodules, or layers within the rock. The primary source of the calcite in limestone is most commonly marine organisms. These organisms secrete shells that settle out of the water column and are deposited on ocean floors as pelagic ooze or alternatively is conglomerated in a coral reef.
Mylonite	Mylonite is a fine-grained, compact rock produced by dynamic crystallization of the constituent minerals resulting in a reduction of the grain size of the rock. It is classified as a metamorphic rock. They can have many different mineralogical compositions, it is a classification based on the textural appearance of the rock.
Gneiss	Gneiss is a common and widely distributed type of rock formed by high-grade regional metamorphic processes from preexisting formations that were originally either igneous or sedimentary rocks. Gneissic rocks are usually medium to coarse foliated and largely recrystallized but do not carry large quantities of micas, chlorite or other platy minerals.
Granulite	Granulite refers to metamorphic rocks that have experienced high temperatures of metamorphism. Many granulites represent samples of the deep continental crust.
Slate	Slate is a fine-grained, homogeneous, metamorphic rock derived from an original shale-type sedimentary rock composed of clay or volcanic ash through low grade regional metamorphism. The result is a foliated rock in which the foliation may not correspond to the original sedimentary layering.
Greenschist	Greenschist is a general field petrologic term applied to metamorphic and/or altered mafic volcanic rock. The green is due to abundant green chlorite, actinolite and epidote minerals that dominate the rock.
Marble	Marble is a metamorphic rock resulting from the metamorphism of limestone, composed mostly of calcite. It is extensively used for sculpture, as a building material, and in many other applications. The word 'marble' is colloquially used to refer to many other stones that are capable of taking a high polish.
Schist	The schist refers to a group of medium-grade metamorphic rocks, chiefly notable for the preponderance of lamellar minerals such as micas, chlorite, talc, hornblende, graphite, and others. Quartz often occurs in drawn-out grains to such an extent that a particular form called quartz schist is produced.
Quartzite	Quartzite is a hard, metamorphic rock which was originally sandstone. Sandstone is converted into quartzite through heating and pressure usually related to tectonic compression within orogenic belts.
Foliated metamorphic rock	Foliated metamorphic rock has penetrative planar fabric present within it. It is common to rocks affected by regional metamorphic compression typical of orogenic belts.
Hornfels	Hornfels is the group designation for a series of contact metamorphic rocks that have been baked and indurated by the heat of intrusive igneous masses and have been rendered massive, hard, splintery, and in some cases exceedingly tough and durable. Most hornfels are fine-grained.
Intrusion	An intrusion is a body of igneous rock that has crystallized from a molten magma below the surface of the Earth.

Equilibrium	Equilibrium is the condition of a system in which competing influences are balanced.
Sedimentary rock	Sedimentary rock is one of the three main rock groups. Sedimentary rock covers 75% of the Earth's land area. Four basic processes are involved in the formation of a clastic sedimentary rock: weathering caused mainly by friction of waves, transportation where the sediment is carried along by a current, deposition and compaction where the sediment is squashed together to form a rock of this kind.
Clay	Clay is a term used to describe a group of hydrous aluminium phyllosilicate minerals, that are typically less than 2 micrometres in diameter. Clay consists of a variety of phyllosilicate minerals rich in silicon and aluminium oxides and hydroxides which include variable amounts of structural water. Clays are generally formed by the chemical weathering of silicate-bearing rocks by carbonic acid but some are formed by hydrothermal activity.
Clay minerals	Clay minerals are hydrous aluminium phyllosilicates, sometimes with variable amounts of iron, magnesium, alkali metals, alkaline earths and other cations. Clays have structures similar to the micas and therefore form flat hexagonal sheets. Clay minerals are common weathering products and low temperature hydrothermal alteration products.
Stratification	In horticulture, stratification is the process of pretreating seeds to simulate natural conditions that a seed must endure before germination.
Vesicle	In cell biology, a vesicle is a relatively small and enclosed compartment, separated from the cytosol by at least one lipid bilayer.
Canadian Shield	The Canadian Shield is a large shield covered by a thin layer of soil that forms the nucleus of the North American craton. It has a deep, common, joined bedrock region in eastern and central Canada and stretches North from the Great Lakes to the Arctic Ocean, covering half the country.
Mount Everest	Mount Everest is the highest mountain on Earth, as measured by the height of its summit above sea level. The mountain, which is part of the Himalaya range in High Asia, is located on the border between Nepal and Tibet, China.
Conglomerate	A conglomerate is a rock consisting of individual stones that have become cemented together. Conglomerates are sedimentary rocks consisting of rounded fragements and are thus differentiated from breccias, which consist of angular clasts. Both conglomerates and breccias are characterized by clasts larger than sand.
Continental crust	The continental crust is the layer of granitic, sedimentary, and metamorphic rocks which form the continents and the areas of shallow seabed close to their shores, known as continental shelves. It is less dense than the material of the Earth's mantle and thus "floats" on top of it. Continental crust is also less dense than oceanic crust, though it is considerably thicker. About 40% of the Earth's surface is now underlain by continental crust.
Mountain range	A mountain range is a group of mountains bordered by lowlands or separated from other mountain ranges by passes or rivers. Individual mountains within the same mountain range do not necessarily have the same geology; they may be a mix of different orogeny, for example volcanoes, uplifted mountains or fold mountains and may, therefore, be of different rock.
Oceanic crust	Oceanic crust is the part of Earth's lithosphere that surfaces in the ocean basins. Oceanic crust is primarily composed of mafic rocks, or sima. It is thinner than continental crust, or sial, generally less than 10 kilometers thick, however it is more dense, having a mean density of about 3.3 grams per cubic centimeter.
Mantle	Earth's mantle is a ~2,900 km thick rocky shell comprizing approximately 70% of Earth's volume. It is predominantly solid and overlies the Earth's iron-rich core, which occupies about 30% of Earth

face=symbol>¢s volume. Past episodes of melting and volcanism at the shallower levels of the mantle have produced a very thin crust of crystallized melt products near the surface, upon which we live.

Chemical reaction	A chemical reaction is a process that results in the interconversion of chemical substances. The substance or substances initially involved in a chemical reaction are called reactants. Chemical reactions are characterized by a chemical change, and they yield one or more products which are, in general, different from the reactants.
Ion	An ion is an atom or group of atoms which have lost or gained one or more electrons, making them negatively or positively charged.
Crystal	A crystal is a solid in which the constituent atoms, molecules, or ions are packed in a regularly ordered, repeating pattern extending in all three spatial dimensions. Most metals encountered in everyday life are polycrystals. Crystals are often symmetrically intergrown to form crystal twins.
Andalusite	Andalusite is an aluminium nesosilicate mineral with the chemical formula Al_2SiO_5.
Migmatite	Migmatite is a rock at the frontier between igneous and metamorphic rocks. They can also be known as diatexite.
Island	An island is any piece of land that is completely surrounded by water, above high tide. There are two main types of islands: continental islands and oceanic islands. There are also artificial islands. A grouping of geographically and/or geologically related islands is called an archipelago.
Subduction	In geology, a subduction zone is an area on Earth where two tectonic plates meet and move towards one another, with one sliding underneath the other and moving down into the mantle, at rates typically measured in centimeters per year. An oceanic plate ordinarily slides underneath a continental plate; this often creates an orogenic zone with many volcanoes and earthquakes.
Continental collision	Continental collision is a phenomenon of the plate tectonics of Earth. Continental collision is a variation on the fundamental process of subduction, whereby the subduction zone is destroyed, mountains produced, and two continents sutured together. Continental collision is known only from this planet and is an interesting example of how our different crusts, oceanic and continental, behave during subduction.
Geology	Geology is the science and study of the solid matter that constitute the Earth. Encompassing such things as rocks, soil, and gemstones, geology studies the composition, structure, physical properties, history, and the processes that shape Earth's components.
Gold	Gold is a highly sought-after precious metal which, for many centuries, has been used as money, a store of value and in jewelery. The metal occurs as nuggets or grains in rocks, underground "veins" and in alluvial deposits. It is one of the coinage metals. Itis dense, soft, shiny and the most malleable and ductile of the known metals.
Rapid	A rapid is a section of a river of relatively steep gradient causing an increase in water flow and turbulence. A rapid is a hydrological feature between a run and a cascade. It is characterized by the river becoming shallower and having some rocks exposed above the flow surface.
Sillimanite	Sillimanite also called Bucholzite is an alumino-silicate mineral with the chemical formula Al_2SiO_5.
Kyanite	Kyanite, whose name derives from the Greek word kyanos, meaning blue, is a typically blue

Go to **Cram101.com** for the Practice Tests for this Chapter.

silicate mineral, commonly found in aluminium-rich metamorphic pegmatites and/or sedimentary rock. Kyanite is a diagnostic mineral of the Blueschist Facies of metamorphic rocks.

Sediment	Sediment is any particulate matter that can be transported by fluid flow and which eventually is deposited as a layer of solid particles on the bed or bottom of a body of water or other liquid.
Thrust	A thrust fault is a particular type of fault, or break in the fabric of the Earth face=symbol>¢s crust with resulting movement of each side against the other, in which a lower stratigraphic position is pushed up and over another. This is the result of compressional forces.
Diamonds	Diamonds are the hardest natural material known to man and the third-hardest known material. Its hardness and high dispersion of light make it useful for industrial applications and jewelry.
Carbon	Carbon is a chemical element. An abundant nonmetallic, tetravalent element, carbon has several allotropic forms. This element is the basis of the chemistry of all known life.
Graphite	Graphite is one of the allotropes of carbon. It holds the distinction of being the most stable form of solid carbon ever discovered. It may be considered to be the highest grade of coal, just above anthracite, although it is not normally used as fuel because it is hard to ignite.
Earths atmosphere	Earths atmosphere is a layer of gases surrounding the planet Earth and retained by the Earth's gravity. This mixture of gases is commonly known as air.
Atmospheric pressure	Atmospheric pressure is the pressure at any point in the Earth's atmosphere.
Metasomatism	Metasomatism is the chemical alteration of a rock by hydrothermal and other fluids. Metasomatism can occur via the action of hydrothermal fluids from an igneous or metamorphic source. In the metamorphic environment, metasomatism is created by mass transfer from a volume of metamorphic rock at higher stress and temperature into a zone with lower stress and temperature, with metamorphic hydrothermal solutions acting as a solvent.
Element	An element is a type of atom that is defined by its atomic number; that is, by the number of protons in its nucleus.
Carbon dioxide	Carbon dioxide is a chemical compound, normally in a gaseous state, and is composed of one carbon and two oxygen atoms. It is often referred to by its formula CO_2. It is present in the Earth's atmosphere at a concentration of approximately .000383 by volume and is an important greenhouse gas due to its ability to absorb many infrared wavelengths of sunlight, and due to the length of time it stays in the atmosphere.
Crystal structure	A crystal structure is a unique arrangement of atoms in a crystal. It is composed of a unit cell, a set of atoms arranged in a particular way, which is periodically repeated in three dimensions on a lattice. The spacing between unit cells in various directions is called its lattice parameters. The symmetry properties of the crystal are embodied in its space group.
Porosity	Porosity is a measure of the void spaces in a material, and is measured as a fraction, between 0–1, or as a percentage between 0–100%.
Ore	An ore is a volume of rock containing components or minerals in a mode of occurrence that renders it valuable for mining.
Quartz	Quartz is the second most common mineral in the Earth's continental crust. It is made up of a lattice of silica tetrahedra. Quartz belongs to the rhombohedral crystal system. In nature quartz crystals are often twinned, distorted, or so intergrown with adjacent crystals of quartz or other minerals as to only show part of this shape, or to lack

obvious crystal faces altogether and appear massive.

Seawater	Seawater is water from a sea or ocean. On average, seawater in the world face=symbol>¢s oceans has a salinity of ~3.5%, or 35 parts per thousand. This means that every 1 kg of seawater has approximately 35 grams of dissolved salts.
Serpentine	Serpentine is a group of common rock-forming hydrous magnesium iron phyllosilicate face=symbol size=2>½Mg, Fe size=2>½$_3$Si$_2$O$_5$ size=2>½OH |$_4$minerals; it may contain minor amounts of other elements including chromium, manganese, cobalt and nickel. There are three important mineral polymorphs of serpentine: antigorite, chrysotile and lizardite.
Talc	Talc is a mineral composed of hydrated magnesium silicate with the chemical formula H2Mg32. It occurs as foliated to fibrous masses, its monoclinic crystals being so rare as to be almost unknown. It has a perfect basal cleavage, and the folia are non-elastic, although slightly flexible.
Silicate	In geology and astronomy, the term silicate is used to denote types of rock that consist predominantly of silicate minerals. Such rocks include a wide range of igneous, metamorphic and sedimentary types. Most of the Earth's mantle and crust are made up of silicate rocks. The same is true of the Moon and the other rocky planets.
Ridge	A ridge is a geological feature that is also known as a Rip in the earth causing magma to flow out and forming an undersea volcano, it also has geological features, a continuous elevational crest for some distance. Ridges are usually termed hills or mountains as well, depending on size.
Olivine	The mineral olivine is a magnesium iron silicate. It is one of the most common minerals on Earth, and has also been identified on the Moon, Mars, and comet Wild 2.
Pyroxenes	The pyroxenes are a group of important rock-forming silicate minerals found in many igneous and metamorphic rocks. They share a common structure comprised of single chains of silica tetrahedra and they crystalise in the monoclinic and orthorhombic system.
Chlorite	A chlorite is a compound that contains this group, with chlorine in oxidation state +3. They are also known as salts of chlorous acid.
Compression	In geology the term compression refers to the system of forces that tend to decrease the volume of or shorten rocks. Compressive strength refers to the maximum compressive stress that can be applied to a material before failure occurs.
Mica	The mica group of sheet silicate minerals includes several closely related materials having highly perfect basal cleavage. All are monoclinic with a tendency towards pseudo-hexagonal crystals and are similar in chemical composition. The highly perfect cleavage, which is the most prominent characteristic of mica, is explained by the hexagonal sheet-like arrangement of its atoms.
Granite	Granite is a common and widely occurring type of intrusive, felsic, igneous rock. Granites are usually medium to coarsely crystalline, occasionally with some individual crystals larger than the groundmass forming a rock known as porphyry. Granites can be pink to dark gray or even black, depending on their chemistry and mineralogy.
Breccia	Breccia is a rock composed of angular fragments of rocks or minerals in a matrix, that is a cementing material, that may be similar or different in composition to the fragments.
Crystallization	Crystallization is the process of formation of solid crystals from a uniform solution. It is also a chemical solid-liquid separation technique, in which mass transfer of a solute from the liquid solution to a pure solid crystalline phase occurs.

Competition	Competition between members of a species is the driving force behind evolution and natural selection; especially for resources such as food, water, territory, and sunlight results in the ultimate survival and dominance of the variation of the species best suited for survival.
Sandstone	Sandstone is a sedimentary rock composed mainly of sand-size mineral or rock grains. Most sandstone is composed of quartz and/or feldspar because these are the most common minerals in the Earth's crust. Like sand, sandstone may be any color, but the most common colors are tan, brown, yellow, red, gray and white.
Shale	Shale is a fine-grained sedimentary rock whose original constituents were clays or muds. It is characterized by thin laminae breaking with an irregular curving fracture, often splintery and usually parallel to the often-indistinguishable bedding plane.
Cleavage	Cleavage, in mineralogy, is the tendency of crystalline materials to split along definite planes, creating smooth surfaces.
Bedding planes	Bedding planes are where one sedimetary deposit ends and another one begins. The rock is prone to breakage at these points because of the weakness between the layers.
Silt	Silt is soil or rock derived granular material of a specific grain size. Silt may occur as a soil or alternatively as suspended sediment in a water column of any surface water body. It may also exist as deposition soil at the bottom of a water body.
Luster	Luster is a description of the way light interacts with the surface of a crystal, rock, or mineral. For example, a diamond is said to have an adamantine luster and pyrite is said to have a metallic luster.
Phyllite	Phyllite is a type of foliated metamorphic rock primarily composed of quartz, sericite mica, and chlorite; the rock represents a gradiation in the degree of metamorphism between slate and mica schist. Minute crystals of graphite, sericite, or chlorite impart a silky, sometimes golden sheen to the surfaces of cleavage.
Muscovite	Muscovite is a phyllosilicate mineral of aluminium and potassium. It has a highly perfect basal cleavage yielding remarkably thin laminae , which are often highly elastic. Sheets of muscovite 5 metres by 3 metres have been found in Nellore, India.
Mafic	In geology, mafic minerals and rocks are silicate minerals, magmas, and volcanic and intrusive igneous rocks that have relatively high concentrations of the heavier elements. The term is a combination of "magnesium" and ferrum.
Feldspar	Feldspar is the name of a group of rock-forming minerals which make up as much as sixty percent of the Earth's crust. Feldspars crystallize from magma in both intrusive and extrusive rocks, and they can also occur as compact minerals, as veins, and are also present in many types of metamorphic rock.
Metaconglomerate	Metaconglomerate is the type of rock which originated from conglomerate after undergoing metamorphism.
Clasts	Clastic sedimentary rocks are rocks composed predominantly of broken pieces or clasts of older weathered and eroded rocks.
Hematite	Hematite is a very common mineral, colored black to steel or silver-gray, brown to reddish brown, or red. It is mined as the main ore of iron. Varieties include kidney ore, martite iron rose and specularite. While the forms of it vary, they all have a rust-red streak. it is harder than pure iron, but much more brittle.
Amphibole	Amphibole defines an important group of generally dark-colored rock-forming inosilicate minerals linked at the vertices and generally containing ions of iron and/or magnesium in their structures. Amphiboles crystallize into two crystal systems, monoclinic and orthorhombic.

Tuff	Tuff is a type of rock consisting of consolidated volcanic ash ejected from vents during a volcanic eruption.
Basalt	Basalt is a common gray to black extrusive volcanic rock. It is usually fine-grained due to rapid cooling of lava on the Earth's surface. It may be porphyritic containing larger crystals in a fine matrix, or vesicular, or frothy scoria.
Biotite	Biotite is a common phyllosilicate mineral within the mica group. Primarily a solid-solution series between the iron-endmember annite, and the magnesium-endmember phlogopite; more aluminous endmembers include siderophyllite.
Aluminum	Aluminum is a silvery and ductile member of the poor metal group of chemical elements. It has the symbol Al and atomic number 13.
Gabbro	Gabbro is a dark, coarse-grained, intrusive igneous rock chemically equivalent to basalt. It is a plutonic rock, formed when molten magma is trapped beneath the Earth face=symbol>¢s surface and cools into a crystalline mass.
Amphibolite	Amphibolite is the name given to a rock consisting mainly of hornblende amphibole, the use of the term being restricted, however, to metamorphic rocks. The modern terminology for a holocrystalline plutonic igneous rocks rock composed primarily of hornblende amphibole is a hornblendite, which are usually crystal cumulates.
Abundance	Abundance is an ecological concept referring to the relative representation of a species in a particular ecosystem. It is usually measured as the mean number of individuals found per sample.
Groundmass	Groundmass rock is the fine-grained mass of material in which larger grains or crystals are embedded. The groundmass of an igneous rock consists of fine-grained, often microscopic, crystals in which larger crystals are embedded. This porphyritic texture is indicative of multi-stage cooling of magma.
Chert	Chert is a fine-grained silica-rich cryptocrystalline sedimentary rock that may contain small fossils. It varies greatly in color from white to black, but most often manifests as gray, brown, grayish brown and light green to rusty red; its color is an expression of trace elements present in the rock, and both red and green are most often related to traces of iron.
Transform fault	A transform fault is a geological fault that is a special case of strike-slip faulting which terminates abruptly, at both ends, at a major transverse geological feature. Also known as a conservative plate boundary.
Fault	Faults are planar rock fractures, which show evidence of relative movement. Large faults within the Earth's crust are the result of shear motion and active fault zones are the causal locations of most earthquakes. Earthquakes are caused by energy release during rapid slippage along faults. The largest examples are at tectonic plate boundaries but many faults occur far from active plate boundaries. Since faults do not usually consist of a single, clean fracture, the term fault zone is used when referring to the zone of complex deformation that is associated with the fault plane.
Oxide	An oxide is a chemical compound containing an oxygen atom and other elements. Most of the earth's crust consists of them. They result when elements are oxidized by air.
Iron	Iron is a chemical element metal. It is a lustrous, silvery soft metal. It and nickel are notable for being the final elements produced by stellar nucleosynthesis, and thus are the heaviest elements which do not require a supernova or similarly cataclysmic event for formation.

Dolostone	Dolostone is a sedimentary carbonate rock that contains a high percentage of the mineral dolomite. It is usually referred to as dolomite rock. In old U.S.G.S. publications it was referred to as magnesian limestone.
Calcite	The carbonate mineral Calcite is a chemical or biochemical calcium carbonate and is one of the most widely distributed minerals on the Earth's surface. It is a common constituent of sedimentary rocks, limestone in particular. It is also the primary mineral in metamorphic marble
Coloration	Coloration is the state of being colored, distinguishing a species due to color arrangement.
Flint	Flint is a hard, sedimentary cryptocrystalline silicate form of the mineral quartz, categorized as a variety of chalcedony and broadly part of the mineral group known as silicas. When struck against steel, it will produce sparks, which when directed onto tinder can be used to start a fire.
Plagioclase	Plagioclase is a very important series of tectosilicate minerals within the feldspar family. Rather than referring to a particular mineral with a specific chemical composition, it is a solid solution series.
Epidote	Epidote is a calcium aluminium iron sorosilicate mineral, crystallizing in the monoclinic system. It occurs in crystalline limestones and schistose rocks of metamorphic origin. It is also a product of hydrothermal alteration of various minerals composing igneous rocks.
Greenstone belts	Greenstone belts are zones of variably metamorphosed mafic to ultramafic volcanic sequences with associated sedimentary rocks that occur within Archaean and Proterozoic cratons between granite and gneiss bodies. The belts have been interpreted as having formed at ancient oceanic spreading centers and island arc terranes.
Lava	Lava is molten rock expelled by a volcano during an eruption. When first extruded from a volcanic vent, it is a liquid at temperatures from 700 °C to 1,200 °C.
Spectrum	A spectrum is a condition or value that is not limited to a specific set of values but can vary infinitely within a continuum. The word saw its first scientific use within the field of optics to describe the rainbow of colors in visible light when separated using a prism; it has since been applied by analogy to many fields.
Parent material	Parent material, in soil science, means the underlying geological material in which soil horizons form. Soils typically get a great deal of structure and minerals from their parent material.
Rhyolite	Rhyolite is an igneous, volcanic rock, of felsic composition. It may have any texture from aphanitic to porphyritic. The mineral assemblage is usually quartz, alkali feldspar and plagioclase. Biotite and pyroxene are common accessory minerals.
Facies	A facies should ideally be a distinctive rock that forms under certain conditions of sedimentation, reflecting a particular process or environment.
Compaction	Compaction is the process of a material being more closely packed together.
Cementation	Cementation is the process of deposition of dissolved mineral components in the interstices of sediments. It is an important factor in the consolidation of coarse-grained clastic sedimentary rocks such as sandstones, conglomerates, or breccias during diagenesis or lithification. Cementing materials may include silica, carbonates, iron oxides, or clay minerals.
Hornblende	Hornblende is a complex inosilicate series of minerals. Hornblende is not a recognized mineral, in its own right but the name is used as a general or field term, to refer to a dark amphibole. It is an isomorphous mixture of three molecules; a calcium-iron-magnesium silicate, an aluminium-iron-magnesium silicate and an iron-magnesium silicate.

Blueschist	Blueschist is a rock that forms by the metamorphism of basalt and rocks with similar composition at high pressures and low temperatures, approximately corresponding to a depth of 15 to 30 kilometers and 200 to ~500 degrees Celsius. The blue color of the rock comes from the presence of the mineral glaucophane.
Blueschist facies	Blueschist facies is determined by the particular Temperature-Pressure conditions required to metamorphose basalt to form blueschist. Felsic rocks and pelitic sediments which are subjected to blueschist facies conditions will form different mineral assemblages than metamorphosed basalt.
Plate tectonics	Plate tectonics is a theory of geology that has been developed to explain the observed evidence for large scale motions of the Earth's lithosphere. The theory encompassed and superseded the older theory of continental drift.
Subduction zone	A subduction zone is an area on Earth where two tectonic plates meet and move towards one another, with one sliding underneath the other and moving down into the mantle, at rates typically measured in centimeters per year. In a sense, subduction zones are the opposite of divergent boundaries, areas where material rises up from the mantle and plates are moving apart.
Rift	In geology, a rift is a place where the Earth's crust and lithosphere are being pulled apart.
Rift zone	A rift zone is a feature of some volcanoes in which a linear series of fissures in the volcanic edifice allows lava to be erupted from the volcano's flank instead of from its summit.
Isostasy	Isostasy is a term used in Geology to refer to the state of gravitational equilibrium between the Earth's lithosphere and asthenosphere such that the tectonic plates "float" at an elevation which depends on their thickness and density. It is invoked to explain how different topographic heights can exist at the Earth's surface.
Seafloor	The seafloor the bottom of the ocean. At the bottom of the continental slope is the continental rise, which is caused by sediment cascading down the continental slope.
Fracture zone	A fracture zone linear oceanic feature--often hundreds, even thousands of kilometers long-- resulting from the action of offset mid-ocean ridge axis segments; they are a consequence of plate tectonics.
Mantle plume	A mantle plume is an upwelling of abnormally hot rock within the Earth face=symbol>¢s mantle. As the heads of mantle plumes can partly melt when they reach shallow depths, they are thought to be the cause of volcanic centers known as hotspots and probably also to have caused flood basalts.
Particle size	Particle size refers to the diameter of individual grains of sediment, or the lithified particles in clastic rocks. The term may also be applied to other granular materials. This is different from the crystallite size, which is the size of a single crystal inside the particles or grains.
Radiometric dating	Radiometric dating is a technique used to date materials based on a knowledge of the decay rates of naturally occurring isotopes, and the current abundances. It is the principal source of information about the age of the Earth and a significant source of information about rates of evolutionary change.
Radiometry	In optics, radiometry is the field that studies the measurement of electromagnetic radiation, including visible light. Note that light is also measured using the techniques of photometry, which deal with brightness as perceived by the human eye, rather than absolute power.

Petrology	Petrology is a field of geology which focuses on the study of rocks and the conditions by which they form. There are three branches of petrology, corresponding to the three types of rocks: igneous, metamorphic, and sedimentary. Petrology utilizes the classical fields of mineralogy, petrography, optical mineralogy, and chemical analyses to describe the composition and texture of rocks.

Crust	In geology, a crust is the outermost layer of a planet, part of its lithosphere. They are generally composed of a less dense material than its deeper layers.Earths face=symbol>¢ is composed mainly of basalt and granite. It is cooler and more rigid than the deeper layers of the mantle and core.
Earthquake	An earthquake is the result from the sudden release of stored energy in the Earth face=symbol>¢s crust that creates seismic waves. At the Earth face=symbol>¢s surface, earthquakes may manifest themselves by a shaking or displacement of the ground. An earthquake is caused by tectonic plates getting stuck and putting a strain on the ground. The strain becomes so great that rocks give way by breaking and sliding along fault planes.
Fault	Faults are planar rock fractures, which show evidence of relative movement. Large faults within the Earth's crust are the result of shear motion and active fault zones are the causal locations of most earthquakes. Earthquakes are caused by energy release during rapid slippage along faults. The largest examples are at tectonic plate boundaries but many faults occur far from active plate boundaries. Since faults do not usually consist of a single, clean fracture, the term fault zone is used when referring to the zone of complex deformation that is associated with the fault plane.
San Andreas fault	The San Andreas Fault is a geological fault that runs a length of roughly 800 miles through western and southern California in the United States. The fault, a right-lateral strike-slip fault, marks a transform boundary between the Pacific Plate and the North American Plate.
Sea level	Mean sea level is the average height of the sea, with reference to a suitable reference surface.
Coast	The coast is defined as the part of the land adjoining or near the ocean. A coastline is properly a line on a map indicating the disposition of a coast, but the word is often used to refer to the coast itself. The adjective coastal describes something as being on, near to, or associated with a coast.
Terrace	In agriculture, a terrace is a leveled section of a hilly cultivated area, designed as a method of soil conservation to slow or prevent the rapid surface runoff of irrigation water
Shoreline	A shoreline is the fringe of land at the edge of a large body of water, such as an ocean, sea, or lake. A strict definition is the strip of land along a water body that is alternately exposed and covered by waves and tides.
Wave	A wave is a disturbance that propagates through space or spacetime, transferring energy and momentum and sometimes angular momentum.
Barnacle	A barnacle is a type of arthropod belonging to infraclass Cirripedia in the subphylum Crustacea and is hence distantly related to crabs and lobsters.
Cliff	A cliff is a significant vertical, or near vertical, rock exposure. Cliffs are categorized as erosion landforms due to the processes of erosion and weathering that produce them. Cliffs are common on coasts, in mountainous areas, escarpments and along rivers. Cliffs are usually formed by rock that is resistant to erosion and weathering.
Continent	A continent is one of several large landmasses on Earth. They are generally identified by convention rather than any strict criteria, but seven areas are commonly reckoned as continents – they are: Asia, Africa, North America, South America, Antarctica, Europe, and Australia.
Geometry	Geometry is a part of mathematics concerned with questions of size, shape, and relative position of figures and with properties of space. Geometry is one of the oldest sciences. Initially a body of practical knowledge concerning lengths, areas, and volumes, in the third century B.C. geometry was put into an axiomatic form by Euclid, whose treatment set a

standard for many centuries to follow.

Mountain	A mountain is a landform that extends above the surrounding terrain in a limited area. A mountain is generally steeper than a hill, but there is no universally accepted standard definition for the height of a mountain or a hill although a mountain usually has an identifiable summit.
Sedimentary	Sedimentary rock is one of the three main rock groups. Rock formed from these covers 75% of the Earth's land area, and includes common types such as chalk, limestone, dolomite, sandstone, and shale.
Sedimentary rock	Sedimentary rock is one of the three main rock groups. Sedimentary rock covers 75% of the Earth's land area. Four basic processes are involved in the formation of a clastic sedimentary rock: weathering caused mainly by friction of waves, transportation where the sediment is carried along by a current, deposition and compaction where the sediment is squashed together to form a rock of this kind.
Thrust	A thrust fault is a particular type of fault, or break in the fabric of the Earth face=symbol>¢s crust with resulting movement of each side against the other, in which a lower stratigraphic position is pushed up and over another. This is the result of compressional forces.
Rocky Mountains	The Rocky Mountains are a broad mountain range in western North America. The Rocky Mountains stretch more than 4,800 kilometers from northernmost British Columbia, in Canada, to New Mexico, in the United States.
Ural Mountains	The Ural Mountains are a mountain range that runs roughly north and south through western Russia. They are sometimes considered as the natural boundary between Europe and Asia. It extends 2,500 km from the Kazakh steppes along the northern border of Kazakhstan to the coast of the Arctic ocean.
Himalayas	The Himalayas are a mountain range in Asia, separating the Indian subcontinent from the Tibetan Plateau. By extension, it is also the name of the massive mountain system which includes the Himalaya proper, the Karakoram, the Hindu Kush, and a host of minor ranges extending from the Pamir Knot.
Alps	The Alps is the name for one of the great mountain range systems of Europe, stretching from Austria and Slovenia in the east, through Italy, Switzerland, Liechtenstein and Germany to France in the west.
Erosion	Erosion is displacement of solids by the agents of ocean currents, wind, water, or ice by downward or down-slope movement in response to gravity or by living organisms.
Tectonics	Tectonics is a field of study within geology concerned generally with the structures within the crust of the Earth, or other planets, and particularly with the forces and movements that have operated in a region to create these structures.
Geological time scale	The geological time scale is used by geologists and other scientists to describe the timing and relationships between events that have occurred during the history of Earth.
Thrust Fault	A thrust fault is a particular type of fault, or break in the fabric of the Earth face=symbol>¢s crust with resulting movement of each side against the other, in which a lower stratigraphic position is pushed up and over another. This is the result of compressional forces.
Fold	The term fold is used in geology when one or a stack of originally flat and planar surfaces, such as sedimentary strata, are bent or curved as a result of plastic, i.e. permanent, deformation.
Syncline	In structural geology, a syncline is a downward-curving fold, with layers that dip toward the

center of the structure. On a geologic map, synclines are recognized by a sequence of rock layers that grow progressively younger, followed by the youngest layer at the fold face=symbol>¢s center or hinge, and by a reverse sequence of the same rock layers on the opposite side of the hinge.

Dome	In geology, a dome is a deformational feature consisting of symmetrically-dipping anticlines; their general outline on a geologic map is circular or oval.
Anticline	An anticline is a fold that is convex up or to the youngest beds. Anticlines are usually recognized by a sequence of rock layers that are progressively older toward the center of the fold because the uplifted core of the fold is preferentially eroded to a deeper stratigraphic level relative to the topographically lower flanks. If an anticline plunges, the surface strata will form Vs that point in the direction of the plunge.
Cohesion	Cohesion in chemistry is the intermolecular attraction between like-molecules. It explains phenomena such as surface tension.
Magma	Magma is molten rock located beneath the surface of the Earth, and which often collects in a magma chamber. Magma is a complex high-temperature fluid substance. Most are silicate solutions. It is capable of intrusion into adjacent rocks or of extrusion onto the surface as lava or ejected explosively as tephra to form pyroclastic rock. Environments of magma formation include subduction zones, continental rift zones, mid-oceanic ridges, and hotspots, some of which are interpreted as mantle plumes.
Swell	A swell, in the context of an ocean, is a formation of long wavelength ocean surface waves on the sea. They are far more stable in their directions and frequency than normal oceanic waves since they are formed by tropical storms and by stable wind systems.
Rapid	A rapid is a section of a river of relatively steep gradient causing an increase in water flow and turbulence. A rapid is a hydrological feature between a run and a cascade. It is characterized by the river becoming shallower and having some rocks exposed above the flow surface.
Mantle	Earth's mantle is a ~2,900 km thick rocky shell comprizing approximately 70% of Earth's volume. It is predominantly solid and overlies the Earth's iron-rich core, which occupies about 30% of Earth face=symbol>¢s volume. Past episodes of melting and volcanism at the shallower levels of the mantle have produced a very thin crust of crystallized melt products near the surface, upon which we live.
Metamorphic rock	Metamorphic rock is the result of the transformation of a pre-existing rock type, the protolith, in a process called metamorphism, which means "change in form". The protolith is subjected to heat and extreme pressure causing profound physical and/or chemical change. The protolith may be sedimentary rock, igneous rock or another older rock.
Metamorphic rocks	Metamorphic rock is the result of the transformation of a pre-existing rock type, the protolith, in a process called metamorphism. The protolith is subjected to heat and extreme pressure causing profound physical and/or chemical change. Metamorphic rocks make up a large part of the Earth's crust. They are formed deep beneath the Earth's surface by great stresses from rocks above and high pressures and temperatures.
Metamorphism	Metamorphism can be defined as the solid state recrystallisation of pre-existing rocks due to changes in heat and/or pressure and/or introduction of fluids. There will be mineralogical, chemical and crystallographic changes. Metamorphism produced with increasing pressure and temperature conditions is known as prograde metamorphism. Conversely, decreasing temperatures and pressure characterize retrograde metamorphism.

Go to **Cram101.com** for the Practice Tests for this Chapter.

Mineral	A mineral is a naturally occurring substance formed through geological processes that has a characteristic chemical composition, a highly ordered atomic structure and specific physical properties. A rock, by comparison, is an aggregate of minerals and need not have a specific chemical composition. Minerals range in composition from pure elements and simple salts to very complex silicates with thousands of known forms.
Compression	In geology the term compression refers to the system of forces that tend to decrease the volume of or shorten rocks. Compressive strength refers to the maximum compressive stress that can be applied to a material before failure occurs.
Blocks	Blocks in meteorology are large scale patterns in the atmospheric pressure field that are nearly stationary, effectively "blocking" or redirecting migratory cyclones. These blocks can remain in place for several days or even weeks, causing the areas affected by them to have the same kind of weather for an extended period of time.
Convergent boundary	In plate tectonics, a convergent boundary is an actively deforming region where two tectonic plates or fragments of lithosphere move towards one another. When two plates move toward one another, they form either a subduction zone or a continental collision.
Bedding planes	Bedding planes are where one sedimetary deposit ends and another one begins. The rock is prone to breakage at these points because of the weakness between the layers.
Alcatraz Island	Alcatraz Island is a small island located in the middle of San Francisco Bay in California, United States. It served as a lighthouse, then a military fortification, then a military prison followed by a federal prison until 1963, when it became a national recreation area.
Ridge	A ridge is a geological feature that is also known as a Rip in the earth causing magma to flow out and forming an undersea volcano, it also has geological features, a continuous elevational crest for some distance. Ridges are usually termed hills or mountains as well, depending on size.
Sandstone	Sandstone is a sedimentary rock composed mainly of sand-size mineral or rock grains. Most sandstone is composed of quartz and/or feldspar because these are the most common minerals in the Earth's crust. Like sand, sandstone may be any color, but the most common colors are tan, brown, yellow, red, gray and white.
Tributary	A tributary is a stream or river which flows into a mainstem river, and which does not flow directly into a sea. In orography, they are ordered from those nearest to the source of the river to those nearest to the mouth of the river.
Stream	A stream is a body of water with a current, confined within a bed and banks. Streams are important as conduits in the water cycle, instruments in aquifer recharge, and corridors for fish and wildlife migration.
Volcanic rock	Volcanic rock is an igneous rock of volcanic origin. They often have a vesicular texture, which is the result voids left by volatiles escaping from the molten lava. Pumice is a rock, which is an example of explosive volcanic eruption. It is so vesicular that it floats in water.
Lava	Lava is molten rock expelled by a volcano during an eruption. When first extruded from a volcanic vent, it is a liquid at temperatures from 700 °C to 1,200 °C.
Weathering	Weathering is the process of breaking down rocks, soils and their minerals through direct contact with the atmosphere. Weathering occurs without movement. Two main classifications of weathering processes exist. Mechanical or physical weathering involves the breakdown of rocks and soils through direct contact with atmospheric conditions. The second classification, chemical weathering, involves the direct effect of atmospheric chemicals in the breakdown of rocks, soils and minerals.

Go to **Cram101.com** for the Practice Tests for this Chapter.

Arches National Park	Arches National Park preserves over 2,000 natural sandstone arches, including the world-famous Delicate Arch, in addition to a variety of unique geological resources and formations.
Crevasse	A crevasse is a fracture in a glacier caused by large tensile stresses at or near the glacier's surface.
Deposition	Deposition is the geological process whereby material is added to a landform. This is the process by which wind and water create a sediment deposit, through the laying down of granular material that has been eroded and transported from another geographical location.
Gold	Gold is a highly sought-after precious metal which, for many centuries, has been used as money, a store of value and in jewelery. The metal occurs as nuggets or grains in rocks, underground "veins" and in alluvial deposits. It is one of the coinage metals. Itis dense, soft, shiny and the most malleable and ductile of the known metals.
Silver	Silver is a soft white lustrous transition metal, it has the highest electrical and thermal conductivity for a metal.
Tungsten	Tungsten is a very hard, heavy, steel-gray to white transition metal, it is found in several ores including wolframite and scheelite and is remarkable for its robust physical properties, especially the fact that it has the highest melting point of all the non-alloyed metals and the second highest of all the elements after carbon. The pure form is used mainly in electrical applications.
Zinc	Zinc is a chemical element in the periodic table that has the symbol Zn and atomic number 30. In some historical and sculptural contexts, it is known as spelter.
Mercury	Mercury is a chemical element in the periodic table that has the symbol Hg and atomic number 80. A heavy, silvery transition metal, mercury is one of five elements that are liquid at or near room temperature and pressure.
Ore	An ore is a volume of rock containing components or minerals in a mode of occurrence that renders it valuable for mining.
Copper	Copper is a ductile metal with excellent electrical conductivity, and finds extensive use as an electrical conductor, heat conductor, as a building material, and as a component of various alloys.
Intrusion	An intrusion is a body of igneous rock that has crystallized from a molten magma below the surface of the Earth.
Igneous	Igneous rocks form when molten rock, magma, cools and solidifies, with or without crystallization, either below the surface as intrusive, plutonic rocks or on the surface as extrusive, volcanic, rocks.
Plucking	Plucking, in the sense relating to glaciers, is when a glacier erodes away chunks of bedrock to be later deposited as glacial erratics. Glacial plucking exploits pre-existing fractures in the bedrock. When the ice comes into contact with a joint, the friction on the ice results in melting of some of the ice.
Quarry	A quarry is a type of open-pit mine from which rock or minerals are extracted. They are generally used for extracting building materials, such as dimension stone. They are usually shallower than other types of open-pit mines.
Canyon	A canyon is a deep valley between cliffs often carved from the landscape by a river. Most were formed by a process of long-time erosion from a plateau level. The cliffs form because harder rock strata that are resistant to erosion and weathering remain exposed on the valley walls.
Creep	Creep, is the slow downward progression of rock and soil down a low grade slope; it can also

refer to slow deformation of such materials as a result of prolonged pressure and stress.

Escarpment	An escarpment is a transition zone between different physiogeographic provinces that involves an elevation differential, often involving high cliffs. Most commonly, an escarpment, is a transition from one series of sedimentary rocks to another series of a different age and composition. In such cases, the escarpment usually represents the line of erosional loss of the newer rock over the older.
Valley	In geology, a valley is a depression with predominant extent in one direction. The terms U-shaped and V-shaped are descriptive terms of geography to characterize the form of valleys. Most valleys belong to one of these two main types or a mixture of them, at least with respect of the cross section of the slopes or hillsides.
Lithosphere	The lithosphere is the solid outermost shell of a rocky planet. On the Earth, the lithosphere includes the crust and the uppermost mantle which is joined to the crust across the Mohorovièiæ discontinuity. Lithosphere is underlain by asthenosphere, the weaker, hotter, and deeper part of the upper mantle.
Rift	In geology, a rift is a place where the Earth's crust and lithosphere are being pulled apart.
Oceanic ridge	A oceanic ridge is an underwater mountain range, formed by plate tectonics. This uplifting of the ocean floor occurs when convection currents rise in the mantle beneath the oceanic crust and create magma where two tectonic plates meet at a divergent boundary.
Basin and Range	The United States Basin and Range is a particular type of topography that covers much of the southwestern United States and northwestern Mexico that is typified by elongate north-south trending arid valleys bounded by mountain ranges which also bound adjacent valleys.
Graben	A graben is a depressed block of land bordered by parallel faults. A graben is the result of a block of land being downthrown producing a valley with a distinct scarp on each side.
Horst	In physical geography and geology, a horst is the raised fault block bounded by normal faults. The raised block is a portion of the Earth's crust that has remained stationary while the land has sunk on either side of it or has been crushed by a mountain range against it.
Mountain range	A mountain range is a group of mountains bordered by lowlands or separated from other mountain ranges by passes or rivers. Individual mountains within the same mountain range do not necessarily have the same geology; they may be a mix of different orogeny, for example volcanoes, uplifted mountains or fold mountains and may, therefore, be of different rock.
Sierra Nevada	The Sierra Nevada is a mountain range that is almost entirely in the eastern portion of the U.S. state of California. The Sierra Nevada stretches 400 miles , from Fredonyer Pass in the north to Tehachapi Pass in the south. It is bounded on the west by California face=symbol>¢s Central Valley, and on the east by the Great Basin.
Alluvial	An alluvial plain is a relatively flat and gently sloping landform found at the base of a range of hills or mountains, formed by the deposition of alluvial soil over a long period of time by one or more rivers coming from the mountains.
Paleozoic	The Paleozoic is the earliest of three geologic eras of the Phanerozoic eon. The Paleozoic is subdivided into six geologic periods; from oldest to youngest they are: the Cambrian, Ordovician, Silurian, Devonian, Carboniferous, and Permian.
Drainage	Drainage is the natural or artificial removal of surface and sub-surface water from a given area. Many agricultural soils need drainage to improve production or to manage water supplies.
Drainage system	A drainage system is the pattern formed by the streams, rivers, and laked in a particular

watershed. They are governed by the topography of the land, whether a particular region is dominated by hard or soft rocks, and the gradient of the land.

Sag pond	A sag pond is a body of water, which forms as water collects in the lowest parts of the depression that forms between two strands of an active strike-slip fault .
Vegetation	Vegetation is a general term for the plant life of a region; it refers to the ground cover provided by plants, and is, by far, the most abundant biotic element of the biosphere. Primeval redwood forests, coastal mangrove stands, sphagnum bogs, desert soil crusts, roadside weed patches, wheat fields, cultivated gardens and lawns; are all encompassed by the term vegetation.
Groundwater	Groundwater is water located beneath the ground surface in soil pore spaces and in the fractures of geologic formations. Groundwater is recharged from, and eventually flows to, the surface naturally; natural discharge often occurs at springs and seeps, streams and can often form oases or wetlands.
Island	An island is any piece of land that is completely surrounded by water, above high tide. There are two main types of islands: continental islands and oceanic islands. There are also artificial islands. A grouping of geographically and/or geologically related islands is called an archipelago.
Geologic map	A geologic map is a special-purpose map made to show geological features. The stratigraphic contour lines are drawn on the surface of a selected deep stratum, so that they can show the topographic trends of the strata under the ground. It is not always possible to properly show this when the strata are extremely fractured, mixed, in some discontinuities, or where they are otherwise disturbed.
Orogeny	Orogeny is the process of building mountains, and may be studied as a tectonic structural event, as a geographical event and a chronological event, in that orogenic events cause distinctive structural phenomena and related tectonic activity, affect certain regions of rocks and crust and happen within a time frame.
Zagros Mountains	The Zagros Mountains make up Iran and Iraq's largest mountain range. They have a total length of 1,500 km from western Iran, specifically South Kurdistan region on the border with Iraq to the southern parts of the Persian Gulf. The mountain range ends at the Straits of Hormuz. The highest points in the Zagros Mountains are Zard Kuh .
Alpine	Alpine climate is the average weather for a region above the tree line. The climate becomes colder at high elevations—this characteristic is described by the lapse rate of air: air will tend to get colder as it rises, since it expands.
Salt dome	A salt dome is formed when a thick bed of evaporite minerals found at depth intrudes vertically into surrounding rock strata, forming a diapir.
Diapir	A diapir is an intrusion caused by buoyancy and pressure differentials. A diapir is any relatively mobile mass that intrudes into preexisting strata. Diapirs commonly intrude vertically upward along fractures or zones of structural weakness through more dense overlying rocks because of density contrast between a less dense, lower rock mass and overlying denser rocks. The density contrast manifests as a force of buoyancy.
Gypsum	Gypsum is a very soft mineral composed of calcium sulfate dihydrate, with the chemical formula $CaSO_4 \cdot 2H_2O$. Gypsum occurs in nature as flattened and often twinned crystals and transparent cleavable masses. It may also occur silky and fibrous. Finally it may also be granular or quite compact.
Ductile materials	Ductile materials have the mechanical property of being capable of sustaining large plastic deformations due to tensile stress without fracture in metals, such as being drawn into a wire. It is characterized by the material flowing under shear stress. It is contrasted with

brittleness.

Black Hills	The Black Hills are a small, isolated mountain range rizing from the Great Plains of North America in western South Dakota and extending into Wyoming, USA. Set off from the main body of the Rocky Mountains, the region is somewhat of a geological anomaly—accurately described as an "island of trees in a sea of grass.
Glacier	A glacier is a large, slow moving river of ice, formed from compacted layers of snow, that slowly deforms and flows in response to gravity. Glacier ice is the largest reservoir of fresh water on Earth, and second only to oceans as the largest reservoir of total water. Glaciers cover vast areas of polar regions but are restricted to the highest mountains in the tropics.
Delta	A delta is a landform where the mouth of a river flows into an ocean, sea, desert, estuary or lake. It builds up sediment outwards into the flat area which the river face=symbol>¢s flow encounters transported by the water and set down as the currents slow.
Subsidence	In geology, engineering, and surveying, subsidence is the motion of a surface as it shifts downward relative to a datum such as sea-level. The opposite of subsidence is uplift, which results in an increase in elevation. In meteorology, subsidence refers to the downward movement of air.
Mesozoic	The Mesozoic is one of three geologic eras of the Phanerozoic eon. The Mesozoic was a time of tectonic, climatic and evolutionary activity, shifting from a state of connectedness into their present configuration. The climate was exceptionally warm throughout the period, also playing an important role in the evolution and diversification of new animal species. By the end of the era, the basis of modern life was in place.
Seafloor	The seafloor the bottom of the ocean. At the bottom of the continental slope is the continental rise, which is caused by sediment cascading down the continental slope.
Element	An element is a type of atom that is defined by its atomic number; that is, by the number of protons in its nucleus.
Fossil	Fossils are the mineralized or otherwise preserved remains or traces of animals, plants, and other organisms. The totality of fossils, both discovered and undiscovered, and their placement in fossiliferous rock formations and sedimentary layers is known as the fossil record.
Limestone	Limestone is a sedimentary rock composed largely of the mineral calcite. Limestone often contains variable amounts of silica in the form of chert or flint, as well as varying amounts of clay, silt and sand as disseminations, nodules, or layers within the rock. The primary source of the calcite in limestone is most commonly marine organisms. These organisms secrete shells that settle out of the water column and are deposited on ocean floors as pelagic ooze or alternatively is conglomerated in a coral reef.
Tension	Tension is a reaction force applied by a stretched string, rope or a similar object on the objects which stretch it. The direction of the force of it is parallel to the string, towards the string.
Structural geology	Structural geology is the study of the three dimensional distribution of rock bodies and their planar or folded surfaces, and their internal fabrics.
Geology	Geology is the science and study of the solid matter that constitute the Earth. Encompassing such things as rocks, soil, and gemstones, geology studies the composition, structure, physical properties, history, and the processes that shape Earth's components.

Geological time scale	The geological time scale is used by geologists and other scientists to describe the timing and relationships between events that have occurred during the history of Earth.
Geology	Geology is the science and study of the solid matter that constitute the Earth. Encompassing such things as rocks, soil, and gemstones, geology studies the composition, structure, physical properties, history, and the processes that shape Earth's components.
Canyon	A canyon is a deep valley between cliffs often carved from the landscape by a river. Most were formed by a process of long-time erosion from a plateau level. The cliffs form because harder rock strata that are resistant to erosion and weathering remain exposed on the valley walls.
Grand Canyon	The Grand Canyon is a very colorful, steep-sided gorge, carved by the Colorado River in the U.S. state of Arizona. It is one of the first national parks in the United States.
Deposition	Deposition is the geological process whereby material is added to a landform. This is the process by which wind and water create a sediment deposit, through the laying down of granular material that has been eroded and transported from another geographical location.
Plateau	A plateau is an area of highland, usually consisting of relatively flat rural area.
Sedimentary	Sedimentary rock is one of the three main rock groups. Rock formed from these covers 75% of the Earth's land area, and includes common types such as chalk, limestone, dolomite, sandstone, and shale.
Sedimentary rock	Sedimentary rock is one of the three main rock groups. Sedimentary rock covers 75% of the Earth's land area. Four basic processes are involved in the formation of a clastic sedimentary rock: weathering caused mainly by friction of waves, transportation where the sediment is carried along by a current, deposition and compaction where the sediment is squashed together to form a rock of this kind.
Sea level	Mean sea level is the average height of the sea, with reference to a suitable reference surface.
Stream	A stream is a body of water with a current, confined within a bed and banks. Streams are important as conduits in the water cycle, instruments in aquifer recharge, and corridors for fish and wildlife migration.
Erosion	Erosion is displacement of solids by the agents of ocean currents, wind, water, or ice by downward or down-slope movement in response to gravity or by living organisms.
Dome	In geology, a dome is a deformational feature consisting of symmetrically-dipping anticlines; their general outline on a geologic map is circular or oval.
Lava	Lava is molten rock expelled by a volcano during an eruption. When first extruded from a volcanic vent, it is a liquid at temperatures from 700 °C to 1,200 °C.
Lava dome	A lava dome is a roughly circular mound-shaped protrusion resulting from the slow eruption of felsic lava from a volcano. The viscosity, or stickiness, of the lava does not allow for the lava to flow very far from its vent before solidifying. Domes may reach heights of several hundred meters, and can grow slowly and steadily for months or years. The sides of these structures are composed of unstable rock debris.
Cinder	A cinder is a fragment of cooled pyroclastic material, lava or magma.
Cinder cones	Cinder cones are steep, conical hills of volcanic fragments that accumulate around and downwind from a volcanic vent. The rock fragments, often called cinders are glassy and contain numerous gas bubbles "frozen" into place as magma exploded into the air and then cooled quickly.

Go to **Cram101.com** for the Practice Tests for this Chapter.

Organism	In biology and ecology, an organism is a living complex adaptive system of organs that influence each other in such a way that they function in some way as a stable whole.
Fossil	Fossils are the mineralized or otherwise preserved remains or traces of animals, plants, and other organisms. The totality of fossils, both discovered and undiscovered, and their placement in fossiliferous rock formations and sedimentary layers is known as the fossil record.
Mineral	A mineral is a naturally occurring substance formed through geological processes that has a characteristic chemical composition, a highly ordered atomic structure and specific physical properties. A rock, by comparison, is an aggregate of minerals and need not have a specific chemical composition. Minerals range in composition from pure elements and simple salts to very complex silicates with thousands of known forms.
Radioactive decay	Radioactive decay is the process in which an unstable atomic nucleus loses energy by emitting radiation in the form of particles or electromagnetic waves.
Relative dating	Before the advent of absolute dating in the 20th century, archaeologists and geologists were largely limited to the use of Relative Dating techniques. Estimates of the order of prehistoric and geological events were determined by using basic stratigraphic rules, and by observing where fossil organisms lay in the geological record, stratified bands of rocks present throughout the world.
Superposition	The basic idea of this is that an object, event or entity can be spanned across multiple realities or universes. When combined, these multiple, unique, pan-dimensional segments of the object, consciousness or event, make up parts or constituents of its superposition.
Radiometric dating	Radiometric dating is a technique used to date materials based on a knowledge of the decay rates of naturally occurring isotopes, and the current abundances. It is the principal source of information about the age of the Earth and a significant source of information about rates of evolutionary change.
Radiometry	In optics, radiometry is the field that studies the measurement of electromagnetic radiation, including visible light. Note that light is also measured using the techniques of photometry, which deal with brightness as perceived by the human eye, rather than absolute power.
Uniformitari-nism	Uniformitarianism refers to the principle that the same processes that shape the universe occurred in the past as they do now, and that the same laws of physics apply in all parts of the knowable universe.
Mountain	A mountain is a landform that extends above the surrounding terrain in a limited area. A mountain is generally steeper than a hill, but there is no universally accepted standard definition for the height of a mountain or a hill although a mountain usually has an identifiable summit.
Tectonic uplift	Tectonic uplift is a geological process most often caused by plate tectonics which increases elevation. The opposite of uplift is subsidence, which results in a decrease in elevation. Uplift may be orogenic or isostatic.
Continent	A continent is one of several large landmasses on Earth. They are generally identified by convention rather than any strict criteria, but seven areas are commonly reckoned as continents – they are: Asia, Africa, North America, South America, Antarctica, Europe, and Australia.
James Hutton	James Hutton was a Scottish geologist, noted for formulating uniformitarianism and the Plutonist School of thought. He is considered by many to be the father of modern geology.
Coast	The coast is defined as the part of the land adjoining or near the ocean. A coastline is properly a line on a map indicating the disposition of a coast, but the word is often used to

	refer to the coast itself. The adjective coastal describes something as being on, near to, or associated with a coast.
Matter	Matter is the substance of which physical objects are composed. Matter can be solid, liquid, plasma or gas. It constitutes the observable universe.
Planet	A planet, as defined by the International Astronomical Union, is a celestial body orbiting a star or stellar remnant that is massive enough to be rounded by its own gravity, not massive enough to cause thermonuclear fusion in its core, and has cleared its neighboring region of planetesimals.
Star	A star is a massive, luminous ball of plasma. Stars group together to form galaxies, and they dominate the visible universe. The nearest star to Earth is the Sun, which is the source of most of the energy on Earth, including daylight. Other stars are visible in the night sky, when they are not outshone by the Sun. A star shines because nuclear fusion in its core releases energy which traverses the star's interior and then radiates into outer space.
Gravitation	Gravitation, in everyday life, is most familiar as the agency that endows objects with weight. Gravitation is responsible for keeping the Earth and the other planets in their orbits around the Sun; for the formation of tides; and for various other phenomena that we observe. Gravitation is also the reason for the very existence of the Earth, the Sun, and most macroscopic objects in the universe; without it, matter would not have coalesced into these large masses, and life, as we know it, would not exist.
Hydrogen	Hydrogen is a chemical element represented by the symbol H and an atomic number of 1. At standard temperature and pressure it is a colorless, odorless, nonmetallic, tasteless, highly flammable diatomic gas . With an atomic mass of 1.00794 g/mol, hydrogen is the lightest element.Hydrogen is the most abundant of the chemical elements, constituting roughly 75% of the universe's elemental mass.
Universe	The Universe is defined as the summation of all particles and energy that exist and the space-time in which all events occur.
Charles Lyell	Charles Lyell was a Scottish lawyer, geologist, and populariser of uniformitarianism. Principles of Geology, his first book, was also his most famous, most influential, and most important. First published in three volumes in 1830-33, it established his credentials as an important geological theorist and introduced the doctrine of uniformitarianism.
Charles Darwin	Charles Darwin was already eminent as an English naturalist when he proposed and provided evidence for the theory that all species have evolved over time from one or a few common ancestors through the process of natural selection. The fact that evolution occurs became accepted by the scientific community and the general public in his lifetime, while his theory of natural selection came to be widely seen as the primary explanation of the process of evolution in the 1930s, and now forms the basis of modern evolutionary theory. In modified form, Darwin's theory remains a cornerstone of biology, as it provides a unifying explanation for the diversity of life.
Magma	Magma is molten rock located beneath the surface of the Earth, and which often collects in a magma chamber. Magma is a complex high-temperature fluid substance. Most are silicate solutions. It is capable of intrusion into adjacent rocks or of extrusion onto the surface as lava or ejected explosively as tephra to form pyroclastic rock. Environments of magma formation include subduction zones, continental rift zones, mid-oceanic ridges, and hotspots, some of which are interpreted as mantle plumes.
Thermal	A thermal column is a column of rizing air in the lower altitudes of the Earth face=symbol>¢s atmosphere. Thermals are created by the uneven heating of the Earth's surface from solar radiation, and are an example of

convection. The Sun warms the ground, which in turn warms the air directly above it.

Igneous	Igneous rocks form when molten rock, magma, cools and solidifies, with or without crystallization, either below the surface as intrusive, plutonic rocks or on the surface as extrusive, volcanic, rocks.
Igneous rock	Igneous rock forms when rock cools and solidifies either below the surface as intrusive rocks or on the surface as extrusive rocks. This magma can be derived from partial melts of pre-existing rocks in either the Earth's mantle or crust. Typically, the melting is caused by one or more of the following processes -- an increase in temperature, a decrease in pressure, or a change in composition.
Swamp	A swamp is a wetland that features temporary or permanent inundation of large areas of land by shallow bodies of water, generally with a substantial number of hummocks, or dry-land protrusions, and covered by aquatic vegetation, or vegetation that tolerates periodical inundation.
Vegetation	Vegetation is a general term for the plant life of a region; it refers to the ground cover provided by plants, and is, by far, the most abundant biotic element of the biosphere. Primeval redwood forests, coastal mangrove stands, sphagnum bogs, desert soil crusts, roadside weed patches, wheat fields, cultivated gardens and lawns; are all encompassed by the term vegetation.
Coal	Coal is a fossil fuel formed in swamp ecosystems where plant remains were saved by water and mud from oxidization and biodegradation. It is a sedimentary rock, but the harder forms, such as anthracite coal, can be regarded as metamorphic rocks because of later exposure to elevated temperature and pressure. It is composed primarily of carbon along with assorted other elements, including sulfur.
Limestone	Limestone is a sedimentary rock composed largely of the mineral calcite. Limestone often contains variable amounts of silica in the form of chert or flint, as well as varying amounts of clay, silt and sand as disseminations, nodules, or layers within the rock. The primary source of the calcite in limestone is most commonly marine organisms. These organisms secrete shells that settle out of the water column and are deposited on ocean floors as pelagic ooze or alternatively is conglomerated in a coral reef.
Seawater	Seawater is water from a sea or ocean. On average, seawater in the world face=symbol>¢s oceans has a salinity of ~3.5%, or 35 parts per thousand. This means that every 1 kg of seawater has approximately 35 grams of dissolved salts.
Lake	A lake is a body of water or other liquid of considerable size contained on a body of land. A vast majority are fresh water, and lie in the Northern Hemisphere at higher latitudes. Most have a natural outflow in the form of a river or stream, but some do not, and lose water solely by evaporation and/or underground seepage.
Climate	Climate is the average and variations of weather over long periods of time. Climate zones can be defined using parameters such as temperature and rainfall.
Unconformity	An unconformity is a buried erosion surface separating two rock masses or strata of different ages, indicating that sediment deposition was not continuous. In general, the older layer was exposed to erosion for an interval of time before deposition of the younger, but the term is used to describe any break in the sedimentary geologic record.
Siccar Point	Siccar Point is a rocky promontory in the county of Berwickshire on the east coast of Scotland. It is famous in the history of geology as a result of a boat trip in 1788 in which James Hutton, with James Hall and John Playfair, observed the angular unconformity which Hutton regarded as conclusive proof of the theory of geological evolution.
Sandstone	Sandstone is a sedimentary rock composed mainly of sand-size mineral or rock grains. Most

Go to **Cram101.com** for the Practice Tests for this Chapter.

sandstone is composed of quartz and/or feldspar because these are the most common minerals in the Earth's crust. Like sand, sandstone may be any color, but the most common colors are tan, brown, yellow, red, gray and white.

John Playfair	Professor John Playfair was a Scottish scientist. He is perhaps best known for his book Illustrations of the Huttonian Theory of the Earth which was a summary of the work of James Hutton.
Old Red Sandstone	The Old Red Sandstone is a huge set of sedimentary rocks dating to the Devonian. It is a marine formation, having been laid down as earlier Silurian rocks uplifted by the formation of Pangaea eroded and slid into a body of fresh water.
Angular unconformity	Angular unconformity is an unconformity where horizontally parallel strata of sedimentary rock are deposited on tilted and eroded layers that may be either vertical or at an angle to the overlying horizontal layers
Stratigraphy	Stratigraphy, a branch of geology, studies rock layers and layering. It is primarily used in the study of sedimentary and layered volcanic rocks. Stratigraphy includes two related subfields: lithologic or lithostratigraphy and biologic stratigraphy or biostratigraphy.
Metamorphism	Metamorphism can be defined as the solid state recrystallisation of pre-existing rocks due to changes in heat and/or pressure and/or introduction of fluids. There will be mineralogical, chemical and crystallographic changes. Metamorphism produced with increasing pressure and temperature conditions is known as prograde metamorphism. Conversely, decreasing temperatures and pressure characterize retrograde metamorphism.
Foliated metamorphic rock	Foliated metamorphic rock has penetrative planar fabric present within it. It is common to rocks affected by regional metamorphic compression typical of orogenic belts.
Granite	Granite is a common and widely occurring type of intrusive, felsic, igneous rock. Granites are usually medium to coarsely crystalline, occasionally with some individual crystals larger than the groundmass forming a rock known as porphyry. Granites can be pink to dark gray or even black, depending on their chemistry and mineralogy.
Intrusion	An intrusion is a body of igneous rock that has crystallized from a molten magma below the surface of the Earth.
Subsidence	In geology, engineering, and surveying, subsidence is the motion of a surface as it shifts downward relative to a datum such as sea-level. The opposite of subsidence is uplift, which results in an increase in elevation. In meteorology, subsidence refers to the downward movement of air.
Metamorphic rock	Metamorphic rock is the result of the transformation of a pre-existing rock type, the protolith, in a process called metamorphism, which means "change in form". The protolith is subjected to heat and extreme pressure causing profound physical and/or chemical change. The protolith may be sedimentary rock, igneous rock or another older rock.
Metamorphic rocks	Metamorphic rock is the result of the transformation of a pre-existing rock type, the protolith, in a process called metamorphism. The protolith is subjected to heat and extreme pressure causing profound physical and/or chemical change. Metamorphic rocks make up a large part of the Earth's crust. They are formed deep beneath the Earth's surface by great stresses from rocks above and high pressures and temperatures.
Sediment	Sediment is any particulate matter that can be transported by fluid flow and which eventually is deposited as a layer of solid particles on the bed or bottom of a body of water or other liquid.

Crust	In geology, a crust is the outermost layer of a planet, part of its lithosphere. They are generally composed of a less dense material than its deeper layers.Earths face=symbol>¢ is composed mainly of basalt and granite. It is cooler and more rigid than the deeper layers of the mantle and core.
Dike	A dike is an intrusion into a cross-cutting fissure, meaning a dike cuts across other pre-existing layers or bodies of rock, this means that a dike is always younger than the rocks that contain it. The thickness is usually much smaller than the other two dimensions. Thickness can vary from sub-centimeter scale to many meters in thickness and the lateral dimensions can extend over many kilometers.
Terrain	Terrain is the third or vertical dimension of land surface. When terrain is described underwater, the term bathymetry is used.
Quarry	A quarry is a type of open-pit mine from which rock or minerals are extracted. They are generally used for extracting building materials, such as dimension stone. They are usually shallower than other types of open-pit mines.
Shale	Shale is a fine-grained sedimentary rock whose original constituents were clays or muds. It is characterized by thin laminae breaking with an irregular curving fracture, often splintery and usually parallel to the often-indistinguishable bedding plane.
Faunal succession	The faunal succession holds that sedimentary rock strata are observed to contain fossilised flora and fauna, and that these fossil forms succeed each other in a specific, reliable order that can be identified over wide distances.
Dinosaurs	Dinosaurs were vertebrate animals that dominated terrestrial ecosystems for over 160 million years, first appearing approximately 230 million years ago. At the end of the Cretaceous Period, approximately 65 million years ago, a catastrophic extinction event ended dinosaurs' dominance on land.
Era	An era is a long period of time with different technical and colloquial meanings, and usages in language. It begins with some beginning event known as an epoch, epochal date, epochal event or epochal moment.
Orogeny	Orogeny is the process of building mountains, and may be studied as a tectonic structural event, as a geographical event and a chronological event, in that orogenic events cause distinctive structural phenomena and related tectonic activity, affect certain regions of rocks and crust and happen within a time frame.
Law of Superposition	The law of superposition is an axiom that forms one of the bases of the sciences of geology, archaeology, and other fields dealing with stratigraphy. In its plainest form, that is: layers are arranged in a time sequence, with the oldest on the bottom and the youngest on the top, unless later processes disturb this arrangement.
Chalk	Chalk is a soft, white, porous sedimentary rock, a form of limestone composed of the mineral calcite. It forms under relatively deep marine conditions from the gradual accumulation of minute calcite plates shed from micro-organisms called coccolithophores. It is common to find flint nodules embedded in it.
Invertebrate	Invertebrate is an English word that describes any animal without a spinal column.
Reptiles	Reptiles are tetrapods and amniotes, animals whose embryos are surrounded by an amniotic membrane, and members of the class Sauropsida.They rely on gathering and losing heat from the environment to regulate their internal temperature, e.g, by moving between sun and shade, or by preferential circulation — moving warmed blood into the body core, while pushing cool blood to the periphery.
Excavation	Excavation is the most commonly used technique within the science of archaeology. It is the

Go to **Cram101.com** for the Practice Tests for this Chapter.

exposure, processing and recording of archaeological remains.

Amphibians	Amphibians are a taxon of animals that include all living tetrapods or four-legged vertebrates, that do not have amniotic eggs, are ectothermic, term for the animals whose body heat is regulated by the external environment; previously known as cold-blooded, and generally spend part of their time on land.
Fault	Faults are planar rock fractures, which show evidence of relative movement. Large faults within the Earth's crust are the result of shear motion and active fault zones are the causal locations of most earthquakes. Earthquakes are caused by energy release during rapid slippage along faults. The largest examples are at tectonic plate boundaries but many faults occur far from active plate boundaries. Since faults do not usually consist of a single, clean fracture, the term fault zone is used when referring to the zone of complex deformation that is associated with the fault plane.
Conglomerate	A conglomerate is a rock consisting of individual stones that have become cemented together. Conglomerates are sedimentary rocks consisting of rounded fragments and are thus differentiated from breccias, which consist of angular clasts. Both conglomerates and breccias are characterized by clasts larger than sand.
Boulders	In geology, boulders are rock s with a grain size of usually no less than 256 mm diameter.
Alcatraz Island	Alcatraz Island is a small island located in the middle of San Francisco Bay in California, United States. It served as a lighthouse, then a military fortification, then a military prison followed by a federal prison until 1963, when it became a national recreation area.
Landform	A landform comprises a geomorphological unit, and is largely defined by its surface form and location in the landscape, as part of the terrain, and as such, is typically an element of topography. They are categorised by features such as elevation, slope, orientation, stratification, rock exposure, and soil type. They include berms, mounds, hills, cliffs, valleys, rivers and numerous other elements.
Electron correlation	Electron correlation refers to the interaction between electrons in a quantum system whose electronic structure is being considered.
Cambrian	The Cambrian is a major division of the geologic timescale. The Cambrian is the earliest period in whose rocks are found numerous large, distinctly fossilizable multicellular organisms that are more complex than sponges or medusoids. During this time, roughly fifty separate major groups of organisms or "phyla" emerged suddenly, in most cases without evident precursors. This radiation of animal phyla is referred to as the Cambrian explosion.
Silurian	The Silurian is a major division of the geologic timescale that extends from the end of the Ordovician period to the beginning of the Devonian period. The base of the Silurian is set at a major extinction event when 60% of marine species were wiped out.
Ordovician	The Ordovician is the second of the six periods of the Paleozoic era. It follows the Cambrian period and is followed by the Silurian period. The Ordovician started at a major extinction called the Cambrian-Ordovician extinction and lasted for about 44.6 million years. It ended with another major extinction event that wiped out 60% of marine genera.
Ridge	A ridge is a geological feature that is also known as a Rip in the earth causing magma to flow out and forming an undersea volcano, it also has geological features, a continuous elevational crest for some distance. Ridges are usually termed hills or mountains as well, depending on size.
Hierarchy	A hierarchy is a system of ranking and organizing things or people, where each element of the system, except for the top element, is subordinate to a single other element.
Precambrian	The Precambrian is an informal name for the eons of the geologic timescale that came before

Go to **Cram101.com** for the Practice Tests for this Chapter.

the current Phanerozoic eon. It spans from the formation of Earth around 4500 Ma to the evolution of abundant macroscopic hard-shelled animals, which marked the beginning of the Cambrian, the first period of the first era of the Phanerozoic eon, some 542 Ma.

Continental crust	The continental crust is the layer of granitic, sedimentary, and metamorphic rocks which form the continents and the areas of shallow seabed close to their shores, known as continental shelves. It is less dense than the material of the Earth's mantle and thus "floats" on top of it. Continental crust is also less dense than oceanic crust, though it is considerably thicker. About 40% of the Earth's surface is now underlain by continental crust.
Volcanic rock	Volcanic rock is an igneous rock of volcanic origin. They often have a vesicular texture, which is the result voids left by volatiles escaping from the molten lava. Pumice is a rock, which is an example of explosive volcanic eruption. It is so vesicular that it floats in water.
Phanerozoic	The Phanerozoic is the current eon in the geologic timescale, and the one during which abundant animal life has existed. It covers roughly 545 million years and goes back to the time when diverse hard-shelled animals first appeared.
Phanerozoic Eon	The Phanerozoic Eon is the current eon in the geologic timescale, and the one during which abundant animal life has existed. It covers roughly 545 million years and goes back to the time when diverse hard-shelled animals first appeared.
Biosphere	The biosphere is the part of the earth, including air, land, surface rocks, and water, within which life occurs, and which biotic processes in turn alter or transform. From the broadest biophysiological point of view, the biosphere is the global ecological system integrating all living beings and their relationships, including their interaction with the elements of the lithosphere, hydrosphere, and atmosphere. This biosphere is postulated to have evolved, beginning through a process of biogenesis or biopoesis, at least some 3.5 billion years ago.
Paleozoic	The Paleozoic is the earliest of three geologic eras of the Phanerozoic eon. The Paleozoic is subdivided into six geologic periods; from oldest to youngest they are: the Cambrian, Ordovician, Silurian, Devonian, Carboniferous, and Permian.
Triassic	Triassic is the first period of the Mesozoic Era. Both the start and end of the Triassic are marked by major extinction events. During the Triassic, both marine and continental life show an adaptive radiation beginning from the starkly impoverished biosphere that followed the Permian-Triassic extinction. Corals of the hexacorallia group made their first appearance. The first flowering plants may have evolved during the Triassic, as did the first flying vertebrates, the pterosaurs.
Mesozoic	The Mesozoic is one of three geologic eras of the Phanerozoic eon. The Mesozoic was a time of tectonic, climatic and evolutionary activity, shifting from a state of connectedness into their present configuration. The climate was exceptionally warm throughout the period, also playing an important role in the evolution and diversification of new animal species. By the end of the era, the basis of modern life was in place.
Cenozoic	The Cenozoic Era meaning "new life", is the most recent of the three classic geological eras. It covers the 65.5 million years since the Cretaceous-Tertiary extinction event at the end of the Cretaceous that marked the demise of the last non-avian dinosaurs and the end of the Mesozoic Era. The Cenozoic era is ongoing.
Cenozoic era	The Cenozoic Era is the most recent of the three classic geological eras. It covers the 65.5 million years since the Cretaceous-Tertiary extinction event at the end of the Cretaceous that marked the demise of the last non-avian dinosaurs and the end of the Mesozoic Era.
Tertiary	The Tertiary covers roughly the time span between the demise of the non-avian dinosaurs and

beginning of the most recent Ice Age. Each epoch of the Tertiary was marked by striking developments in mammalian life. The earliest recognizable hominoid relatives of humans appeared. Tectonic activity continued as Gondwana finally split completely apart.

Uranium	Uranium is approximately 70% more dense than lead and is weakly radioactive. It occurs naturally in low concentrations in soil, rock and water.
Rotation	A rotation is a movement of an object in a circular motion. A two-dimensional object rotates around a center of rotation. A three-dimensional object rotates around a line called an axis. A circular motion about an external point, e.g. the Earth about the Sun, is called an orbit or more properly an orbital revolution.
Neutron	In physics, the neutron is a subatomic particle with no net electric charge.
Isotopes	Isotopes are any of the several different forms of an element each having different atomic mass. Isotopes of an element have nuclei with the same number of protons but different numbers of neutrons.
Proton	In physics, the proton is a subatomic particle with an electric charge of one positive fundamental unit a diameter of about 1.5×10 class="unicode">−15 m, and a mass of 938.27231(28) MeV/c2 (1.6726 class="unicode">× 10^{-27} kg), 1.007 276 466 88(13) u or about 1836 times the mass of an electron.
Element	An element is a type of atom that is defined by its atomic number; that is, by the number of protons in its nucleus.
Atoms	Atoms are the fundamental building blocks of chemistry, and are conserved in chemical reactions.
Electron	The electron is a fundamental subatomic particle that carries a negative electric charge.
Chemical reaction	A chemical reaction is a process that results in the interconversion of chemical substances. The substance or substances initially involved in a chemical reaction are called reactants. Chemical reactions are characterized by a chemical change, and they yield one or more products which are, in general, different from the reactants.
Closed system	In thermodynamics, a closed system can exchange heat and work, but not matter, with its surroundings.
Crystal	A crystal is a solid in which the constituent atoms, molecules, or ions are packed in a regularly ordered, repeating pattern extending in all three spatial dimensions. Most metals encountered in everyday life are polycrystals. Crystals are often symmetrically intergrown to form crystal twins.
Carbon	Carbon is a chemical element. An abundant nonmetallic, tetravalent element, carbon has several allotropic forms. This element is the basis of the chemistry of all known life.
Potassium	Potassium is a chemical element. It is a soft silvery-white metallic alkali metal that occurs naturally bound to other elements in seawater and many minerals. It oxidizes rapidly in air and is very reactive, especially towards water. In many respects, it and sodium are chemically similar, although organisms in general, and animal cells in particular, treat them very differently.
Zircon	Zircon is a mineral belonging to the group of nesosilicates. Its chemical name is zirconium silicate and its corresponding chemical formula is ZrSiO4.
Ion	An ion is an atom or group of atoms which have lost or gained one or more electrons, making them negatively or positively charged.
Sensitive High	The Sensitive High Resolution Ion Microprobe is a large-diameter, double focusing secondary

Resolution Ion Microprobe	ion mass spectrometer. This tool is primarily used for geological and geochemical applications. It can rapidly measure the isotopic and elemental abundances in minerals at a scale as small as 5 μm.
Crater	A crater is an approximately circular depression in the surface of a planet, moon or other solid body in the Solar System, formed by the hyper-velocity impact of a smaller body with the surface. Impact craters typically have raised rims, and they range from small, simple, bowl-shaped depressions to large, complex, multi-ringed, impact basins.
Crystallization	Crystallization is the process of formation of solid crystals from a uniform solution. It is also a chemical solid-liquid separation technique, in which mass transfer of a solute from the liquid solution to a pure solid crystalline phase occurs.
Gneiss	Gneiss is a common and widely distributed type of rock formed by high-grade regional metamorphic processes from preexisting formations that were originally either igneous or sedimentary rocks. Gneissic rocks are usually medium to coarse foliated and largely recrystallized but do not carry large quantities of micas, chlorite or other platy minerals.
Mica	The mica group of sheet silicate minerals includes several closely related materials having highly perfect basal cleavage. All are monoclinic with a tendency towards pseudo-hexagonal crystals and are similar in chemical composition. The highly perfect cleavage, which is the most prominent characteristic of mica, is explained by the hexagonal sheet-like arrangement of its atoms.
Feldspar	Feldspar is the name of a group of rock-forming minerals which make up as much as sixty percent of the Earth's crust. Feldspars crystallize from magma in both intrusive and extrusive rocks, and they can also occur as compact minerals, as veins, and are also present in many types of metamorphic rock.
Amphibole	Amphibole defines an important group of generally dark-colored rock-forming inosilicate minerals linked at the vertices and generally containing ions of iron and/or magnesium in their structures. Amphiboles crystallize into two crystal systems, monoclinic and orthorhombic.
Decay product	In nuclear physics, a decay product, also known as a daughter product, daughter isotope or daughter nuclide, is a nuclide resulting from the radioactive decay of a parent isotope or precursor nuclide. The daughter product may be stable or it may decay to form a daughter product of its own. The daughter of a daughter product is sometimes called a granddaughter product.
Noble gas	The noble gas is the element in group 18 of the periodic table. It is also called helium family or neon family. Chemically, they are very stable due to having the maximum number of valence electrons their outer shell can hold. A thorough explanation requires an understanding of electronic configuration, with references to quantum mechanics.
Clastic	Clastic rocks are rocks formed from fragments of pre-existing rock.
Atmosphere	An atmosphere is a layer of gases that may surround a material body of sufficient mass. The gases are attracted by the gravity of the body, and are retained for a longer duration if gravity is high and the atmosphere's temperature is low. Some planets consist mainly of various gases, and thus have very deep atmospheres.
Carbon dioxide	Carbon dioxide is a chemical compound, normally in a gaseous state, and is composed of one carbon and two oxygen atoms. It is often referred to by its formula $CO2$. It is present in the Earth's atmosphere at a concentration of approximately .000383 by volume and is an important greenhouse gas due to its ability to absorb many infrared wavelengths of sunlight, and due to the length of time it stays in the atmosphere.
Earths	Earths atmosphere is a layer of gases surrounding the planet Earth and retained by the

atmosphere	Earth's gravity. This mixture of gases is commonly known as air.
Atmospheric carbon dioxide	Atmospheric carbon dioxide is present in a low concentration on earth. It is essential to photosynthesis in plants and other photoautotrophs, and is also a prominent greenhouse gas.
Photosynthesis	Photosynthesis generally, is the synthesis of triose phosphates from sunlight, carbon dioxide and water.
Organic matter	Organic matter is matter that has come from a recently living organism; is capable of decay, or the product of decay; or is composed of organic compounds. The definition of organic matter varies upon the subject it is being used for.
Magnetic field	In physics, a magnetic field is a solenoidal vector field in the space surrounding moving electric charges and magnetic dipoles, such as those in electric currents and magnets.
Magnetism	Magnetism is one of the phenomena by which materials exert attractive or repulsive forces on other materials. Some well known materials that exhibit easily detectable magnetic properties are nickel, iron, some steels, and the mineral magnetite; however, all materials are influenced to greater or lesser degree by the presence of a magnetic field.
Weathering	Weathering is the process of breaking down rocks, soils and their minerals through direct contact with the atmosphere. Weathering occurs without movement. Two main classifications of weathering processes exist. Mechanical or physical weathering involves the breakdown of rocks and soils through direct contact with atmospheric conditions. The second classification, chemical weathering, involves the direct effect of atmospheric chemicals in the breakdown of rocks, soils and minerals.
Daughter product	In nuclear physics, a decay product, also known as a daughter product, daughter isotope or daughter nuclide, is a nuclide resulting from the radioactive decay of a parent isotope or precursor nuclide. The daughter product may be stable or it may decay to form a daughter product of its own. The daughter of a daughter product is sometimes called a granddaughter product.
Varve	A varve is an annual layer of sediment or sedimentary rock. The word Varve is derived from the Swedish word varv whose meanings and connotations include revolution, in layers, and circle.
Climate change	Climate change refers to the variation in the Earth's global climate or in regional climates over time. It describes changes in the variability or average state of the atmosphere over time scales ranging from decades to millions of years. These changes can be caused by processes internal to the Earth, external forces or, more recently, human activities.
Season	A season is one of the major divisions of the year, generally based on yearly periodic changes in weather. They are recognized as: spring, summer, autumn, and winter.
Meltwater	Meltwater is the water released by the melting of snow or ice, including glacial ice. Meltwater provides drinking water for a large proportion of the world face=symbol>¢s population, as well as providing water for irrigation and hydroelectric plants.
Baltic Sea	The Baltic Sea is located in Northern Europe. It is bounded by the Scandinavian Peninsula, the mainland of Europe, and the Danish islands. It drains into the Kattegat by way of the Øresund, the Great Belt and the Little Belt. The Kattegat continues through the Skagerrak into the North Sea and the Atlantic Ocean.
Greenland	Greenland is a self-governed Danish territory lying between the Arctic and Atlantic Oceans. Though geographically and ethnically an Arctic island nation associated with the continent of North America, politically and historically Greenland is closely tied to Europe. It is the

largest island in the world that is not also considered a continent.

Ice sheet	An ice sheet is a mass of glacier ice that covers surrounding terrain and is greater than 19,305 mile². The only current ice sheets are in Antarctica and Greenland. Ice sheets are bigger than ice shelves or glaciers. Masses of ice covering less than 50,000 km² are termed an ice cap. An ice cap will typically feed a series of glaciers around its periphery. Although the surface is cold, the base of an ice sheet is generally warmer. This process produces fast-flowing channels in the ice sheet.
Antarctic Ice Sheet	The Antarctic ice sheet covers about 98% of the Antarctic continent and is the largest single mass of ice on Earth.
Ash fall	Ash fall consists of very fine rock and mineral particles less than 2 mm in diameter that are ejected from a volcanic vent. The very fine particles may be carried for many miles, settling out as a dust-like layer across the landscape
Greenland Ice Sheet	The Greenland Ice Sheet is a vast body of ice covering roughly 80% of the surface of Greenland. It's almost 2,400 kilometres long in a north-south direction, and its greatest width is 1,100 kilometres at a latitude of 77 class="unicode">° N, near its northern margin. It sheet covers 1.71 million km², or roughly 80% of the surface of Greenland. The thickness is generally more than 2 km and over 3 km at its thickest point.
Ice cap	An ice cap is a dome-shaped ice mass that covers less than 50,000 km² of land area. Masses of ice covering more than 50,000 km² are termed an ice sheet.
Oxide	An oxide is a chemical compound containing an oxygen atom and other elements. Most of the earth's crust consists of them. They result when elements are oxidized by air.
Copper	Copper is a ductile metal with excellent electrical conductivity, and finds extensive use as an electrical conductor, heat conductor, as a building material, and as a component of various alloys.
Industrial Revolution	The Industrial Revolution was a major shift of technological, socioeconomic, and cultural conditions that occurred in the late 18th century and early 19th century in some Western countries. It began in Britain and spread throughout the world, a process that continues.
Glacier	A glacier is a large, slow moving river of ice, formed from compacted layers of snow, that slowly deforms and flows in response to gravity. Glacier ice is the largest reservoir of fresh water on Earth, and second only to oceans as the largest reservoir of total water. Glaciers cover vast areas of polar regions but are restricted to the highest mountains in the tropics.
Geochronology	Geochronology is the science of determining the absolute age of rocks, fossils, and sediments, within a certain degree of uncertainty inherent within the method used. A variety of dating methods are used by geologists to achieve this.
Colorado River	The Colorado River is a river in the southwestern United States and northwestern Mexico, approximately 1,450 mi long, draining a part of the arid regions on the western slope of the Rocky Mountains. The natural course of the river flows into the Gulf of California, but the heavy use of the river as an irrigation source for the Imperial Valley has desiccated the lower course of the river in Mexico such that it no longer consistently reaches the sea.
Blocks	Blocks in meteorology are large scale patterns in the atmospheric pressure field that are nearly stationary, effectively "blocking" or redirecting migratory cyclones. These blocks can remain in place for several days or even weeks, causing the areas affected by them to have the same kind of weather for an extended period of time.

Schist	The schist refers to a group of medium-grade metamorphic rocks, chiefly notable for the preponderance of lamellar minerals such as micas, chlorite, talc, hornblende, graphite, and others. Quartz often occurs in drawn-out grains to such an extent that a particular form called quartz schist is produced.
Shoreline	A shoreline is the fringe of land at the edge of a large body of water, such as an ocean, sea, or lake. A strict definition is the strip of land along a water body that is alternately exposed and covered by waves and tides.
Evaporation	Evaporation is the process by which molecules in a liquid state become a gas.
Air mass	In meteorology, an air mass is a large volume of air having fairly uniform characteristics of temperature, atmospheric pressure, and water vapor content.
Equilibrium	Equilibrium is the condition of a system in which competing influences are balanced.
Glaciers	A glacier is a large, slow moving river of ice, formed from compacted layers of snow, that slowly deforms and flows in response to gravity. Glacier ice is the largest reservoir of fresh water on Earth, and second only to oceans as the largest reservoir of total water. Glaciers cover vast areas of polar regions but are restricted to the highest mountains in the tropics.

Go to **Cram101.com** for the Practice Tests for this Chapter.

Planet	A planet, as defined by the International Astronomical Union, is a celestial body orbiting a star or stellar remnant that is massive enough to be rounded by its own gravity, not massive enough to cause thermonuclear fusion in its core, and has cleared its neighboring region of planetesimals.
Atmosphere	An atmosphere is a layer of gases that may surround a material body of sufficient mass. The gases are attracted by the gravity of the body, and are retained for a longer duration if gravity is high and the atmosphere's temperature is low. Some planets consist mainly of various gases, and thus have very deep atmospheres.
Climate	Climate is the average and variations of weather over long periods of time. Climate zones can be defined using parameters such as temperature and rainfall.
Water vapor	Water vapor is the gas phase of water. Water vapor is one state of the water cycle within the hydrosphere. Water vapor can be produced from the evaporation of liquid water or from the sublimation of ice. Under normal atmospheric conditions, water vapor is continuously evaporating and condensing.
Vapor	Vapor is the gas phase component of a another state of matter which does not completely fill its container. It is distinguished from the pure gas phase by the presence of the same substance in another state of matter. Hence when a liquid has completely evaporated, it is said that the system has been completely transformed to the gas phase.
Wave	A wave is a disturbance that propagates through space or spacetime, transferring energy and momentum and sometimes angular momentum.
Evaporation	Evaporation is the process by which molecules in a liquid state become a gas.
Gulf of Mexico	The Gulf of Mexico is the ninth largest body of water in the world. It is an ocean basin largely surrounded by the North American continent and the island of Cuba. It is bounded on the northeast, north and northwest by the Gulf Coast of the United States, on the southwest and south by Mexico, and on the southeast by Cuba.
Season	A season is one of the major divisions of the year, generally based on yearly periodic changes in weather. They are recognized as: spring, summer, autumn, and winter.
Weather	The weather is the set of all extant phenomena in a given atmosphere at a given time. The term usually refers to the activity of these phenomena over short periods, as opposed to the term climate, which refers to the average atmospheric conditions over longer periods of time.
Precipitation	Precipitation is any product of the condensation of atmospheric water vapor that is deposited on the earth's surface. It occurs when the atmosphere becomes saturated with water vapour and the water condenses and falls out of solution. Air becomes saturated via two processes, cooling and adding moisture.
Sediment	Sediment is any particulate matter that can be transported by fluid flow and which eventually is deposited as a layer of solid particles on the bed or bottom of a body of water or other liquid.
Desert	In geography, a desert is a landscape form or region that receives very little precipitation. They are defined as areas that receive an average annual precipitation of less than 250 mm. A desert where vegetation cover is exceedingly sparse correspond to the face=symbol>¢hyperarid' regions of the earth, where rainfall is exceedingly rare and infrequent.
Ocean current	An ocean current is any more or less continuous, directed movement of ocean water that flows in one of the Earth's oceans.Ocean Currents are rivers of hot or cold water within the ocean. The currents are generated from the forces acting upon the water like the earth's rotation, the wind, the temperature and salinity

Go to **Cram101.com** for the Practice Tests for this Chapter.

differences and the gravitation of the moon. The depth contours, the shoreline and other currents influence the current's direction and strength.

Ice cap	An ice cap is a dome-shaped ice mass that covers less than 50,000 km² of land area. Masses of ice covering more than 50,000 km² are termed an ice sheet.
Currents	Ocean currents are any more or less continuous, directed movement of ocean water that flows in one of the Earth's oceans.They are rivers of hot or cold water within the ocean. They are generated from the forces acting upon the water like the earth's rotation, the wind, the temperature and salinity differences and the gravitation of the moon.
Seawater	Seawater is water from a sea or ocean. On average, seawater in the world face=symbol>¢s oceans has a salinity of ~3.5%, or 35 parts per thousand. This means that every 1 kg of seawater has approximately 35 grams of dissolved salts.
Solar radiation	Solar radiation is radiant energy emitted by the sun from a nuclear fusion reaction that creates electromagnetic energy. The spectrum of solar radiation is close to that of a black body with a temperature of about 5800 K. About half of the radiation is in the visible short-wave part of the electromagnetic spectrum. The other half is mostly in the near-infrared part, with some in the ultraviolet part of the spectrum.
Radiation	Radiation as used in physics, is energy in the form of waves or moving subatomic particles.
Humidity	Humidity is a term used to describe the amount of water vapor in air. Absolute humidity, relative humidity, and specific humidity are different ways to express the water content in a parcel of air.
Sea ice	Sea ice is formed from ocean water that freezes. Because the oceans consist of saltwater, this occurs at about -1.8 °C.
Carbonate	In organic chemistry, a carbonate is a salt of carbonic acid.
Salinity	Salinity is the saltiness or dissolved salt content of a body of water. In oceanography, it has been traditional to express halinity not as percent, but as parts per thousand, which is approximately grams of salt per liter of solution.
Carbon	Carbon is a chemical element. An abundant nonmetallic, tetravalent element, carbon has several allotropic forms. This element is the basis of the chemistry of all known life.
Carbon dioxide	Carbon dioxide is a chemical compound, normally in a gaseous state, and is composed of one carbon and two oxygen atoms. It is often referred to by its formula $CO2$. It is present in the Earth's atmosphere at a concentration of approximately .000383 by volume and is an important greenhouse gas due to its ability to absorb many infrared wavelengths of sunlight, and due to the length of time it stays in the atmosphere.
Mountain	A mountain is a landform that extends above the surrounding terrain in a limited area. A mountain is generally steeper than a hill, but there is no universally accepted standard definition for the height of a mountain or a hill although a mountain usually has an identifiable summit.
Tectonics	Tectonics is a field of study within geology concerned generally with the structures within the crust of the Earth, or other planets, and particularly with the forces and movements that have operated in a region to create these structures.
Volcano	A volcano is an opening, or rupture, in the Earth's surface or crust, which allows hot, molten rock, ash and gases to escape from deep below the surface.
Continent	A continent is one of several large landmasses on Earth. They are generally identified by convention rather than any strict criteria, but seven areas are commonly reckoned as

Go to **Cram101.com** for the Practice Tests for this Chapter.

continents – they are: Asia, Africa, North America, South America, Antarctica, Europe, and Australia.

Climate change	Climate change refers to the variation in the Earth's global climate or in regional climates over time. It describes changes in the variability or average state of the atmosphere over time scales ranging from decades to millions of years. These changes can be caused by processes internal to the Earth, external forces or, more recently, human activities.
Global warming	Global warming is the increase in the average temperature of the Earth face=symbol>¢s near-surface air and oceans in recent decades and its projected continuation. An increase in global temperatures can in turn cause other changes, including sea level rise, and changes in the amount and pattern of precipitation resulting in floods and drought. There may also be changes in the frequency and intensity of extreme weather events.
Earths atmosphere	Earths atmosphere is a layer of gases surrounding the planet Earth and retained by the Earth's gravity. This mixture of gases is commonly known as air.
Fossil	Fossils are the mineralized or otherwise preserved remains or traces of animals, plants, and other organisms. The totality of fossils, both discovered and undiscovered, and their placement in fossiliferous rock formations and sedimentary layers is known as the fossil record.
Fossil fuel	Fossil fuels are hydrocarbons, primarily coal and petroleum, formed from the fossilized remains of dead plants and animals. In common parlance, the term fossil fuel also includes hydrocarbon-containing natural resources that are not derived from animal or plant sources. Fossil fuels have made large-scale industrial development possible and have largely supplanted water-driven mills, as well as the combustion of wood or peat for heat.
Fossil fuels	Fossil fuels are hydrocarbons, primarily coal and petroleum, formed from the fossilized remains of dead plants and animals by exposure to heat and pressure in the Earth face=symbol>¢s crust over hundreds of millions of years. The burning of fossil fuels by humans is the largest source of emissions of carbon dioxide, which is one of the greenhouse gases that enhances radiative forcing and contributes to global warming.
Solar system	The Solar System consists of the Sun and the other celestial objects gravitationally bound to it: the eight planets, their 165 known moons, three currently identified dwarf planets and their four known moons, and billions of small bodies.
Nitrogen	Nitrogen is a chemical element which has the symbol N and atomic number 7. Elemental nitrogen is a colorless, odourless, tasteless and mostly inert diatomic gas at standard conditions, constituting 78.1% by volume of Earth's atmosphere.
Troposphere	The troposphere is the lowest portion of Earth's atmosphere. It is the densest layer of the atmosphere and contains approximately 75% of the mass of the atmosphere and almost all the water vapor and aerosols.
Tropics	The tropics are the geographic region of the Earth centered on the equator and limited in latitude by the Tropic of Cancer in the northern hemisphere, at approximately 23 class="unicode">°30′ N latitude, and the Tropic of Capricorn in the southern hemisphere at 23°30′ S latitude.
Groundwater	Groundwater is water located beneath the ground surface in soil pore spaces and in the fractures of geologic formations. Groundwater is recharged from, and eventually flows to, the surface naturally; natural discharge often occurs at springs and seeps, streams and can often form oases or wetlands.

Glacier	A glacier is a large, slow moving river of ice, formed from compacted layers of snow, that slowly deforms and flows in response to gravity. Glacier ice is the largest reservoir of fresh water on Earth, and second only to oceans as the largest reservoir of total water. Glaciers cover vast areas of polar regions but are restricted to the highest mountains in the tropics.
Weathering	Weathering is the process of breaking down rocks, soils and their minerals through direct contact with the atmosphere. Weathering occurs without movement. Two main classifications of weathering processes exist. Mechanical or physical weathering involves the breakdown of rocks and soils through direct contact with atmospheric conditions. The second classification, chemical weathering, involves the direct effect of atmospheric chemicals in the breakdown of rocks, soils and minerals.
Mineral	A mineral is a naturally occurring substance formed through geological processes that has a characteristic chemical composition, a highly ordered atomic structure and specific physical properties. A rock, by comparison, is an aggregate of minerals and need not have a specific chemical composition. Minerals range in composition from pure elements and simple salts to very complex silicates with thousands of known forms.
Ore	An ore is a volume of rock containing components or minerals in a mode of occurrence that renders it valuable for mining.
Chemical reaction	A chemical reaction is a process that results in the interconversion of chemical substances. The substance or substances initially involved in a chemical reaction are called reactants. Chemical reactions are characterized by a chemical change, and they yield one or more products which are, in general, different from the reactants.
Abundance	Abundance is an ecological concept referring to the relative representation of a species in a particular ecosystem. It is usually measured as the mean number of individuals found per sample.
Solar power	Solar power is Solar Radiation emitted from our sun. It has been used in many traditional technologies for centuries, and has come into widespread use where other power supplies are absent, such as in remote locations and in space.
Oxide	An oxide is a chemical compound containing an oxygen atom and other elements. Most of the earth's crust consists of them. They result when elements are oxidized by air.
Ozone	Ozone is a triatomic molecule, consisting of three oxygen atoms. It is an allotrope of oxygen that is much less stable than the diatomic species O2. Ground-level ozone is an air pollutant with harmful effects on the respiratory systems of animals. On the other hand, ozone in the upper atmosphere protects living organisms by preventing damaging ultraviolet light from reaching the Earth's surface.
Methane	Methane is a chemical compound with the molecular formula CH_4. It is the simplest alkane, and the principal component of natural gas. Burning one molecule of methane in the presence of oxygen releases one molecule. Methane's relative abundance and clean burning process makes it a very attractive fuel.
Thermal	A thermal column is a column of rizing air in the lower altitudes of the Earth face=symbol>¢s atmosphere. Thermals are created by the uneven heating of the Earth's surface from solar radiation, and are an example of convection. The Sun warms the ground, which in turn warms the air directly above it.
Nuclear fusion	In physics and nuclear chemistry, nuclear fusion is the process by which multiple atomic particles join together to form a heavier nucleus. It is accompanied by the release or absorption of energy.

Go to **Cram101.com** for the Practice Tests for this Chapter.

Altitude	Altitude is the elevation of an object from a known level or datum. Common datums are mean sea level and the surface of the World Geodetic System geoid, used by Global Positioning System. In aviation, altitude is measured in feet. For non-aviation uses, altitude may be measured in other units such as metres or miles.
Storm	A storm is any disturbed state of an astronomical body's atmosphere, especially affecting its surface, and strongly implying severe weather. It may be marked by strong wind, thunder and lightning, heavy precipitation, such as ice, or wind transporting some substance through the atmosphere.
Condensation	Condensation is the change in matter of a substance to a denser phase, such as a gas to a liquid. Condensation commonly occurs when a vapor is cooled to a liquid, but can also occur if a vapor is compressed into a liquid, or undergoes a combination of cooling and compression.
Stratosphere	The stratosphere is the second layer of Earth's atmosphere, just above the troposphere, and below the mesosphere. It is stratified in temperature, with warmer layers higher up and cooler layers farther down.
Molecule	In chemistry, a molecule is defined as a sufficiently stable electrically neutral group of at least two atoms in a definite arrangement held together by strong chemical bonds.
Venus	Venus is the second-closest planet to the Sun, orbiting it every 224.7 Earth days. It is the brightest natural object in the night sky, except for the Moon, reaching an apparent magnitude of −4.6. Because Venus is an inferior planet, from Earth it never appears to venture far from the Sun: its elongation reaches a maximum of 47.8°.
Mars	Mars the fourth planet from the Sun in the Solar System. The planet is named after Mars, the Roman god of war. It is also referred to as the "Red Planet" because of its reddish appearance as seen from Earth.
Jupiter	Jupiter is the fifth planet from the Sun and the largest planet within the solar system. It is two and a half times as massive as all of the other planets in our solar system combined. Jupiter, along with Saturn, Uranus and Neptune, is classified as a gas giant.
Hydrogen	Hydrogen is a chemical element represented by the symbol H and an atomic number of 1. At standard temperature and pressure it is a colorless, odorless, nonmetallic, tasteless, highly flammable diatomic gas . With an atomic mass of 1.00794 g/mol, hydrogen is the lightest element. Hydrogen is the most abundant of the chemical elements, constituting roughly 75% of the universe's elemental mass.
Electromagnetic radiation	Electromagnetic radiation is a self-propagating wave in space with electric and magnetic components.
Electromagnetism	Electromagnetism is the physics of the electromagnetic field: a field which exerts a force on particles that possess the property of electric charge, and is in turn affected by the presence and motion of those particles. The magnetic field is produced by the motion of electric charges, i.e. electric current.
Ozone layer	The ozone layer is the part of the Earth's atmosphere which contains relatively high concentrations of ozone . "Relatively high" means a few parts per million—much higher than the concentrations in the lower atmosphere but still small compared to the main components of the atmosphere.
Ultraviolet	Ultraviolet light is electromagnetic radiation with a wavelength shorter than that of visible light, but longer than soft X-rays. The color violet has the shortest wavelength in the visible spectrum. UV light has a shorter wavelength than that of violet light.
Temperate	In geography, temperate latitudes of the globe lie between the tropics and the polar circles.

	The changes in these regions between summer and winter are generally subtle: warm or cool, rather than extreme hot or cold.
Thermosphere	The thermosphere is the layer of the earth's atmosphere directly above the mesosphere and directly below the exosphere. Within this layer, ultraviolet radiation causes ionization. It is the fourth atmospheric layer from earth.
Mesosphere	The mesosphere is the layer of the Earth's atmosphere that is directly above the stratosphere and directly below the thermosphere. The mesosphere is located from about 50 km to 80-90 km altitude above Earth's surface. Within this layer, temperature decreases with increasing altitude. The main dynamical features in this region are atmospheric tides, internal atmospheric gravity waves and planetary waves.
Magnetosphere	A magnetosphere is the region around an astronomical object in which phenomena are dominated or organized by its magnetic field. Earth is surrounded by a magnetosphere, as are the magnetized planets Jupiter, Saturn, Uranus and Neptune. The term magnetosphere has also been used to describe regions dominated by the magnetic fields of celestial objects, e.g. pulsar magnetospheres.
Sea level	Mean sea level is the average height of the sea, with reference to a suitable reference surface.
Atmospheric pressure	Atmospheric pressure is the pressure at any point in the Earth's atmosphere.
Mount Everest	Mount Everest is the highest mountain on Earth, as measured by the height of its summit above sea level. The mountain, which is part of the Himalaya range in High Asia, is located on the border between Nepal and Tibet, China.
Greenhouse	A greenhouse is a building where plants are cultivated.
Air mass	In meteorology, an air mass is a large volume of air having fairly uniform characteristics of temperature, atmospheric pressure, and water vapor content.
Buoyancy	In physics, buoyancy is the upward force on an object produced by the surrounding fluid in which it is fully or partially immersed, due to the pressure difference of the fluid between the top and bottom of the object. The net upward buoyancy force is equal to the magnitude of the weight of fluid displaced by the body.
Equator	The equator is an imaginary line on the Earth's surface equidistant from the North Pole and South Pole. It thus divides the Earth into a Northern Hemisphere and a Southern Hemisphere.
Latitude	Latitude gives the location of a place on Earth north or south of the equator. Lines of Latitude are the horizontal lines shown running east-to-west on maps. Technically, Latitude is an angular measurement in degrees ranging from 0° at the Equator to 90° at the poles.
Climatology	Climatology is the study of climate, scientifically defined as weather conditions averaged over a period of time. Basic knowledge of climate can be used within shorter term weather forecasting using analog techniques such as teleconnections and climate indices.
Winter	Winter is one of the four seasons of temperate zones. Almost all English-language calendars, going by astronomy, state that winter begins on the winter solstice, and ends on the spring equinox. Calculated more by the weather, it begins and ends earlier and is the season with the shortest days and the lowest temperatures.
Weather satellite	A weather satellite is primarily used to monitor the weather and climate of the Earth. These meteorological satellites, however, see more than clouds and cloud systems. Weather satellite

Go to **Cram101.com** for the Practice Tests for this Chapter.

	images helped in monitoring the volcanic ash cloud from Mount St. Helens and activity from other volcanoes such as Mount Etna.
National Oceanic and Atmospheric Administration	The National Oceanic and Atmospheric Administration is a scientific agency of the United States Department of Commerce focused on the conditions of the oceans and the atmosphere. National Oceanic and Atmospheric Administration warns of dangerous weather, charts seas and skies, guides the use and protection of ocean and coastal resources, and conducts research to improve understanding and stewardship of environment.
Sunlight	Sunlight in the broad sense is the total spectrum of the electromagnetic radiation given off by the Sun. On Earth, it is filtered through the atmosphere, and the solar radiation is obvious as daylight when the Sun is above the horizon.
Equilibrium	Equilibrium is the condition of a system in which competing influences are balanced.
Atmospheric circulation	Atmospheric circulation is the large-scale movement of air, and the means by which heat is distributed on the surface of the Earth. The large-scale structure of the atmospheric circulation varies from year to year, but the basic structure remains fairly constant.
Sinkhole	A sinkhole is a natural depression or hole in the surface topography caused by the removal of soil or bedrock, often both, by water. They may vary in size from less than a meter to several hundred meters both in diameter and depth, and vary in form from soil-lined bowls to bedrock-edged chasms.
Trough	In geology, a trough generally refers to a linear structural depression that extends laterally over a distance, while being less steep than a trench. It can be a narrow basin or a geologic rift. In meteorolology a trough is an elongated region of relatively low atmospheric pressure, often associated with fronts.
Polar cells	Polar cells are part of a three cell movement involving Hadley Cells and Ferrel Cells which show atmospheric circulation and Surface winds.Cold dense air descends over the Poles which creates high pressure , this cold air moves along the surface to lower latitudes.
Rotation	A rotation is a movement of an object in a circular motion. A two-dimensional object rotates around a center of rotation. A three-dimensional object rotates around a line called an axis. A circular motion about an external point, e.g. the Earth about the Sun, is called an orbit or more properly an orbital revolution.
Prevailing winds	The prevailing winds are the trends in speed and direction of wind over a particular point on the earth's surface. A region's prevailing winds often reflect global patterns of movement in the earth's atmosphere.
Horse latitudes	Horse latitudes are subtropical latitudes between 30 and 35 degrees both north and south. This region is an area of variable winds mixed with calm, and it owes its name to the fact that the confused sea, muggy heat, and rolling and pitching of waves often slowed colonial ships for days to weeks due to lack of propulsion.
Trade winds	The trade winds are a pattern of wind that are found in bands around the Earth face=symbol>¢s equatorial region. The trade winds are the prevailing winds in the tropics, blowing from the high-pressure area in the horse latitudes towards the low-pressure area around the equator. The trade winds blow predominantly from the northeast in the northern hemisphere and from the southeast in the southern hemisphere.
Polar front	In meteorology, a polar front is the boundary between the polar cell and the Ferrel cell in each hemisphere. At this boundary a sharp gradient in temperature occurs between these two air masses, each at very different temperatures.
Forest	A forest is an area with a high density of trees, historically, a wooded area set aside for

Go to **Cram101.com** for the Practice Tests for this Chapter.

hunting. These plant communities cover large areas of the globe and function as animal habitats, hydrologic flow modulators, and soil conservers, constituting one of the most important aspects of the Earth's biosphere.

Coriolis effect	The Coriolis effect is the apparent deflection of objects from a straight path if the objects are viewed from a rotating frame of reference.
Convection	Convection in the most general terms refers to the movement of currents within fluids. Convection is one of the major modes of Heat and mass transfer. In fluids, convective heat and mass transfer take place through both diffusion and by advection, in which matter or heat is transported by the larger-scale motion of currents in the fluid.
Convection cell	A convection cell is a phenomenon of fluid dynamics which occurs in situations where there are temperature differences within a body of liquid or gas.
Global Telecommunic-tions System	The Global telecommunications system is a global network for the transmission of meteorological data from weather stations, satellites and numerical weather prediction centres. The system consists of an integrated network of point-to-point circuits, and multi-point circuits which interconnect meteorological telecommunication centres.
Element	An element is a type of atom that is defined by its atomic number; that is, by the number of protons in its nucleus.
Stream	A stream is a body of water with a current, confined within a bed and banks. Streams are important as conduits in the water cycle, instruments in aquifer recharge, and corridors for fish and wildlife migration.
Jet stream	A Jet stream is a fast flowing, relatively narrow air currents found in the atmosphere at around 11 kilometres above the surface of the Earth, just under the tropopause. It forms at the boundaries of adjacent air masses with significant differences in temperature, such as of the polar region and the warmer air to the south. The jet stream is mainly found in the Stratosphere.
Terrestrial	Terrestrial refers to things having to do with the land or with the planet Earth.
Amazon River	The Amazon River of South America is the largest river in the world by volume, with greater total river flow than the next eight largest rivers combined, and with the largest drainage basin in the world. Because of its vast dimensions it is sometimes called The River Sea.
Ganges	The Ganges is a major river in the Indian subcontinent flowing east through the eponymous plains of northern India into Bangladesh. It is held sacred by Hindus and is worshipped in its personified form as the goddess Ganga.
Carbonate minerals	Carbonate minerals are those minerals containing the carbonate ion: $CO_3 2-$.
Limestone	Limestone is a sedimentary rock composed largely of the mineral calcite. Limestone often contains variable amounts of silica in the form of chert or flint, as well as varying amounts of clay, silt and sand as disseminations, nodules, or layers within the rock. The primary source of the calcite in limestone is most commonly marine organisms. These organisms secrete shells that settle out of the water column and are deposited on ocean floors as pelagic ooze or alternatively is conglomerated in a coral reef.
Metamorphism	Metamorphism can be defined as the solid state recrystallisation of pre-existing rocks due to changes in heat and/or pressure and/or introduction of fluids. There will be mineralogical, chemical and crystallographic changes. Metamorphism produced with increasing pressure and temperature conditions is known as prograde metamorphism. Conversely, decreasing temperatures and pressure characterize retrograde metamorphism.
Seafloor	The seafloor the bottom of the ocean. At the bottom of the continental slope is the

Go to **Cram101.com** for the Practice Tests for this Chapter.

continental rise, which is caused by sediment cascading down the continental slope.

Oceanic crust	Oceanic crust is the part of Earth's lithosphere that surfaces in the ocean basins. Oceanic crust is primarily composed of mafic rocks, or sima. It is thinner than continental crust, or sial, generally less than 10 kilometers thick, however it is more dense, having a mean density of about 3.3 grams per cubic centimeter.
Crust	In geology, a crust is the outermost layer of a planet, part of its lithosphere. They are generally composed of a less dense material than its deeper layers.Earths face=symbol>¢ is composed mainly of basalt and granite. It is cooler and more rigid than the deeper layers of the mantle and core.
Biosphere	The biosphere is the part of the earth, including air, land, surface rocks, and water, within which life occurs, and which biotic processes in turn alter or transform. From the broadest biophysiological point of view, the biosphere is the global ecological system integrating all living beings and their relationships, including their interaction with the elements of the lithosphere, hydrosphere, and atmosphere. This biosphere is postulated to have evolved, beginning through a process of biogenesis or biopoesis, at least some 3.5 billion years ago.
Melting	Melting is the process of heating a solid substance to a point where it turns into a liquid. An object that has melted is molten.
Fresh water	Fresh water contains low concentrations of dissolved salts and other total dissolved solids. It is an important renewable resource, necessary for the survival of most terrestrial organisms, and required by humans for drinking and agriculture, among many other uses.
Surface water	Water collecting on the ground or in a stream, river, lake, or wetland is called surface water; as opposed to groundwater. Surface water is naturally replenished by precipitation and naturally lost through discharge to the oceans, evaporation, and sub-surface seepage into the groundwater. Surface water is the largest source of fresh water.
Shale	Shale is a fine-grained sedimentary rock whose original constituents were clays or muds. It is characterized by thin laminae breaking with an irregular curving fracture, often splintery and usually parallel to the often-indistinguishable bedding plane.
Seep	A seep is a wet place where a liquid, usually groundwater, has oozed from the ground to the surface. They are usually not flowing, with the liquid sourced only from underground. The term may also refer to the movement of liquid hydrocarbons to the surface through fractures and fissures in the rock and between geological layers. It may be a significant source of pollution.
Arctic	The Arctic is the region around the Earth's North Pole, opposite the Antarctic region around the South Pole. In the northern hemisphere, the Arctic includes the Arctic Ocean and parts of Canada, Greenland, Russia, the United States, Iceland, Norway, Sweden and Finland. The word Arctic comes from the Greek word arktos, which means bear. This is due to the location of the constellation Ursa Major, the "Great Bear", above the Arctic region.
Ice sheet	An ice sheet is a mass of glacier ice that covers surrounding terrain and is greater than 19,305 mile². The only current ice sheets are in Antarctica and Greenland. Ice sheets are bigger than ice shelves or glaciers. Masses of ice covering less than 50,000 km² are termed an ice cap. An ice cap will typically feed a series of glaciers around its periphery. Although the surface is cold, the base of an ice sheet is generally warmer. This process produces fast-flowing channels in the ice sheet.
Arctic Ocean	The Arctic Ocean, located in the northern hemisphere and mostly in the Arctic north polar region, is the smallest of the world's five major oceanic divisions and the shallowest.

Blocks	Blocks in meteorology are large scale patterns in the atmospheric pressure field that are nearly stationary, effectively "blocking" or redirecting migratory cyclones. These blocks can remain in place for several days or even weeks, causing the areas affected by them to have the same kind of weather for an extended period of time.
Ridge	A ridge is a geological feature that is also known as a Rip in the earth causing magma to flow out and forming an undersea volcano, it also has geological features, a continuous elevational crest for some distance. Ridges are usually termed hills or mountains as well, depending on size.
Pacific Ocean	The Pacific Ocean is the largest of the Earth's oceanic divisions. It extends from the Arctic in the north to the Antarctic in the south, bounded by Asia and Australia on the west and the Americas on the east. At 169.2 million square kilometres in area, this largest division of the World Ocean – and, in turn, the hydrosphere – covers about 46% of the Earth's water surface and about 32% of its total surface area, making it larger than all of the Earth's land area combined.
Abyssal	The abyssal is the pelagic zone that contains the very deep benthic communities near the bottom of oceans.
Anomalous	An anomalous phenomenon is an observed event which deviates from what is expected according to existing rules or scientific theory.
Gulf stream	The Gulf Stream is a powerful, warm, and swift Atlantic ocean current that originates in the Gulf of Mexico, exits through the Strait of Florida, and follows the eastern coastlines of the United States and Newfoundland before crossing the Atlantic Ocean. It then splits in two, with the northern stream crossing to northern Europe and the southern stream recirculating off West Africa. The Gulf Stream influences the climate of the east coast of North America from Florida to Newfoundland, and the west coast of Europe.
Atlantic Ocean	The Atlantic Ocean is the second-largest of the world's oceanic divisions; with a total area of about 106.4 million square kilometres , it covers approximately one-fifth of the Earth's surface. The Atlantic Ocean occupies an elongated, S-shaped basin extending longitudinally between the Americas to the west, and Eurasia and Africa to the east.
Antarctic Circumpolar Current	The Antarctic Circumpolar Current is an ocean current that flows from west to east around Antarctica.
Humboldt Current	The Humboldt Current is a cold, low salinity ocean current that extends along the West Coast of South America from Northern Peru to the southern tip of Chile. The waters of the Humboldt Current system flow in the direction of the Equator and can extend 1,000 kilometers offshore.
Velocity	In physics, velocity is defined as the rate of change of displacement or the rate of displacement. Simply put, it is distance per units of time.
Los Alamos National Laboratory	Los Alamos National Laboratory is a United States Department of Energy national laboratory, managed and operated by Los Alamos National Security, LLC, located in Los Alamos, New Mexico. The laboratory is one of the largest multidisciplinary institutions in the world.
Food web	Food web refers to describe the feeding relationships between species in an ecological community. Typically a food web refers to a graph where only connections are recorded, and a food web or ecosystem network refers to a network where the connections are given weights representing the quantity of nutrients or energy being transferred.
Shoreline	A shoreline is the fringe of land at the edge of a large body of water, such as an ocean, sea, or lake. A strict definition is the strip of land along a water body that is alternately exposed and covered by waves and tides.

Upwelling	Upwelling is an oceanographic phenomenon that involves wind-driven motion of dense, cooler, and usually nutrient-rich water towards the ocean surface, replacing the warmer, usually nutrient-deplete surface water.
Coast	The coast is defined as the part of the land adjoining or near the ocean. A coastline is properly a line on a map indicating the disposition of a coast, but the word is often used to refer to the coast itself. The adjective coastal describes something as being on, near to, or associated with a coast.
Phosphate	A phosphate, in inorganic chemistry, is a salt of phosphoric acid. In organic chemistry it is an ester of phosphoric acid.
Organism	In biology and ecology, an organism is a living complex adaptive system of organs that influence each other in such a way that they function in some way as a stable whole.
Drought	A drought is an extended period of months or years when a region notes a deficiency in its water supply. Generally, this occurs when a region receives consistently below average precipitation.
El Nino	El Nino is a global coupled ocean-atmosphere phenomenon. The Pacific ocean signatures, are important temperature fluctuations in surface waters of the tropical Eastern Pacific Ocean.
Turbulent flow	Turbulent flow is a flow regime characterized by chaotic, stochastic property changes. This includes low momentum diffusion, high momentum convection, and rapid variation of pressure and velocity in space and time.
Ocean basins	Ocean basins are large geologic basins that are below sea level. Geologically, there are other undersea geomorphological features such as the continental shelves, the deep ocean trenches, and the undersea mountain rangeswhich are not considered to be part of the ocean basins.
Canopy	The canopy is the uppermost level of a forest, formed by the tree crowns.
Terrain	Terrain is the third or vertical dimension of land surface. When terrain is described underwater, the term bathymetry is used.
Topographic maps	Topographic maps are a variety of maps characterized by large-scale detail and quantitative representation of relief, usually using contour lines in modern mapping, but historically using a variety of methods.
Antarctic Bottom Water	The Antarctic Bottom Water is a type of water mass in the seas surrounding Antarctica with temperatures ranging from 0 to -0.8∘ C, salinities from 34.6 to 34.7, and a density near 27.88.
Indian Ocean	The Indian Ocean is the third largest of the world's oceanic divisions, covering about 20% of the Earth's water surface. It is bounded on the north by Asia on the west by Africa; on the east by the Malay Peninsula, the Sunda Islands, and Australia; and on the south by the Southern Ocean.
Bering Sea	The Bering Sea is a body of water in the Pacific Ocean that comprises a deep water basin which rises through a narrow slope into the shallower water above the continental shelves.
Heat transfer	In thermal physics, heat transfer is the passage of thermal energy from a hot to a cold body. When a physical body, e.g. an object or fluid, is at a different temperature than its surroundings or another body, transfer of thermal energy, also known as heat transfer, occurs in such a way that the body and the surroundings reach thermal equilibrium.

Weathering	Physical and chemical processes in which solid rock exposed at earth's surface is changed to separate solid particles and dissolved material, which can then be moved to another place as sediment is referred to as weathering.
Stone	Stone refers to quarried or artificially broken rock for use in construction.
Cement	Minerals such as silica and carbonate that are chemically precipitated in the pores of sediments, binding the grain are referred to as cement.
Pile	A long substantial pole of wood, concrete or metal, driven into the earth or sea bed to serve as a support or protection is called pile.
Rock	Any material that makes up a large, natural, continuous part of earth's crust is rock.
Atmosphere	Atmosphere refers to the whole mass of air surrounding the earth.
Mass	The amount of material in an object is the mass.
Sedimentary rock	Rock that forms from the accumulated products of erosion and in some cases from the compacted shells, skeletons, and other remains of dead organisms is sedimentary rock. Compare with igneous rock, metamorphic roc is a sedimentary rock.
Bedding	The layering of rocks, especially sedimentary rooks is called bedding.
Weather	Weather refers to short-term changes in the temperature, barometric pressure, humidity, precipitation, sunshine, cloud cover, wind direction and speed, and other conditions in the troposphere at a given place and time. Compare with climate.
Rocks	An aggregate of one or more minerals rather large in area are rocks. The three classes of rocks are the following: Igneous rock - crystalline rocks formed from molten material. Sedimentary rock - A rock resulting from the consolidation of loose sediment that has accumulated in layers. Metamorphic rock - Rock that has formed from preexisting rock as a result of heat or pressure.
Point	Point refers to the extreme end of a cape, or the outer end of any land area protruding into the water, usually less prominent than a cape. A low profile shoreline promontory of more or less triangular shape, the top of which extends seaward.
Bedrock	Solid rock lying beneath loose soil or unconsolidated sediment is called bedrock.
Erosion	Erosion refers to wearing away of the land by natural forces. On a beach, the carrying away of beach material by wave action, tidal currents or by DEFLATION. The wearing away of land by the action of natural forces.
Soil	Soil refers to a layer of weathered, unconsolidated material on top of bedrock; often also defined as containing organic matter and being capable of supporting plant growth.
Sand	Sand refers to an unconsolidated mixture of inorganic soil consisting of small but easily distinguishable grains ranging in size from about.062 mm to 2.0 mm.
Gravel	Gravel refers to loose, rounded fragments of rock, larger than sand, but smaller than cobbles. Small stones and pebbles, or a mixture of these with sand.
Clay	Clay refers to a fine grained sediment with a typical grain size less than 0.004 mm. Possesses electromagnetic properties which bind the grains together to give a bulk strength or cohesion.
Ore	Part of a metal-yielding material that can be economically extracted at a given time is ore. An ore typically contains two parts: the ore mineral, which contains the desired metal, and waste mineral material.
Chemical	One of the millions of different elements and compounds found naturally or synthesized by

human is referred to as chemical.

Hydrosphere	The earth's liquid water earth's frozen water, and small amounts of water vapor in the atmosphere are hydrosphere.
Stress	Stress refers to force per unit area. May be compression, tension, or shear.
Chemical weathering	The decomposition of rocks under attack of base- or acidladen waters is called chemical weathering.
Dissolution	The act of dissolving a solid is called dissolution.
Acid	A substance that releases a hydrogen ion in solution is an acid.
Oxidation	Oxidation refers to a combination with oxygen. In fire, oxygen combines with organic matter, in rust, oxygen combines with iron.
Depth	Depth refers to vertical distance from still-water level to the bottom.
Mixture	Mixture refers to combination of two or more elements and compounds.
Clay minerals	Clay minerals refers to layered, platy minerals composed of hydrous aluminum silicates. Examples include kaolinite and chlorite.
Matter	Matter refers to anything that has mass and takes up space. On earth, where gravity is present, we weigh an object to determine its mass.
Climate	Physical properties of the troposphere of an area based on analysis of its weather records over a long period. The two main factors determining an area's climate are temperature, with its seasonal variations, and the amount and distri.
Debris	Debris refers to any accumulation of rock fragments; detritus.
Wind	Wind refers to the mass movement of air.
Metamorphism	The changes in minerals and rock textures that occur with the elevated temperatures and pressures below the Earth's surface are referred to as metamorphism.
Crust	Solid outer zone of the earth. It consists of oceanic crust and continental crust. Compare core, mantle.
Environment	Environment refers to all external conditions and factors, living and nonliving, that affect an organism or other specified system during its lifetime; the earth's life-support systems for us and for all other forms of life-another term for solar capita.
Equilibrium	A point of rest. A system that does not tend to undergo any change of its own accord but remains in a single, fixed condition is said to be in equilibrium. Compare with steady state.
Temperature	Temperature refers to a measure of the average speed of motion of the atoms, ions, or molecules in a substance or combination of substances at a given moment. Compare with heat.
Conditions	Conditions refers to physical or chemical attributes of the environment that, while not being consumed, influence biological processes and population growth. Examples are temperature, salinity, and acidity. Compare resources.
Fall	Fall refers to a mass moving nearly vertical and downward under the influence of gravity.
Base	A substance that combines with a hydrogen ion in solution is called the base.
Cliff	A high steep face of rock is referred to as cliff.
Gravity	The attraction between bodies of matter is a gravity.
Range	Land used for grazing is referred to as the range.

Go to **Cram101.com** for the Practice Tests for this Chapter.

Accumulation	Buildup of matter, energy, or information in a system is referred to as accumulation.
Granite	A light-colored, coarsegrained, intrusive igneous rock composed mainly of quartz and feldspar and that typifies the continental crust are called the granite.
Freezing point	The temperature at which a liquid is transformed into a solid is referred to as freezing point.
Shore	That strip of ground bordering any body of water which is alternately exposed, or covered by tides and/or waves is a shore. A shore of unconsolidated material is usually called a beach.
Lake	Large natural body of standing fresh water formed when water from precipitation, land runoff, or groundwater flow fills a depression in the earth created by glaciation, earth movement, volcanic activity, or a giant meteorit are called the lake.
Groundwater	Groundwater refers to water that sinks into the soil and is stored in slowly flowing and slowly renewed underground reservoirs called aquifers; underground water in the zone of saturation, below the water table. Compare runoff, surface water.
Valley	An elongated depression, usually with an outlet, between bluffs or between ranges of hills or mountains is a valley.
Evaporation	Evaporation refers to conversion of a liquid into a gas.
Sea	Sea refers to the ocean. A large body of salt water, second in rank to an ocean, more or less landlocked and generally part of, or connected with, an ocean or a larger sea. State of the ocean or lake surface, in regard to waves.
Slumping	The sliding of large, cohering blocks of sediment or rock downslope under the influence of gravity is called slumping.
Plants	Eukaryotic, mostly multicelled organisms such as algae, mosses, ferns, flowers, cacti, grasses, beans, wheat, rice, and trees are plants. These organisms use photosynthesis to produce organic nutrients for themselves and for other organisms is referred to as plants.
Nutrients	Nutrients refer to chemicals such as phosphorus and nitrogen that, when released into water sources, may cause pollution events such as eutrophication.
Ion	Ion refers to atom or group of atoms with one or more positive or negative electrical charges. Compare atone, molecule.
Recent	A synonym of Holocene is called recent.
Thermal	The energy of the random motion of atoms and molecules is referred to as thermal.
Avalanche	A large mass of snow, ice, soil, or rock that moves rapidly downslope under the pull of gravity is referred to as an avalanche.
Ice age	Ice age refers to one of several periods of low temperature during the last million years.
Mineral	A naturally occurring, inorganic, crystalline solid that has a definite chemical composition and possesses characteristic physical properties is a mineral.
Halite	A mineral composed of sodium chloride and commonly referred to as rock salt are called the halite.
Gypsum	An evaporite deposit composed of hydrous calcium sulfat is a gypsum.
Surface water	Precipitation that does not infiltrate the ground or return to the atmosphere by evaporation or transpiration is called surface water.
Molecule	Molecule refers to a combination of two or more atoms of the same chemical element or different chemical elements held together by chemical bonds. Compare with atom, ion.

Atoms	Atoms refers to minute units made of subatomic particles that are the basic building blocks of all chemical elements and thus all matter; the smallest unit of an element that can exist and still have the unique characteristics of that element. Compare to ion, molecule.
Atom	The smallest component of an element, comprising neutrons, protons, and electrons is an atom.
Concentration	Amount of a chemical in a particular volume or weight of air, water, soil, or other medium is referred to as concentration.
Carbonic acid	Carbonic acid refers to a common but weak acid formed by carbon dioxide dissolving in water. carnivore A flesh-eating animal.
Carbon dioxide	Molecule of carbon and oxygen present in the atmosphere at approximately 350 ppm. Emissions of carbon dioxide resulting from burning of fossil fuels are thought to be co:.itributing to potential global warming through an enhanced greenhouse effect.
Bacteria	Bacteria refer to prokaryotic, one-celled organisms. Some transmit diseases. Most act as decomposers and get the nutrients they need by breaking down complex organic compounds in the tissues of living or dead organisms into simpler inorganic nutrient compounds.
Coal	Solid, combustible mixture of organic compounds with 30-98% carbon by weight, mixed with various amounts of water and small amounts of sulfur and nitrogen compounds. It is formed in several stages as the remains of plants are subjected to heat and press is a coal.
Biota	A general term for all the organisms of all species living in an area or region up to and including the biosphere, as in 'the biota of the Mojave Desert' or 'the biota in that aquarium'.
Silicate minerals	The most important group of rock-forming minerals is called silicate minerals.
Sedimentary rocks	Sedimentary rocks refer to rocks that have formed by the compaction and cementation of sediment.
Residual	The components of water level not attributable to astronomical effects are called residual.
Silica	A compound with a composition, such as quartz in granite and opal in the shells of radiolaria is referred to as silica.
Site	A factor considering the summation of all environmental features of a location that influences the placement of a city is a site.
Ocean	The great body of salt water which occupies two-thirds of the surface of the Earth, or one of its major subdivisions is called an ocean.
Absorption	Conversion of sound or light energy into heat is called absorption.
Electron	Tiny particle moving around outside the nucleus of an atom. Each electron has one unit of negative charge and almost no mass. Compare neutron, proton.
Mud	Mud refers to a mixture of silt and clay sized particles.
State	State refers to an expression of the internal form of matter. Water exists in three states: solid, liquid, and gas. A solid has a fixed volume and fixed shape; a liquid has a fixed volume but no fixed shape; and a gas has neither fixed volume nor fixed shape.
Planet	A smaller, usually nonluminous body orbiting a star is a planet.
Sediment	Loose, fragments of rocks, minerals or organic material which are transported from their source for varying distances and deposited by air, wind, ice and water are sediment. Other sediments are precipitated from the overlying water or form chemically, in place. Sediment includes all the unconsolidated materials on the sea floor. The fine-grained material

	deposited by water or wind.
Outcrop	A surface exposure of bare rock, not covered by soil or vegetation is called outcrop.
Limestone	A sedimentary rock composed dominantly of calcium carbonate, either precipitated from seawater or deposited as shell debris are called the limestone.
Porosity	Porosity refers to a percentage of space in rock or soil occupied by voids, whether the voids are isolated or connected. Compare with permeability.
Permeability	Permeability refers to the property of bulk material which permits movement of water through its pores.
Precipitation	Water in the form of rain, sleet, hail, and snow that falls from the atmosphere onto the land and bodies of water is called precipitation.
Runoff	Fresh water from precipitation and melting ice that flows on the earth's surface into nearby streams, lakes, wetlands, and reservoir is referred to as runoff.
Chemical equilibrium	In seawater, the condition in which the proportion and amounts of dissolved salts per unit volume of ocean are nearly constant is a chemical equilibrium.
Development	Development refers to change from a society that is largely rural, agricultural, illiterate, and poor, with a rapidly growing population, to one that is mostly urban, industrial, educated, and wealthy, with a slowly growing or stationary population.
River	A natural stream of water larger than a brook or creek is a river.
Basalt	Basalt refers to a dark, fine-grained igneous rock composed of minerals enriched in ferromagnesian silicates.
System	A set of components that function and interact in some regular and theoretically predictable manner is called a system.
Network	Network refers to a set consisting of stations for which geometric relationships have been determined and which are so related that removal of one station from the set will affect the relationships between the other stations; and lines connecting the stations to show this interdependence.
Lead	Lead refers to a heavy metal that is an important constituent of automobile batteries and other industrial products. A toxic metal capable of causing environmental disruption and producing a health problem to people and other living organisms.
Igneous rocks	Igneous rocks refers to rocks formed from the solidification of magma. They are extrusive if they crystallize on the surface of the Earth and intrusive if they crystallize beneath the surface.
Dike	Dike refers to sometimes written as dyke; earth structure along a sea or river in order to protect LITTORAL lands from flooding by high water; DIKES along rivers are sometimes called levees.
Zone	Division or province of the ocean with homogeneous characteristics is referred to as a zone.
Metamorphic rock	Metamorphic rock refers to rock produced when a preexisting rock is subjected to high temperatures, high pressures, chemically active fluids, or a combination of these agents. Compare with igneous rock, sedimentary rock.
Plates	Various-sized areas of earth's lithosphere that move slowly around with the mantle's flowing asthenosphere are plates. Most earthquakes and volcanoes occur around the boundaries of these plates.
Precision	Precision refers to a measure of reproducibility, or how closely a series of measurements of

the same quantity agree with one another. Compare with accuracy.

Stream	Stream refers to any flow of water; a current. A course of water flowing along a bed in the earth.
Front	The boundary between two air masses with different temperatures and densitie is referred to as front.
Horizon	Horizon refers to the line or circle which forms the apparent boundary between Earth and sky. A plane in rock strata characterized by particular features, as occurrence of distinctive fossil species. One of the series of distinctive layers found in a vertical cross-section of any well-developed soil.
Soil profile	Cross-sectional view of the horizons in a soil is referred to as the soil profile.
Forest	Forest refers to biome with enough average annual precipitation to support growth of various species of trees and smaller forms of vegetation. Compare desert, grassland.
Ore deposits	Earth materials in which metals are concentrated in high concentrations, sufficient to be mined are called ore deposits.
Map	Map refers to a representation of Earth's surface usually depicting mostly land areas.
Degree	An arbitrary measure of temperature. One degree Celsius _ 1.8 degrees Fahrenheit.
Nutrient	Any atom, ion, or molecule an organism that needs to live, grow, or reproduce is a nutrient.
Dunes	Dunes refers to accumulations of windblown sand on the BACKSHORE, usually in the form of small hills or ridges, stabilized by vegetation or control structures. A type of bed form indicating significant sediment transport over a sandy seabed.
Permafrost	Permafrost refers to a perennially frozen layer of the soil that forms when the water there freezes. It is found in arctic tundra.
Well	Well refers to a hole, generally cylindrical and usually walled or lined with pipe, that is dug or drilled into the ground to penetrate an aquifer below the zone of saturation.
Swell	Waves that have traveled a long distance from their generating area and have been sorted out by travel into long waves of the same approximate period are referred to as a swell.
Relief	The difference in elevation between the highest and lowest points in an area is called relief.
Fertilizer	Fertilizer refers to substance that adds inorganic or organic plant nutrients to soil and improves its ability to grow crops, trees, or other vegetation.
Topography	Topography refers to the form of the features of the actual surface of the Earth in a particular region considered collectively.
Lava	Molten rock that is extruded out of volcanoes is called lava.
Slope	The degree of inclination to the horizontal is the slope.
Evaporite	A type of sediment precipitated from an aqueous solution, usually by the evaporation of water from a basin with restricted circulation is the evaporite.
Infiltration	Infiltration refers to downward movement of water through soil.
Annual	Annual refers to plant that grows, sets seed, and dies in one growing season. Compare to perennial.
Desert	Desert refers to biome in which evaporation exceeds precipitation and the average amount of precipitation is less than 25 centimeters a year. Such areas have little vegetation or have widely spaced, mostly low vegetation. Compare forest, grassland.

Pore	Pore refers to an opening or void space in; oil or rock.
Latitude	Latitude refers to distance from the equator. Compare altitude.
Ash	The loose debris that is ejected from an erupting volcano is called ash.
Volcano	Vent or fissure in the earth's surface through which magma, liquid lava, and gases are released into the environment is a volcano.
Core	Core refers to a cylindrical sample extracted from a beach or seabed to investigate the types and DEPTHS of sediment layers. An inner, often much less permeable portion of a BREAKWATER, or BARRIER beach.
Reach	Reach refers to an arm of the ocean extending into the land. A straight section of restricted waterway of considerable extent; may be similar to a narrows, except much longer in extent.
Cross section	A two-dimensional drawing showing features in the vertical plane as in a canvon wall or road cut is referred to as cross section.
Observations	Information obtained through one or more of the five senses or through instruments that extend the senses are observations.
Key	Key refers to a low, insular BANK of sand, coral, etc., as one of the islets off the southern coast of Florida.
Geochemical cycles	Geochemical cycles refers to the pathways of chemical elements in geologic processes, including the chemistry of the lithosphere, atmosphere, and hydrosphere.
Geology	Geology refers to the science which treats of the origin, history and structure of the Earth, as recorded in rocks; together with the forces and processes now operating to modify rocks.
Resources	Resources refer to substances that can be consumed by an organism and, as a result, become unavailable to other organisms.
Feedback	A kind of system response that occurs when output of the system also serves as input leading to changes in the system is called feedback.
Slide	In mass wasting, movement of a descending mass along a plane approximately parallel to the slope of the surface is referred to as slide.

Slope	The degree of inclination to the horizontal is the slope.
Stream	Stream refers to any flow of water; a current. A course of water flowing along a bed in the earth.
Erosion	Erosion refers to wearing away of the land by natural forces. On a beach, the carrying away of beach material by wave action, tidal currents or by DEFLATION. The wearing away of land by the action of natural forces.
Wave	An oscillatory movement in a body of water manifested by an alternate rise and fall of the surface is called a wave. Disturbances of the surface of a liquid body, as the ocean, in the form of a ridge, swell or hump. The term wave by itself usually refers to the term surface gravity wave.
Lava	Molten rock that is extruded out of volcanoes is called lava.
Gravity	The attraction between bodies of matter is a gravity.
Creep	Creep refers to very slow, continuous downslope movement of soil or debris.
Soil	Soil refers to a layer of weathered, unconsolidated material on top of bedrock; often also defined as containing organic matter and being capable of supporting plant growth.
Crater	Crater refers to an abrupt basin commonly rimmed by ejected material. In volcanoes, craters form by outwarcc explosion, are commonly less than 2 km diameter, and occur at the summit of a volcanic cone. Similar rimmed basins form by impacts with meteorites, asteroids, and.
Sea	Sea refers to the ocean. A large body of salt water, second in rank to an ocean, more or less landlocked and generally part of, or connected with, an ocean or a larger sea. State of the ocean or lake surface, in regard to waves.
Mass	The amount of material in an object is the mass.
Rock	Any material that makes up a large, natural, continuous part of earth's crust is rock.
Ocean	The great body of salt water which occupies two-thirds of the surface of the Earth, or one of its major subdivisions is called an ocean.
Front	The boundary between two air masses with different temperatures and densitie is referred to as front.
Force	Force refers to a push or pull that affects motion. The product of mass and acceleration of a material.
Weathering	Physical and chemical processes in which solid rock exposed at earth's surface is changed to separate solid particles and dissolved material, which can then be moved to another place as sediment is referred to as weathering.
Subsidence	Sinking or down warping of a part of the earth's surface is called subsidence.
Hurricane	A cyclonic storm, usually of tropic origin, covering an extensive area, and containing winds in excess of 75 miles per hour is the hurricane.
Debris	Debris refers to any accumulation of rock fragments; detritus.
Stability	Ability of a living system to withstand or recover from externally imposed changes or stresses is called stability.
Saturation	A chemical state whereby the maximum amount of solute is dissolved under the given conditions is saturation.
Viscous	Ease of flow is viscous. The more viscous a substance, the less readily it flows.
Lahar	Lahar refers to a volcanic inudflow composed of unconsolidated volcanic debris and water.

Debris flow	Debris flow refers to a general term applied to the rapid downslope flow of unconsolidated debris; an exam.
Compaction	The decrease in volume and porosity of a sediment via burial is referred to as compaction.
Bedrock	Solid rock lying beneath loose soil or unconsolidated sediment is called bedrock.
System	A set of components that function and interact in some regular and theoretically predictable manner is called a system.
Undercutting	Undercutting refers to erosion of material at the foot of a cliff or bank; a sea cliff, or river bank on the outside of a meander. Ultimately, the overhang collapses, and the process is repeated.
Angle of repose	The maximum slope at which soils and loose materials on the banks of canals, rivers or embankments stay stable is called the angle of repose.
Equilibrium	A point of rest. A system that does not tend to undergo any change of its own accord but remains in a single, fixed condition is said to be in equilibrium. Compare with steady state.
Sand	Sand refers to an unconsolidated mixture of inorganic soil consisting of small but easily distinguishable grains ranging in size from about.062 mm to 2.0 mm.
Sorting	Sorting refers to a process of selection and separation of sediment grains according to their grain size.
Rocks	An aggregate of one or more minerals rather large in area are rocks. The three classes of rocks are the following: Igneous rock - crystalline rocks formed from molten material. Sedimentary rock - A rock resulting from the consolidation of loose sediment that has accumulated in layers. Metamorphic rock - Rock that has formed from preexisting rock as a result of heat or pressure.
Cohesion	Attachment of water molecules to each other by hydrogen bonds is a cohesion.
Unconsolidated	Unconsolidated with regards to sediment grains refers to loose, separate, or unattached to one another.
Storm	Local or regional atmospheric disturbance characterized by strong winds often accompanied by precipitation is referred to as a storm.
Earthquake	Shaking of the ground resulting either from the fracturing and displacement of rock, producing a fault, or from subsequent movement along the fault is the earthquake.
Shoreline	Shoreline refers to the intersection of a specified plane of water with the shore. All of the water areas of the state, including reservoirs and their associated uplands, together with the lands underlying them.
Deforestation	Removal of trees from a forested area without adequate replanting is a deforestation.
Recent	A synonym of Holocene is called recent.
River	A natural stream of water larger than a brook or creek is a river.
Degree	An arbitrary measure of temperature. One degree Celsius _ 1.8 degrees Fahrenheit.
Bedding	The layering of rocks, especially sedimentary rooks is called bedding.
Dam	Dam refers to structure built in rivers or estuaries, basically to separate water at both sides and/or to retain water at one side.
Catastrophe	Catastrophe refers to a situation or event that causes significant damage to people and property, such that recovery and/or rehabilitation is a long and involved process. Examples of natural catastrophes include hurricanes, volcanic eruptions, large wildfires, and floods.

Go to **Cram101.com** for the Practice Tests for this Chapter.

Slide	In mass wasting, movement of a descending mass along a plane approximately parallel to the slope of the surface is referred to as slide.
Friction	The resistance to motion of two bodies in contact is a friction.
Sediment	Loose, fragments of rocks, minerals or organic material which are transported from their source for varying distances and deposited by air, wind, ice and water are sediment. Other sediments are precipitated from the overlying water or form chemically, in place. Sediment includes all the unconsolidated materials on the sea floor. The fine-grained material deposited by water or wind.
Valley	An elongated depression, usually with an outlet, between bluffs or between ranges of hills or mountains is a valley.
Event	Event refers to an occurrence meeting specified conditions, e.g. damage, a threshold wave height or a threshold water level.
Flood	Period when tide level is rising; often taken to mean the flood current which occurs during this period. A flow above the CARRYING CAPACITY of a CHANNEL.
Conditions	Conditions refers to physical or chemical attributes of the environment that, while not being consumed, influence biological processes and population growth. Examples are temperature, salinity, and acidity. Compare resources.
Limestone	A sedimentary rock composed dominantly of calcium carbonate, either precipitated from seawater or deposited as shell debris are called the limestone.
Clay	Clay refers to a fine grained sediment with a typical grain size less than 0.004 mm. Possesses electromagnetic properties which bind the grains together to give a bulk strength or cohesion.
Water level	Elevation of a particular point or small patch on the surface of a body of water above a specific point or surface, averaged over a period of time sufficiently long to remove the effects of short period disturbances is referred to as water level.
Cross section	A two-dimensional drawing showing features in the vertical plane as in a canyon wall or road cut is referred to as cross section.
Sedimentary rocks	Sedimentary rocks refer to rocks that have formed by the compaction and cementation of sediment.
Map	Map refers to a representation of Earth's surface usually depicting mostly land areas.
Magnitude	An assessment of the size of an event is a magnitude. Magnitude scales exist for earthquakes, volcanic eruptions. hurricanes, and tornadoes. For earthquakes, different magnitudes are calculated for the same earthquake when different types of seismic waves are used.
Frequency	Number of events in a given time interval. For earthquakes, it is the number of cycles of seismic waves that pass in a second; frequency = l/period.
Reduce	With respect to waste management, reduce refers to practices that will reduce the amount of waste we produce.
Migration	A term that refers to the habit of some animals is a migration.
Mixture	Mixture refers to combination of two or more elements and compounds.
Mud	Mud refers to a mixture of silt and clay sized particles.
Slump	In mass wasting, movement along a curved surface in which the upper part moves vertically downward while the lower part moves outward is a slump.
Bedding plane	A surface parallel to the surface of deposition, which may or may not have a physical

expression is called a bedding plane. The original attitude of a bedding plane should not be assumed to have been horizontal.

Outcrop	A surface exposure of bare rock, not covered by soil or vegetation is called outcrop.
Temperature	Temperature refers to a measure of the average speed of motion of the atoms, ions, or molecules in a substance or combination of substances at a given moment. Compare with heat.
Freezing point	The temperature at which a liquid is transformed into a solid is referred to as freezing point.
Point	Point refers to the extreme end of a cape, or the outer end of any land area protruding into the water, usually less prominent than a cape. A low profile shoreline promontory of more or less triangular shape, the top of which extends seaward.
Pore	Pore refers to an opening or void space in; oil or rock.
Pebble	Pebble refers to a rock fragment between 4 and 64 mm in diameter.
Precipitation	Water in the form of rain, sleet, hail, and snow that falls from the atmosphere onto the land and bodies of water is called precipitation.
Desiccation	Desiccation refers to drying.
Plants	Eukaryotic, mostly multicelled organisms such as algae, mosses, ferns, flowers, cacti, grasses, beans, wheat, rice, and trees are plants. These organisms use photosynthesis to produce organic nutrients for themselves and for other organisms is referred to as plants.
Runoff	Fresh water from precipitation and melting ice that flows on the earth's surface into nearby streams, lakes, wetlands, and reservoir is referred to as runoff.
Disturbance	Disturbance refers to a discrete event in time that disrupts an ecosystem or community. Examples of natural disturbances include fires, hurricanes, tornadoes, droughts, and floods. Examples of humancaused disturbances include deforestation, overgrazing, and plowing.
Groundwater	Groundwater refers to water that sinks into the soil and is stored in slowly flowing and slowly renewed underground reservoirs called aquifers; underground water in the zone of saturation, below the water table. Compare runoff, surface water.
Permafrost	Permafrost refers to a perennially frozen layer of the soil that forms when the water there freezes. It is found in arctic tundra.
Spring	A place where groundwater flows out onto the surface is a spring.
Impermeable	Impermeable refers to impervious; the condition of rock that does not allow fluids to flow through it.
Zone	Division or province of the ocean with homogeneous characteristics is referred to as a zone.
Coast	A strip of land of indefinite length and width that extends from the SEASHORE inland to the first major change in terrain features is referred to as coast.
Range	Land used for grazing is referred to as the range.
Pyroclastic	Pertaining to magma and volcanic rock blasted up into the air is referred to as pyroclastic.
Volcano	Vent or fissure in the earth's surface through which magma, liquid lava, and gases are released into the environment is a volcano.
Heat	Total kinetic energy of all the randomly moving atoms, ions, or molecules within a given substance, excluding the overall motion of the whole object. This form of kinetic energy flows from one body to another when there is a temperature difference betwe is referred to as heat.

Lake	Large natural body of standing fresh water formed when water from precipitation, land runoff, or groundwater flow fills a depression in the earth created by glaciation, earth movement, volcanic activity, or a giant meteorit are called the lake.
Reach	Reach refers to an arm of the ocean extending into the land. A straight section of restricted waterway of considerable extent; may be similar to a narrows, except much longer in extent.
High water	High water refers to maximum height reached by a rising tide. The height may be solely due to the periodic tidal forces or it may have superimposed upon it the effects of prevailing meteorological conditions. Nontechnically, also called the HIGH TIDE.
Density	Density refers to the ratio of a mass to a unit volume specified as grams per cubic centimeter.
Ash	The loose debris that is ejected from an erupting volcano is called ash.
Fracture	Fracture refers to a general term for any breaks in rock. Fractures include faults, joints, and crack,_.
Development	Development refers to change from a society that is largely rural, agricultural, illiterate, and poor, with a rapidly growing population, to one that is mostly urban, industrial, educated, and wealthy, with a slowly growing or stationary population.
Harbor	A water area nearly surrounded by land, sea walls, BREAKWATERS or artificial dikes, forming a safe anchorage for ships is referred to as harbor.
Matter	Matter refers to anything that has mass and takes up space. On earth, where gravity is present, we weigh an object to determine its mass.
Flooding	The natural process whereby waters emerge from their stream channel to cover part of the floodplain. Natural flooding is not a problem until people choose to build homes and other structures on floodplains.
Lead	Lead refers to a heavy metal that is an important constituent of automobile batteries and other industrial products. A toxic metal capable of causing environmental disruption and producing a health problem to people and other living organisms.
Stress	Stress refers to force per unit area. May be compression, tension, or shear.
Dip	The angle of inclination measured in degrees from the horizontal is called dip.
Coal	Solid, combustible mixture of organic compounds with 30-98% carbon by weight, mixed with various amounts of water and small amounts of sulfur and nitrogen compounds. It is formed in several stages as the remains of plants are subjected to heat and press is a coal.
Avalanche	A large mass of snow, ice, soil, or rock that moves rapidly downslope under the pull of gravity is referred to as an avalanche.
Fall	Fall refers to a mass moving nearly vertical and downward under the influence of gravity.
Cliff	A high steep face of rock is referred to as cliff.
Base	A substance that combines with a hydrogen ion in solution is called the base.
Depth	Depth refers to vertical distance from still-water level to the bottom.
Desert	Desert refers to biome in which evaporation exceeds precipitation and the average amount of precipitation is less than 25 centimeters a year. Such areas have little vegetation or have widely spaced, mostly low vegetation. Compare forest, grassland.
Risk	Risk refers to the probability that something undesirable will happen from deliberate or accidental exposure.

Elevation	Elevation refers to the distance of a point above a specified surface of constant potential; the distance is measured along the direction of gravity between the point and the surface.
Topographic map	A map on which elevations are shown by means of contour lines is referred to as a topographic
Topography	Topography refers to the form of the features of the actual surface of the Earth in a particular region considered collectively.
Theories	Scientific models that offer broad, fundamental explanations of related phenomena and are supported by consistent and extensive evidence are theories.
Acoustic fluidization	A theorized process where sound waves trapped inside a dry, fallen mass lessen internal friction to enable fluid like flow is referred to as acoustic fluidization.
Glacier	Large masses of moving ice on land derived by the recrystallization of snow into ice under pressure are referred to as glacier.
Climate	Physical properties of the troposphere of an area based on analysis of its weather records over a long period. The two main factors determining an area's climate are temperature, with its seasonal variations, and the amount and distri.
Continental slope	Continental slope refers to the declivity from the offshore border of the CONTINENTAL SHELF to oceanic depths. It is characterized by a marked increase in slope.
Turbidity	Turbidity refers to a condition of a liquid due to fine visible material in suspension, which may not be of sufficient size to be seen as individual particles by the naked eye but which prevents the passage of light through the liquid. A measure of fine suspended matter in liquids.
Abyssal plain	A flat area on the deep-sea floor having a very gentle slope of less than one meter per kilometer, and consisting chiefly of graded terrigenous sediments known as turbidites is an abyssal plain.
Shore	That strip of ground bordering any body of water which is alternately exposed, or covered by tides and/or waves is a shore. A shore of unconsolidated material is usually called a beach.
Oil	The liquid form of petroleum consisting of a complex mixture of large hydrocarbon molecules is referred to as oil.
Cement	Minerals such as silica and carbonate that are chemically precipitated in the pores of sediments, binding the grain are referred to as cement.
Methane hydrate	A white ice-like compound made up of molecules of methane gas trapped in 'cages' of frozen water in the sediments of the deep seafloor is referred to as methane hydrate.
Planet	A smaller, usually nonluminous body orbiting a star is a planet.
Tsunami	Tsunami refers to a large, high-velocity wave generated by displacement of the sea floor; also called seismic sea wave. Commonly misnamed tidal wave.
Coral	Coral refers to any of more than 6,000 species of small cnidarians, many of which are capable of generating hard calcareous skeletons.
Fault	A fracture in rock along which there has been an observable amount of displacement. Faults are rarely single planar units; normally they occur as parallel to sub-parallel sets of planes along which movement has taken place to a greater or lesser extent. Such sets are called fault or fracture-zones.
Escarpment	A more or less continuous line of CLIFFS or steep slopes facing in one general direction which are caused by EROSION or faulting, also called SCARP is an escarpment.
Magma	Molten rock below the earth's surface is referred to as magma.

Go to **Cram101.com** for the Practice Tests for this Chapter.

Mantle	Zone of the earth's interior between its core and its crust. Compare with core, crust is the mantle.
Seamount	Conical mountain rising 1000 m or more above the sea floor is a seamount.
Relief	The difference in elevation between the highest and lowest points in an area is called relief.
Head	A comparatively high promontory with either a CLIFF or steep face. It extends into a large body of water, such as a sea or lake. An unnamed HEAD is usually called a headland. The section of RIP CURRENT which has widened out seaward of the BREAKERS, also called head of
Subduction zone	Elongate region in which the sea floor slides beneath a continent or island arc is referred to as the subduction zone.
Roller	Roller refers to an indefinite term, sometimes considered to denote one of a series of long-crested waves which roll in upon a coast, as after a storm. Long, high swell, also called a ground swell.
Dissolution	The act of dissolving a solid is called dissolution.
Peat	Peat refers to an organic deposit consisting predominantly of partly decayed plant matter.
Bacteria	Bacteria refer to prokaryotic, one-celled organisms. Some transmit diseases. Most act as decomposers and get the nutrients they need by breaking down complex organic compounds in the tissues of living or dead organisms into simpler inorganic nutrient compounds.
Carbon dioxide	Molecule of carbon and oxygen present in the atmosphere at approximately 350 ppm. Emissions of carbon dioxide resulting from burning of fossil fuels are thought to be co:.itributing to potential global warming through an enhanced greenhouse effect.
Subsurface mining	Subsurface mining refers to the extraction of a metal ore or fuel resource such as coal from a deep underground deposit. Compare with surface mining.
Recharge	The addition of new water to an aquifer or to the zone of saturation is called a recharge.
Delta	ALLUVIAL DEPOSIT, usually triangular, at the mouth of a river of other stream. It is normally built up only where there is no tidal or CURRENT action capable of removing the sediment as fast as it is deposited, and hence the DELTA builds forward from the COASTLINE. A TIDAL DELTA is a similar deposit at the mouth of a tidal INLET, put there by TIDAL CURRENTS. A WAVE DELTA is a deposit made by large waves which run over the top of a SPIT or BAR beach and down the landward side.
Annual	Annual refers to plant that grows, sets seed, and dies in one growing season. Compare to perennial.
Chemical weathering	The decomposition of rocks under attack of base- or acidladen waters is called chemical weathering.
Energy	Capacity to do work by performing mechanical, physical, chemical, or electrical tasks or to cause a heat transfer between two objects at different temperatures is an energy.
Output	Output refers to matter, energy, or information leaving a system. Compare with input, throughput.
Network	Network refers to a set consisting of stations for which geometric relationships have been determined and which are so related that removal of one station from the set will affect the relationships between the other stations; and lines connecting the stations to show this interdependence.
Observations	Information obtained through one or more of the five senses or through instruments that extend the senses are observations.

Go to **Cram101.com** for the Practice Tests for this Chapter.

Key	Key refers to a low, insular BANK of sand, coral, etc., as one of the islets off the southern coast of Florida.
Boulder	A rounded rock on a beach, greater than 256 mm in diameter, larger than a cobbl is a boulder.
Geomorphology	Geomorphology refers to that branch of physical geography which deals with the form of the Earth, the general configuration of its surface, the distribution of the land, water, etc. The investigation of the history of geologic changes through the interpretation of topographic forms.
Geology	Geology refers to the science which treats of the origin, history and structure of the Earth, as recorded in rocks; together with the forces and processes now operating to modify rocks.
Environment	Environment refers to all external conditions and factors, living and nonliving, that affect an organism or other specified system during its lifetime; the earth's life-support systems for us and for all other forms of life-another term for solar capita.
Mitigation	Process that identifies actions to avoid, lessen, or compensate for anticipated adverse environmental impacts is called mitigation.
Resources	Resources refer to substances that can be consumed by an organism and, as a result, become unavailable to other organisms.
Feedback	A kind of system response that occurs when output of the system also serves as input leading to changes in the system is called feedback.

River	A natural stream of water larger than a brook or creek is a river.
System	A set of components that function and interact in some regular and theoretically predictable manner is called a system.
Network	Network refers to a set consisting of stations for which geometric relationships have been determined and which are so related that removal of one station from the set will affect the relationships between the other stations; and lines connecting the stations to show this interdependence.
Ocean	The great body of salt water which occupies two-thirds of the surface of the Earth, or one of its major subdivisions is called an ocean.
Rock	Any material that makes up a large, natural, continuous part of earth's crust is rock.
Debris	Debris refers to any accumulation of rock fragments; detritus.
Erosion	Erosion refers to wearing away of the land by natural forces. On a beach, the carrying away of beach material by wave action, tidal currents or by DEFLATION. The wearing away of land by the action of natural forces.
Planet	A smaller, usually nonluminous body orbiting a star is a planet.
Matter	Matter refers to anything that has mass and takes up space. On earth, where gravity is present, we weigh an object to determine its mass.
Power	Power refers to the time rate of doing work.
Range	Land used for grazing is referred to as the range.
Flood	Period when tide level is rising; often taken to mean the flood current which occurs during this period. A flow above the CARRYING CAPACITY of a CHANNEL.
Basin	A large submarine depression of a generally circular, elliptical or oval shape is called a basin.
Energy	Capacity to do work by performing mechanical, physical, chemical, or electrical tasks or to cause a heat transfer between two objects at different temperatures is an energy.
Gravity	The attraction between bodies of matter is a gravity.
Open system	Open system refers to a system, such as a living organism, in which both matter and energy are exchanged between the system and the environment. Compare with closed system.
Sediment	Loose, fragments of rocks, minerals or organic material which are transported from their source for varying distances and deposited by air, wind, ice and water are sediment. Other sediments are precipitated from the overlying water or form chemically, in place. Sediment includes all the unconsolidated materials on the sea floor. The fine-grained material deposited by water or wind.
Precipitation	Water in the form of rain, sleet, hail, and snow that falls from the atmosphere onto the land and bodies of water is called precipitation.
Groundwater	Groundwater refers to water that sinks into the soil and is stored in slowly flowing and slowly renewed underground reservoirs called aquifers; underground water in the zone of saturation, below the water table. Compare runoff, surface water.
Climate change	Refers to any long-term trend in MEAN SEA LEVEL, wave height, wind speed, drift rate etc are called the climate change.
Well	Well refers to a hole, generally cylindrical and usually walled or lined with pipe, that is dug or drilled into the ground to penetrate an aquifer below the zone of saturation.

Go to **Cram101.com** for the Practice Tests for this Chapter.

Stream	Stream refers to any flow of water; a current. A course of water flowing along a bed in the earth.
Discharge	The volume of water flowing in a stream per unit of time is the discharge.
Gradient	A measure of slope in meters of rise or fall per meter of horizontal distance. More general, a change of a value per unit of distance, e.g. the GRADIENT in longshore transport causes EROSION or ACCRETION. With reference to winds or currents, the rate of increase or decrease in speed, usually in the vertical; or the curve that represents this rate.
Load	The quantity of sediment transported by a current. It includes the suspended load of small particles in the water, and the bedload of large particles that move along the bottom.
Base level	The level below, which a stream cannot erode, usually sea level, is referred to as base level.
State	State refers to an expression of the internal form of matter. Water exists in three states: solid, liquid, and gas. A solid has a fixed volume and fixed shape; a liquid has a fixed volume but no fixed shape; and a gas has neither fixed volume nor fixed shape.
Equilibrium	A point of rest. A system that does not tend to undergo any change of its own accord but remains in a single, fixed condition is said to be in equilibrium. Compare with steady state.
Abrasion	Frictional erosion by material transported by wind and waves is referred to as abrasion.
Point	Point refers to the extreme end of a cape, or the outer end of any land area protruding into the water, usually less prominent than a cape. A low profile shoreline promontory of more or less triangular shape, the top of which extends seaward.
Lake	Large natural body of standing fresh water formed when water from precipitation, land runoff, or groundwater flow fills a depression in the earth created by glaciation, earth movement, volcanic activity, or a giant meteorit are called the lake.
Delta	ALLUVIAL DEPOSIT, usually triangular, at the mouth of a river of other stream. It is normally built up only where there is no tidal or CURRENT action capable of removing the sediment as fast as it is deposited, and hence the DELTA builds forward from the COASTLINE. A TIDAL DELTA is a similar deposit at the mouth of a tidal INLET, put there by TIDAL CURRENTS. A WAVE DELTA is a deposit made by large waves which run over the top of a SPIT or BAR beach and down the landward side.
Wind	Wind refers to the mass movement of air.
Desert	Desert refers to biome in which evaporation exceeds precipitation and the average amount of precipitation is less than 25 centimeters a year. Such areas have little vegetation or have widely spaced, mostly low vegetation. Compare forest, grassland.
Plate tectonics	Plate tectonics refer to the theory of geophysical processes that explains the movements of lithospheric plates and the processes that occur at their boundaries.
Tectonics	The study of the major structural features of the Earth's crust or the broad structure of a region is referred to as tectonics.
Geology	Geology refers to the science which treats of the origin, history and structure of the Earth, as recorded in rocks; together with the forces and processes now operating to modify rocks.
Drainage basin	Total area drained by a stream and its tributaries is referred to as drainage basin.
Ridge	Ridge refers to the volcanic mountain ranges that lie along the spreading centers on the floors of the oceans.
Surface runoff	Water flowing off the land into bodies of surface water is called surface runoff.

Go to **Cram101.com** for the Practice Tests for this Chapter.

Runoff	Fresh water from precipitation and melting ice that flows on the earth's surface into nearby streams, lakes, wetlands, and reservoir is referred to as runoff.
Map	Map refers to a representation of Earth's surface usually depicting mostly land areas.
Resolution	Resolution in general, refers to a measure of the finest detail distinguishable in an object or phenomenon. In particular, a measure of the finest detail distinguishable in an image.
Radar	An instrument for determining the distance and direction to an object by measuring the time needed for radio signals to travel from the instrument to the object and back, and by measuring the angle through which the instrument's antenna has traveled is referred to as radar.
Topography	Topography refers to the form of the features of the actual surface of the Earth in a particular region considered collectively.
Alluvial fan	A cone-shaped deposit of sediment that forms at the mouth of steep mountain streams in desert regions is an alluvial fan.
Valley	An elongated depression, usually with an outlet, between bluffs or between ranges of hills or mountains is a valley.
Slope	The degree of inclination to the horizontal is the slope.
Degree	An arbitrary measure of temperature. One degree Celsius _ 1.8 degrees Fahrenheit.
Development	Development refers to change from a society that is largely rural, agricultural, illiterate, and poor, with a rapidly growing population, to one that is mostly urban, industrial, educated, and wealthy, with a slowly growing or stationary population.
Force	Force refers to a push or pull that affects motion. The product of mass and acceleration of a material.
Cross section	A two-dimensional drawing showing features in the vertical plane as in a canyon wall or road cut is referred to as cross section.
Fall	Fall refers to a mass moving nearly vertical and downward under the influence of gravity.
Frictional drag	Frictional drag refers to the retarding force associated with the surface of a body moving through a fluid or gas, or with a fluid or gas moving across a body surface. Sometimes called surface drag.
Drag	The resistance to movement of an organism induced by the fluid through which it swims is called drag.
Zone	Division or province of the ocean with homogeneous characteristics is referred to as a zone.
Migration	A term that refers to the habit of some animals is a migration.
Meander	Loop-like bends or curves in the flow path of a current are a meander.
Bank	The rising ground bordering a lake, river or sea is referred to as a bank.
Weathering	Physical and chemical processes in which solid rock exposed at earth's surface is changed to separate solid particles and dissolved material, which can then be moved to another place as sediment is referred to as weathering.
Bed	The bottom of a watercourse, or any body of water is called a bed.
Viscosity	Resistance to flow is called viscosity.
Lava	Molten rock that is extruded out of volcanoes is called lava.
Current	Current refers to the flowing of water, or other liquid or gas. That portion of a stream of

water which is moving with a velocity much greater than the average or in which the progress of the water is principally concentrated. Ocean currents can be classified in a number of different ways.

Turbulent flow	Turbulent flow refers to any flow, which is not laminar, i.e., the stream lines of the fluid, instead of remaining parallel, become confused and intermingled.
Suspended load	Suspended load refers to the finest of the beach sediments, light enough in weight to remain lifted indefinitely above the bottom by water turbulence.
Saltation	Saltation refers to a term used to describe the movement of a particle being transported by wind or water, which is too heavy to remain in suspension. The particle is rolled forward by the current, generates lift and rises, loses the forward momentum supplying the lift and settles to the floor, where the process is repeated. The size of the particles, which can be saltated, depends upon the velocity of the current and its density.
Bed load	Bed load refers to heavy or large sediment particles that travel near or on the bed.
Mud	Mud refers to a mixture of silt and clay sized particles.
Spring	A place where groundwater flows out onto the surface is a spring.
Sand	Sand refers to an unconsolidated mixture of inorganic soil consisting of small but easily distinguishable grains ranging in size from about.062 mm to 2.0 mm.
Gravel	Gravel refers to loose, rounded fragments of rock, larger than sand, but smaller than cobbles. Small stones and pebbles, or a mixture of these with sand.
Bedrock	Solid rock lying beneath loose soil or unconsolidated sediment is called bedrock.
Chemical	One of the millions of different elements and compounds found naturally or synthesized by human is referred to as chemical.
Mineral	A naturally occurring, inorganic, crystalline solid that has a definite chemical composition and possesses characteristic physical properties is a mineral.
Parts per million	Parts per million refers to the number of parts of a chemical found in one million parts of a particular gas, liquid, or solid.
Threshold velocity	The maximum orbital velocity at which the sediment on the bed begins to move as waves approach shallow water is called threshold velocity.
Elevation	Elevation refers to the distance of a point above a specified surface of constant potential; the distance is measured along the direction of gravity between the point and the surface.
Sediment transport	Sediment transport refers to the main agencies by which sedimentary materials are moved. Running water and wind are the most widespread transporting agents. In both cases, three mechanisms operate, although the particle size of the transported material involved is very different, owing to the differences in density and viscosity of air and water.
Depth	Depth refers to vertical distance from still-water level to the bottom.
Density	Density refers to the ratio of a mass to a unit volume specified as grams per cubic centimeter.
Sea	Sea refers to the ocean. A large body of salt water, second in rank to an ocean, more or less landlocked and generally part of, or connected with, an ocean or a larger sea. State of the ocean or lake surface, in regard to waves.
Unconformity	A surface that represents a break in the geologic record, with the rock unit immediately above it being considerably younger than the rock beneath is called unconformity.
Graded stream	An equilibrium stream with evenly sloping bottom adjusted to efficiently handle water flow

241

	and sediment transport is referred to as graded stream.
Fault	A fracture in rock along which there has been an observable amount of displacement. Faults are rarely single planar units; normally they occur as parallel to sub-parallel sets of planes along which movement has taken place to a greater or lesser extent. Such sets are called fault or fracture-zones.
Creek	Creek refers to a stream, less predominant than a river, and generally tributary to a river. A small tidal CHANNEL through a coastal MARSH.
Dam	Dam refers to structure built in rivers or estuaries, basically to separate water at both sides and/or to retain water at one side.
Earthquake	Shaking of the ground resulting either from the fracturing and displacement of rock, producing a fault, or from subsequent movement along the fault is the earthquake.
Cliff	A high steep face of rock is referred to as cliff.
Forest	Forest refers to biome with enough average annual precipitation to support growth of various species of trees and smaller forms of vegetation. Compare desert, grassland.
Annual	Annual refers to plant that grows, sets seed, and dies in one growing season. Compare to perennial.
Coastline	Coastline refers to technically, the line that forms the boundary between the COAST and the SHORE. Commonly, the line that forms the boundary between land and the water. The line where terrestrial processes give way to marine processes, TIDAL CURRENTS, wind waves, etc.
Wave	An oscillatory movement in a body of water manifested by an alternate rise and fall of the surface is called a wave. Disturbances of the surface of a liquid body, as the ocean, in the form of a ridge, swell or hump. The term wave by itself usually refers to the term surface gravity wave.
Front	The boundary between two air masses with different temperatures and densitie is referred to as front.
Aquatic	Aquatic refers to pertaining to water.
Food chain	Food chain refers to series of organisms in which each eats or decomposes the preceding one. Compare food web.
Recent	A synonym of Holocene is called recent.
Plankton	Plankton refers to small plant organisms and animal organisms that float in aquatic ecosystems. Compare with benthos, nekton.
Soil	Soil refers to a layer of weathered, unconsolidated material on top of bedrock; often also defined as containing organic matter and being capable of supporting plant growth.
Nutrients	Nutrients refer to chemicals such as phosphorus and nitrogen that, when released into water sources, may cause pollution events such as eutrophication.
Fertilizer	Fertilizer refers to substance that adds inorganic or organic plant nutrients to soil and improves its ability to grow crops, trees, or other vegetation.
Ecology	Ecology refers to study of the interactions of living organisms with one another and with their nonliving environment of matter and energy; study of the structure and functions of nature.
Salinity	Salinity refers to number of grams of salt per thousand grams of seawater, usually expressed in parts per thousand.
Population	Group of individual organisms of the same species living within a particular area is referred

Go to **Cram101.com** for the Practice Tests for this Chapter.

to as a population.

Parasite	Consumer organism that lives on or in and feeds on a living plant or animal, known as the host, over an extended period of time is a parasite. The parasite draws nourishment from and gradually weakens its host; it may or may not kill the host.
Infiltration	Infiltration refers to downward movement of water through soil.
Conditions	Conditions refers to physical or chemical attributes of the environment that, while not being consumed, influence biological processes and population growth. Examples are temperature, salinity, and acidity. Compare resources.
Surface water	Precipitation that does not infiltrate the ground or return to the atmosphere by evaporation or transpiration is called surface water.
Evaporation	Evaporation refers to conversion of a liquid into a gas.
Transpiration	Process in which water is absorbed by the root systems of plants, moves up through the plants, passes through pores in their leaves or other parts, and then evaporates into the atmosphere as water vapor is called transpiration.
Plants	Eukaryotic, mostly multicelled organisms such as algae, mosses, ferns, flowers, cacti, grasses, beans, wheat, rice, and trees are plants. These organisms use photosynthesis to produce organic nutrients for themselves and for other organisms is referred to as plants.
Storm	Local or regional atmospheric disturbance characterized by strong winds often accompanied by precipitation is referred to as a storm.
Flooding	The natural process whereby waters emerge from their stream channel to cover part of the floodplain. Natural flooding is not a problem until people choose to build homes and other structures on floodplains.
Frequency	Number of events in a given time interval. For earthquakes, it is the number of cycles of seismic waves that pass in a second; frequency = l/period.
Sedimentary rocks	Sedimentary rocks refer to rocks that have formed by the compaction and cementation of sediment.
Rocks	An aggregate of one or more minerals rather large in area are rocks. The three classes of rocks are the following: Igneous rock - crystalline rocks formed from molten material. Sedimentary rock - A rock resulting from the consolidation of loose sediment that has accumulated in layers. Metamorphic rock - Rock that has formed from preexisting rock as a result of heat or pressure.
Permafrost	Permafrost refers to a perennially frozen layer of the soil that forms when the water there freezes. It is found in arctic tundra.
Glacier	Large masses of moving ice on land derived by the recrystallization of snow into ice under pressure are referred to as glacier.
Stone	Stone refers to quarried or artificially broken rock for use in construction.
Pebbles	Beach material usually well-rounded and between about 4 mm to 64 mm diameter are pebbles.
Depression	Depression refers to a general term signifying any depressed or lower area in the ocean floor.
Limestone	A sedimentary rock composed dominantly of calcium carbonate, either precipitated from seawater or deposited as shell debris are called the limestone.
Mass	The amount of material in an object is the mass.
Reduce	With respect to waste management, reduce refers to practices that will reduce the amount of

waste we produce.

Reach	Reach refers to an arm of the ocean extending into the land. A straight section of restricted waterway of considerable extent; may be similar to a narrows, except much longer in extent.
Head	A comparatively high promontory with either a CLIFF or steep face. It extends into a large body of water, such as a sea or lake. An unnamed HEAD is usually called a headland. The section of RIP CURRENT which has widened out seaward of the BREAKERS, also called head of
Sheet flow	Sheet flow refers to sediment grains under high sheer stress moving as a layer that extends from the bed surface to some distance below. Grains are transported in the direction of fluid flow.
Escarpment	A more or less continuous line of CLIFFS or steep slopes facing in one general direction which are caused by EROSION or faulting, also called SCARP is an escarpment.
Uplift	The rising of one part of the Earth's crust relative to another part is referred to as uplift.
Upland	Upland refers to the dry land area above and landward of the ordinary high water mark.
Offshore	Offshore in beach terminology refers to the comparatively flat zone of variable width, extending from the shoreface to the edge of the continental shelf. It is continually submerged. The direction seaward from the shore. The zone beyond the nearshore zone where sediment motion induced by waves alone effectively ceases and where the influence of the sea bed on wave action is small in comparison with the effect of wind. The breaker zone directly seaward of the low tide line.
Gulf	A relatively large portion of sea, partly enclosed by land is called gulf.
Shoreline	Shoreline refers to the intersection of a specified plane of water with the shore. All of the water areas of the state, including reservoirs and their associated uplands, together with the lands underlying them.
Shore	That strip of ground bordering any body of water which is alternately exposed, or covered by tides and/or waves is a shore. A shore of unconsolidated material is usually called a beach.
High tide	The high-water position corresponding to a tidal crest are called the high tide.
Tide	The periodic rising and falling of the water that results from gravitational attraction of the moon and sun acting upon the rotating earth is the tide. Although the accompanying horizontal movement of the water resulting from the same cause is also sometimes called the tide, it is preferable to designate the latter as tidal current, reserving the name tide for the vertical movement.
Morphology	Morphology refers to river/estuary/lake/seabed form and its change with time.
Sedimentary rock	Rock that forms from the accumulated products of erosion and in some cases from the compacted shells, skeletons, and other remains of dead organisms is sedimentary rock. Compare with igneous rock, metamorphic roc is a sedimentary rock.
Undercutting	Undercutting refers to erosion of material at the foot of a cliff or bank; a sea cliff, or river bank on the outside of a meander. Ultimately, the overhang collapses, and the process is repeated.
Point bar	Point bar refers to a stream bar deposited on the inside of a curve in the stream, where the water velocity is low.
Bar	An offshore ridge or mound of sand, gravel, or other unconsolidated material which is submerged, especially at the mouth of a river or estuary, or lying parallel to, and a short distance from, the beach is referred to as a bar.

Go to **Cram101.com** for the Practice Tests for this Chapter.

Loop	That part of a STANDING WAVE where the vertical motion is greatest and the horizontal velocities are least is a loop.
Bluff	A high, steep BANK or CLIFF is referred to as bluff.
Neck	Neck refers to the narrow strip of land which connects a peninsula with the mainland, or connects two ridges. The narrow band of water flowing seaward through the surf.
Key	Key refers to a low, insular BANK of sand, coral, etc., as one of the islets off the southern coast of Florida.
Embankment	Embankment refers to an artificial BANK, mound, DIKE, or the like, built to hold back water or to carry a roadway.
Site	A factor considering the summation of all environmental features of a location that influences the placement of a city is a site.
Hydrology	The study of surface and subsurface water is called hydrology.
Levee	Levee refers to eMBANKMENT to prevent inundation. A large DIKE or EMBANKMENT, often having an access road along the top, which is designed as part of a system to protect land from floods.
Braided stream	An overloaded stream so full of sediment that water flow is forced to divide and recombine in a braided pattern is referred to as braided stream.
Climatic change	Change in mean annual temperature and other aspects of climate over periods of time ranging from decades to hundreds of years to several million years is referred to as the climatic change.
Ice age	Ice age refers to one of several periods of low temperature during the last million years.
Tides	The periodic rise and fall of the Earth's water surface as a consequence of the gravitational attraction of the Moon and the Sun, which are called tides.
Input	Input refers to matter, energy, or information entering a system. Compare output, throughput.
Tidal flats	Tidal flats refer to marshy or muddy areas covered and uncovered by the rise and fall of the tide. A tidal marsh. Marshy or muddy areas of the seabed, which are covered and uncovered by the rise and fall of tidal water.
Coast	A strip of land of indefinite length and width that extends from the SEASHORE inland to the first major change in terrain features is referred to as coast.
Tidal current	The alternating horizontal movement of water associated with the rise and fall of the tide caused by astronomical tide-producing forces is called tidal current.
Water level	Elevation of a particular point or small patch on the surface of a body of water above a specific point or surface, averaged over a period of time sufficiently long to remove the effects of short period disturbances is referred to as water level.
Event	Event refers to an occurrence meeting specified conditions, e.g. damage, a threshold wave height or a threshold water level.
High water	High water refers to maximum height reached by a rising tide. The height may be solely due to the periodic tidal forces or it may have superimposed upon it the effects of prevailing meteorological conditions. Nontechnically, also called the HIGH TIDE.
Drought	Condition in which an area does not get enough water because of lower-than-normal precipitation, higher-than-normal temperatures that increase evaporation, or both is called drought.
Weather	Weather refers to short-term changes in the temperature, barometric pressure, humidity,

precipitation, sunshine, cloud cover, wind direction and speed, and other conditions in the troposphere at a given place and time. Compare with climate.

Saturation

A chemical state whereby the maximum amount of solute is dissolved under the given conditions is saturation.

Lightning

A flashing of light as atmospheric electricity flows between clouds or between cloud and ground is a lightning.

Submarine canyon

V-shaped valleys that run across the continental shelf and down the continental slope are referred to as submarine canyon.

Peninsula

An elongated portion of land nearly surrounded by water and connected to a larger body of land, usually by a neck or an isthmus is called a peninsula.

Relief

The difference in elevation between the highest and lowest points in an area is called relief.

Convergence

Resemblance among species belonging to different taxonomic groups as the result from adaptation to similar environments is the convergence.

Plates

Various-sized areas of earth's lithosphere that move slowly around with the mantle's flowing asthenosphere are plates. Most earthquakes and volcanoes occur around the boundaries of these plates.

Crust

Solid outer zone of the earth. It consists of oceanic crust and continental crust. Compare core, mantle.

Rift

Rift refers to the valley created at a pull-apart zone.

Pangaea

Pangaea refers to the megacontinent of the Mesozoic Era that consisted of all of the present-day continents joined together into a single unit.

Basalt

Basalt refers to a dark, fine-grained igneous rock composed of minerals enriched in ferromagnesian silicates.

Subsidence

Sinking or down warping of a part of the earth's surface is called subsidence.

Continent

Continent refers to lower-density masses of rock, exposed as about 40 percent of the Earth's surface: 29 percent as land and I 1 percent as the floor of shallow seas.

Tethys Sea

An immense seaway that separated Gondwanaland from Laurasia during the Mesozoic Era is called Tethys Sea.

Probability

A mathematical statement about how likely it is that something will happen is probability.

Continental margin

The drowned edges of continents consisting of the continental shelf, the continental slope, and the continental rise is called continental margin.

Observations

Information obtained through one or more of the five senses or through instruments that extend the senses are observations.

Terrace

A horizontal or nearly horizontal natural or artificial topographic feature interrupting a steeper slop, sometimes occurring in a series is called a terrace.

Passive continental margin

A subsiding continental margin situated in a nontectonic setting away from a lithospheric plate boundary is a passive continental margin.

Geomorphology

Geomorphology refers to that branch of physical geography which deals with the form of the Earth, the general configuration of its surface, the distribution of the land, water, etc.
The investigation of the history of geologic changes through the interpretation of topographic forms.

Go to **Cram101.com** for the Practice Tests for this Chapter.

Resources	Resources refer to substances that can be consumed by an organism and, as a result, become unavailable to other organisms.
Feedback	A kind of system response that occurs when output of the system also serves as input leading to changes in the system is called feedback.
Slide	In mass wasting, movement of a descending mass along a plane approximately parallel to the slope of the surface is referred to as slide.

Pore	Pore refers to an opening or void space in; oil or rock.
Rocks	An aggregate of one or more minerals rather large in area are rocks. The three classes of rocks are the following: Igneous rock - crystalline rocks formed from molten material. Sedimentary rock - A rock resulting from the consolidation of loose sediment that has accumulated in layers. Metamorphic rock - Rock that has formed from preexisting rock as a result of heat or pressure.
Groundwater	Groundwater refers to water that sinks into the soil and is stored in slowly flowing and slowly renewed underground reservoirs called aquifers; underground water in the zone of saturation, below the water table. Compare runoff, surface water.
System	A set of components that function and interact in some regular and theoretically predictable manner is called a system.
Resource	Resource refers to anything obtained from the living and nonliving environment to meet human needs and wants. It can also be applied to other species.
Desert	Desert refers to biome in which evaporation exceeds precipitation and the average amount of precipitation is less than 25 centimeters a year. Such areas have little vegetation or have widely spaced, mostly low vegetation. Compare forest, grassland.
Surface runoff	Water flowing off the land into bodies of surface water is called surface runoff.
Runoff	Fresh water from precipitation and melting ice that flows on the earth's surface into nearby streams, lakes, wetlands, and reservoir is referred to as runoff.
River	A natural stream of water larger than a brook or creek is a river.
Discharge	The volume of water flowing in a stream per unit of time is the discharge.
Zone	Division or province of the ocean with homogeneous characteristics is referred to as a zone.
Recharge	The addition of new water to an aquifer or to the zone of saturation is called a recharge.
Surface water	Precipitation that does not infiltrate the ground or return to the atmosphere by evaporation or transpiration is called surface water.
Permeability	Permeability refers to the property of bulk material which permits movement of water through its pores.
Porosity	Porosity refers to a percentage of space in rock or soil occupied by voids, whether the voids are isolated or connected. Compare with permeability.
Aquifer	A geologic formation that is water-bearing, and which transmits water from one point to another is called an aquifer.
Lake	Large natural body of standing fresh water formed when water from precipitation, land runoff, or groundwater flow fills a depression in the earth created by glaciation, earth movement, volcanic activity, or a giant meteorit are called the lake.
Ocean	The great body of salt water which occupies two-thirds of the surface of the Earth, or one of its major subdivisions is called an ocean.
Fall	Fall refers to a mass moving nearly vertical and downward under the influence of gravity.
Spring	A place where groundwater flows out onto the surface is a spring.
Rock	Any material that makes up a large, natural, continuous part of earth's crust is rock.
Topography	Topography refers to the form of the features of the actual surface of the Earth in a particular region considered collectively.
Sand	Sand refers to an unconsolidated mixture of inorganic soil consisting of small but easily

	distinguishable grains ranging in size from about.062 mm to 2.0 mm.
Water table	The upper surface of a zone of saturation, where the body of groundwater is not confined by an overlying impermeable formation is a water table. Where an overlying confining formation exists, the aquifer in question has no water table.
Zone of saturation	Area where all available pores in soil and rock in the earth's crust are filled by water is a zone of saturation.
Saturation	A chemical state whereby the maximum amount of solute is dissolved under the given conditions is saturation.
Impermeable	Impermeable refers to impervious; the condition of rock that does not allow fluids to flow through it.
Erosion	Erosion refers to wearing away of the land by natural forces. On a beach, the carrying away of beach material by wave action, tidal currents or by DEFLATION. The wearing away of land by the action of natural forces.
Precipitation	Water in the form of rain, sleet, hail, and snow that falls from the atmosphere onto the land and bodies of water is called precipitation.
Sedimentary rocks	Sedimentary rocks refer to rocks that have formed by the compaction and cementation of sediment.
Pollution	Pollution refers to an undesirable change in the physical, chemical, or biological characteristics of air, water, soil, or food that can adversely affect the health, survival, or activities of humans or other living organisms.
Subsidence	Sinking or down warping of a part of the earth's surface is called subsidence.
Gravity	The attraction between bodies of matter is a gravity.
Force	Force refers to a push or pull that affects motion. The product of mass and acceleration of a material.
Open system	Open system refers to a system, such as a living organism, in which both matter and energy are exchanged between the system and the environment. Compare with closed system.
Well	Well refers to a hole, generally cylindrical and usually walled or lined with pipe, that is dug or drilled into the ground to penetrate an aquifer below the zone of saturation.
Soil	Soil refers to a layer of weathered, unconsolidated material on top of bedrock; often also defined as containing organic matter and being capable of supporting plant growth.
Mineral	A naturally occurring, inorganic, crystalline solid that has a definite chemical composition and possesses characteristic physical properties is a mineral.
Gravel	Gravel refers to loose, rounded fragments of rock, larger than sand, but smaller than cobbles. Small stones and pebbles, or a mixture of these with sand.
Granite	A light-colored, coarsegrained, intrusive igneous rock composed mainly of quartz and feldspar and that typifies the continental crust are called the granite.
Limestone	A sedimentary rock composed dominantly of calcium carbonate, either precipitated from seawater or deposited as shell debris are called the limestone.
Bedding	The layering of rocks, especially sedimentary rooks is called bedding.
Basalt	Basalt refers to a dark, fine-grained igneous rock composed of minerals enriched in ferromagnesian silicates.
Conglomerate	A sedimentary rock dominated by gravel is the conglomerate.

Go to **Cram101.com** for the Practice Tests for this Chapter.

Lava	Molten rock that is extruded out of volcanoes is called lava.
Base	A substance that combines with a hydrogen ion in solution is called the base.
Viscosity	Resistance to flow is called viscosity.
Hydrostatic pressure	Hydrostatic pressure refers to the pressure at a specified water depth that is the result of the weight of the overlying column of water.
Degree	An arbitrary measure of temperature. One degree Celsius _ 1.8 degrees Fahrenheit.
Turbulent flow	Turbulent flow refers to any flow, which is not laminar, i.e., the stream lines of the fluid, instead of remaining parallel, become confused and intermingled.
Percolation	Percolation refers to passage of a liquid through the spaces of a porous material such as soil.
Zone of aeration	Zone of aeration refers to the zone or layer above the water table in which some water may be suspended or moving in a downward migration toward the water table or laterally toward a discharge point, which is often called the vadose zone.
Depth	Depth refers to vertical distance from still-water level to the bottom.
Climate	Physical properties of the troposphere of an area based on analysis of its weather records over a long period. The two main factors determining an area's climate are temperature, with its seasonal variations, and the amount and distri.
Tension	A state of stress that tends to pull the body apart is called tension.
Isotopes	Two or more forms of a chemical element that have the same number of protons but different mass numbers because of different numbers of neutrons in their nuclei is referred to as isotopes.
Valley	An elongated depression, usually with an outlet, between bluffs or between ranges of hills or mountains is a valley.
Outcrop	A surface exposure of bare rock, not covered by soil or vegetation is called outcrop.
Reach	Reach refers to an arm of the ocean extending into the land. A straight section of restricted waterway of considerable extent; may be similar to a narrows, except much longer in extent.
Bed	The bottom of a watercourse, or any body of water is called a bed.
Stream	Stream refers to any flow of water; a current. A course of water flowing along a bed in the earth.
Elevation	Elevation refers to the distance of a point above a specified surface of constant potential; the distance is measured along the direction of gravity between the point and the surface.
Head	A comparatively high promontory with either a CLIFF or steep face. It extends into a large body of water, such as a sea or lake. An unnamed HEAD is usually called a headland. The section of RIP CURRENT which has widened out seaward of the BREAKERS, also called head of
Till	Poorly sorted sediment that is deposited by glaciers is referred to as till.
Unconsolidated	Unconsolidated with regards to sediment grains refers to loose, separate, or unattached to one another.
Volcanic rock	Rock formed by solidification of magma at the Earth's surface is a volcanic rock.
Map	Map refers to a representation of Earth's surface usually depicting mostly land areas.
Point	Point refers to the extreme end of a cape, or the outer end of any land area protruding into the water, usually less prominent than a cape. A low profile shoreline promontory of more or

less triangular shape, the top of which extends seaward.

Bedrock	Solid rock lying beneath loose soil or unconsolidated sediment is called bedrock.
Infiltration	Infiltration refers to downward movement of water through soil.
Conditions	Conditions refers to physical or chemical attributes of the environment that, while not being consumed, influence biological processes and population growth. Examples are temperature, salinity, and acidity. Compare resources.
Fault	A fracture in rock along which there has been an observable amount of displacement. Faults are rarely single planar units; normally they occur as parallel to sub-parallel sets of planes along which movement has taken place to a greater or lesser extent. Such sets are called fault or fracture-zones.
Natural recharge	Natural recharge refers to natural replenishment of an aquifer by precipitation that percolates downward through soil and rock.
State	State refers to an expression of the internal form of matter. Water exists in three states: solid, liquid, and gas. A solid has a fixed volume and fixed shape; a liquid has a fixed volume but no fixed shape; and a gas has neither fixed volume nor fixed shape.
Power	Power refers to the time rate of doing work.
Site	A factor considering the summation of all environmental features of a location that influences the placement of a city is a site.
Energy	Capacity to do work by performing mechanical, physical, chemical, or electrical tasks or to cause a heat transfer between two objects at different temperatures is an energy.
Cone of depression	A cone-shaped depression in the water table around a well caused by withdrawal by pumping of water at rates greater than the rates at which the water can he replenished by natural groundwater flow is called cone of depression.
Depression	Depression refers to a general term signifying any depressed or lower area in the ocean floor.
Plants	Eukaryotic, mostly multicelled organisms such as algae, mosses, ferns, flowers, cacti, grasses, beans, wheat, rice, and trees are plants. These organisms use photosynthesis to produce organic nutrients for themselves and for other organisms is referred to as plants.
Migration	A term that refers to the habit of some animals is a migration.
Fracture	Fracture refers to a general term for any breaks in rock. Fractures include faults, joints, and crack,_.
Recharge area	Any area of land allowing water to pass through it and into an aquifer is a recharge area.
Front	The boundary between two air masses with different temperatures and densitie is referred to as front.
Gulf	A relatively large portion of sea, partly enclosed by land is called gulf.
Coast	A strip of land of indefinite length and width that extends from the SEASHORE inland to the first major change in terrain features is referred to as coast.
Clay	Clay refers to a fine grained sediment with a typical grain size less than 0.004 mm. Possesses electromagnetic properties which bind the grains together to give a bulk strength or cohesion.
Dip	The angle of inclination measured in degrees from the horizontal is called dip.
Basin	A large submarine depression of a generally circular, elliptical or oval shape is called a

Go to **Cram101.com** for the Practice Tests for this Chapter.

261

basin.

Thermal	The energy of the random motion of atoms and molecules is referred to as thermal.
Recent	A synonym of Holocene is called recent.
Magma	Molten rock below the earth's surface is referred to as magma.
Igneous rocks	Igneous rocks refers to rocks formed from the solidification of magma. They are extrusive if they crystallize on the surface of the Earth and intrusive if they crystallize beneath the surface.
Development	Development refers to change from a society that is largely rural, agricultural, illiterate, and poor, with a rapidly growing population, to one that is mostly urban, industrial, educated, and wealthy, with a slowly growing or stationary population.
Geyser	A hot spring that gushes magmaheated water and steam is called geyser.
Temperature	Temperature refers to a measure of the average speed of motion of the atoms, ions, or molecules in a substance or combination of substances at a given moment. Compare with heat.
Boil	An upward flow of water in a sandy formation due to an unbalanced hydrostatic pressure resulting from a rise in a nearby stream, or from removing the overburden in making excavations is a boil.
Heat	Total kinetic energy of all the randomly moving atoms, ions, or molecules within a given substance, excluding the overall motion of the whole object. This form of kinetic energy flows from one body to another when there is a temperature difference betwe is referred to as heat.
Geothermal energy	Geothermal energy refers to heat transferred from the earth's underground concentrations of dry steam, wet steam, or hot water trapped in fractured or porous rock.
Forcing	With respect to global change, processes capable of changing global temperature, such as changes in solar energy emitted from the sun, or volcanic activity, we have forcing.
Mud	Mud refers to a mixture of silt and clay sized particles.
Gypsum	An evaporite deposit composed of hydrous calcium sulfat is a gypsum.
Matter	Matter refers to anything that has mass and takes up space. On earth, where gravity is present, we weigh an object to determine its mass.
Network	Network refers to a set consisting of stations for which geometric relationships have been determined and which are so related that removal of one station from the set will affect the relationships between the other stations; and lines connecting the stations to show this interdependence.
Sinkhole	Over long periods of time, groundwater may dissolve large areas of limestone rock, slowly enlarging underground caverns and eroding support for the land above, which is called a sinkhole. There may be no evidence of what is taking place until the ground collapses abruptly into a void below, producing a sinkhole.
Acid	A substance that releases a hydrogen ion in solution is an acid.
Carbonic acid	Carbonic acid refers to a common but weak acid formed by carbon dioxide dissolving in water. carnivore A flesh-eating animal.
Carbon dioxide	Molecule of carbon and oxygen present in the atmosphere at approximately 350 ppm. Emissions of carbon dioxide resulting from burning of fossil fuels are thought to be co:.itributing to potential global warming through an enhanced greenhouse effect.
Atmosphere	Atmosphere refers to the whole mass of air surrounding the earth.

Go to **Cram101.com** for the Practice Tests for this Chapter.

263

Coal	Solid, combustible mixture of organic compounds with 30-98% carbon by weight, mixed with various amounts of water and small amounts of sulfur and nitrogen compounds. It is formed in several stages as the remains of plants are subjected to heat and press is a coal.
Peat	Peat refers to an organic deposit consisting predominantly of partly decayed plant matter.
Dissolution	The act of dissolving a solid is called dissolution.
Chemical	One of the millions of different elements and compounds found naturally or synthesized by human is referred to as chemical.
Oil	The liquid form of petroleum consisting of a complex mixture of large hydrocarbon molecules is referred to as oil.
Natural gas	Underground deposits of gases consisting of 50-90% by weight methane gas and small amounts of heavier gaseous hydrocarbon compounds such as propane and butane are called natural gas.
Accumulation	Buildup of matter, energy, or information in a system is referred to as accumulation.
Planet	A smaller, usually nonluminous body orbiting a star is a planet.
Entrance	The entrance to a navigable BAY, HARBOR or CHANNEL, INLET or mouth separating the ocean from an inland water body.
Cavern	A large cave is a cavern.
Debris	Debris refers to any accumulation of rock fragments; detritus.
Corridors	Long areas of land that connect habitat that would otherwise become fragmented are referred to as corridors.
Residual	The components of water level not attributable to astronomical effects are called residual.
Wind	Wind refers to the mass movement of air.
Forest	Forest refers to biome with enough average annual precipitation to support growth of various species of trees and smaller forms of vegetation. Compare desert, grassland.
Relief	The difference in elevation between the highest and lowest points in an area is called relief.
Solubility	A measure of how readily one chemical substance can dissolve another is a solubility.
Peninsula	An elongated portion of land nearly surrounded by water and connected to a larger body of land, usually by a neck or an isthmus is called a peninsula.
Cement	Minerals such as silica and carbonate that are chemically precipitated in the pores of sediments, binding the grain are referred to as cement.
Equilibrium	A point of rest. A system that does not tend to undergo any change of its own accord but remains in a single, fixed condition is said to be in equilibrium. Compare with steady state.
Diffusion	The dispersal of material by random molecular movement from regions of high concentration to those of lower concentration is a diffusion.
Evaporation	Evaporation refers to conversion of a liquid into a gas.
Range	Land used for grazing is referred to as the range.
Silica	A compound with a composition, such as quartz in granite and opal in the shells of radiolaria is referred to as silica.
Atoms	Atoms refers to minute units made of subatomic particles that are the basic building blocks of all chemical elements and thus all matter; the smallest unit of an element that can exist and still have the unique characteristics of that element. Compare to ion, molecule.

Weathering	Physical and chemical processes in which solid rock exposed at earth's surface is changed to separate solid particles and dissolved material, which can then be moved to another place as sediment is referred to as weathering.
Sediment	Loose, fragments of rocks, minerals or organic material which are transported from their source for varying distances and deposited by air, wind, ice and water are sediment. Other sediments are precipitated from the overlying water or form chemically, in place. Sediment includes all the unconsolidated materials on the sea floor. The fine-grained material deposited by water or wind.
Mass	The amount of material in an object is the mass.
Fossil	The remains or traces of organisms preserved in rocks or ancient sediment are referred to as fossil.
Resources	Resources refer to substances that can be consumed by an organism and, as a result, become unavailable to other organisms.
Mineral resources	Elements, chemical compounds, minerals, or rocks concentrated in a form that can be extracted to obtain a usable commodity are referred to as mineral resources.
Reserves	Reserves refer to resources that have been identified from which a usable mineral can be extracted profitably at present prices with current mining.
Ore deposits	Earth materials in which metals are concentrated in high concentrations, sufficient to be mined are called ore deposits.
Lead	Lead refers to a heavy metal that is an important constituent of automobile batteries and other industrial products. A toxic metal capable of causing environmental disruption and producing a health problem to people and other living organisms.
Marble	Marble refers to metamorphosed limestone.
Opal	Amorphous silica secreted by organisms such as radiolaria and diatoms is referred to as opal.
Sea	Sea refers to the ocean. A large body of salt water, second in rank to an ocean, more or less landlocked and generally part of, or connected with, an ocean or a larger sea. State of the ocean or lake surface, in regard to waves.
Slope	The degree of inclination to the horizontal is the slope.
Concentration	Amount of a chemical in a particular volume or weight of air, water, soil, or other medium is referred to as concentration.
Leaching	Process in which various chemicals in upper layers of soil are dissolved and carried to lower layers and, in some cases, to groundwater are called leaching.
Environment	Environment refers to all external conditions and factors, living and nonliving, that affect an organism or other specified system during its lifetime; the earth's life-support systems for us and for all other forms of life-another term for solar capita.
Interface	The surface or boundary at which two different substances are in contact, for example, the air-sea interface.
Dispersion	Dispersion refers to act of dispersing, or state of being dispersed. The separation of waves by virtue of their differing rates of travel.
Bacteria	Bacteria refer to prokaryotic, one-celled organisms. Some transmit diseases. Most act as decomposers and get the nutrients they need by breaking down complex organic compounds in the tissues of living or dead organisms into simpler inorganic nutrient compounds.
Host	Plant or animal on which a parasite feeds is referred to as host.

Go to **Cram101.com** for the Practice Tests for this Chapter.

Risk	Risk refers to the probability that something undesirable will happen from deliberate or accidental exposure.
Plume	Plume refers to an arm of magna rising upward from the mantle.
Chlorinated hydrocarbons	The most abundant and dangerous class of halogenated hydrocarbons, synthetic organic chemicals hazardous to the marine environment are called chlorinated hydrocarbons.
Hydrocarbons	Compounds containing only carbon and hydrogen are a large group of organic compounds, including petroleum products, such as crude oil and natural gas are hydrocarbons.
Core	Core refers to a cylindrical sample extracted from a beach or seabed to investigate the types and DEPTHS of sediment layers. An inner, often much less permeable portion of a BREAKWATER, or BARRIER beach.
Contamination	Presence of undesirable material that makes something unfit for a particular use is a contamination.
Key	Key refers to a low, insular BANK of sand, coral, etc., as one of the islets off the southern coast of Florida.
Shallow water	Shallow water refers to water of such depths that surface waves are noticeably affected by bottom topography. Typically this implies water depths equivalent to less than half the wave length.
Gradient	A measure of slope in meters of rise or fall per meter of horizontal distance. More general, a change of a value per unit of distance, e.g. the GRADIENT in longshore transport causes EROSION or ACCRETION. With reference to winds or currents, the rate of increase or decrease in speed, usually in the vertical; or the curve that represents this rate.
Ecology	Ecology refers to study of the interactions of living organisms with one another and with their nonliving environment of matter and energy; study of the structure and functions of nature.
Flooding	The natural process whereby waters emerge from their stream channel to cover part of the floodplain. Natural flooding is not a problem until people choose to build homes and other structures on floodplains.
Compaction	The decrease in volume and porosity of a sediment via burial is referred to as compaction.
High water	High water refers to maximum height reached by a rising tide. The height may be solely due to the periodic tidal forces or it may have superimposed upon it the effects of prevailing meteorological conditions. Nontechnically, also called the HIGH TIDE.
Marsh	Marsh refers to a tract of soft, wet land, usually vegetated by reeds, grasses and occasionally small shrubs. Soft, wet area periodically or continuously flooded to a shallow depth, usually characterized by a particular subclass of grasses, cattails and other low plants.
Drought	Condition in which an area does not get enough water because of lower-than-normal precipitation, higher-than-normal temperatures that increase evaporation, or both is called drought.
Climatic change	Change in mean annual temperature and other aspects of climate over periods of time ranging from decades to hundreds of years to several million years is referred to as the climatic change.
Radar	An instrument for determining the distance and direction to an object by measuring the time needed for radio signals to travel from the instrument to the object and back, and by measuring the angle through which the instrument's antenna has traveled is referred to as radar.

Seismic	Referring to vibrations in the Earth produced by earthquakes is referred to as seismic.
Harbor	A water area nearly surrounded by land, sea walls, BREAKWATERS or artificial dikes, forming a safe anchorage for ships is referred to as harbor.
Ash	The loose debris that is ejected from an erupting volcano is called ash.
Dam	Dam refers to structure built in rivers or estuaries, basically to separate water at both sides and/or to retain water at one side.
Observations	Information obtained through one or more of the five senses or through instruments that extend the senses are observations.
Geomorphology	Geomorphology refers to that branch of physical geography which deals with the form of the Earth, the general configuration of its surface, the distribution of the land, water, etc. The investigation of the history of geologic changes through the interpretation of topographic forms.
Geology	Geology refers to the science which treats of the origin, history and structure of the Earth, as recorded in rocks; together with the forces and processes now operating to modify rocks.
Feedback	A kind of system response that occurs when output of the system also serves as input leading to changes in the system is called feedback.
Slide	In mass wasting, movement of a descending mass along a plane approximately parallel to the slope of the surface is referred to as slide.

Glacier	Large masses of moving ice on land derived by the recrystallization of snow into ice under pressure are referred to as glacier.
Event	Event refers to an occurrence meeting specified conditions, e.g. damage, a threshold wave height or a threshold water level.
Recent	A synonym of Holocene is called recent.
Ice age	Ice age refers to one of several periods of low temperature during the last million years.
Valley	An elongated depression, usually with an outlet, between bluffs or between ranges of hills or mountains is a valley.
Debris	Debris refers to any accumulation of rock fragments; detritus.
River	A natural stream of water larger than a brook or creek is a river.
Precipitation	Water in the form of rain, sleet, hail, and snow that falls from the atmosphere onto the land and bodies of water is called precipitation.
Ocean	The great body of salt water which occupies two-thirds of the surface of the Earth, or one of its major subdivisions is called an ocean.
Sea	Sea refers to the ocean. A large body of salt water, second in rank to an ocean, more or less landlocked and generally part of, or connected with, an ocean or a larger sea. State of the ocean or lake surface, in regard to waves.
Hydrology	The study of surface and subsurface water is called hydrology.
Topography	Topography refers to the form of the features of the actual surface of the Earth in a particular region considered collectively.
Crust	Solid outer zone of the earth. It consists of oceanic crust and continental crust. Compare core, mantle.
Stream	Stream refers to any flow of water; a current. A course of water flowing along a bed in the earth.
Development	Development refers to change from a society that is largely rural, agricultural, illiterate, and poor, with a rapidly growing population, to one that is mostly urban, industrial, educated, and wealthy, with a slowly growing or stationary population.
Planet	A smaller, usually nonluminous body orbiting a star is a planet.
Abrasion	Frictional erosion by material transported by wind and waves is referred to as abrasion.
Sediment	Loose, fragments of rocks, minerals or organic material which are transported from their source for varying distances and deposited by air, wind, ice and water are sediment. Other sediments are precipitated from the overlying water or form chemically, in place. Sediment includes all the unconsolidated materials on the sea floor. The fine-grained material deposited by water or wind.
Pleistocene	An epoch of the Quaternary Period characterized by several glacial ages is referred to as the Pleistocene.
Erosion	Erosion refers to wearing away of the land by natural forces. On a beach, the carrying away of beach material by wave action, tidal currents or by DEFLATION. The wearing away of land by the action of natural forces.
Fall	Fall refers to a mass moving nearly vertical and downward under the influence of gravity.
Lithosphere	Lithosphere refers to outer shell of the earth, composed of the crust and the rigid, outermost part of the mantle outside of the asthenosphere; material found in the earth's

plates.

Migration	A term that refers to the habit of some animals is a migration.
Extinction	Extinction refers to complete disappearance of a species from the earth. This happens when a species cannot adapt and successfully reproduce under new environmental conditions or when it evolves into one or more new species. Compare speciation.
Species	Group of organisms that resemble one another in appearance, behavior, chemical makeup and processes, and genetic structure is a species. Organisms that reproduce sexually are classified as members of the same species only if they can breed with one another and produce offspring.
Plate tectonics	Plate tectonics refer to the theory of geophysical processes that explains the movements of lithospheric plates and the processes that occur at their boundaries.
Tectonics	The study of the major structural features of the Earth's crust or the broad structure of a region is referred to as tectonics.
Open system	Open system refers to a system, such as a living organism, in which both matter and energy are exchanged between the system and the environment. Compare with closed system.
System	A set of components that function and interact in some regular and theoretically predictable manner is called a system.
Compaction	The decrease in volume and porosity of a sediment via burial is referred to as compaction.
Evaporation	Evaporation refers to conversion of a liquid into a gas.
Accumulation	Buildup of matter, energy, or information in a system is referred to as accumulation.
Mass	The amount of material in an object is the mass.
Groundwater	Groundwater refers to water that sinks into the soil and is stored in slowly flowing and slowly renewed underground reservoirs called aquifers; underground water in the zone of saturation, below the water table. Compare runoff, surface water.
Sea ice	Frozen seawater as opposed to glacial ice on land is referred to as sea ice.
Perennial	Perennial refers to plant that can live for more than 2 years. Compare with annual.
Metamorphic rock	Metamorphic rock refers to rock produced when a preexisting rock is subjected to high temperatures, high pressures, chemically active fluids, or a combination of these agents. Compare with igneous rock, sedimentary rock.
Rock	Any material that makes up a large, natural, continuous part of earth's crust is rock.
Mineral	A naturally occurring, inorganic, crystalline solid that has a definite chemical composition and possesses characteristic physical properties is a mineral.
Zone	Division or province of the ocean with homogeneous characteristics is referred to as a zone.
Annual	Annual refers to plant that grows, sets seed, and dies in one growing season. Compare to perennial.
Temperature	Temperature refers to a measure of the average speed of motion of the atoms, ions, or molecules in a substance or combination of substances at a given moment. Compare with heat.
Gravity	The attraction between bodies of matter is a gravity.
Pore	Pore refers to an opening or void space in; oil or rock.
Front	The boundary between two air masses with different temperatures and densitie is referred to as front.

Continent	Continent refers to lower-density masses of rock, exposed as about 40 percent of the Earth's surface: 29 percent as land and I 1 percent as the floor of shallow seas.
Electron	Tiny particle moving around outside the nucleus of an atom. Each electron has one unit of negative charge and almost no mass. Compare neutron, proton.
Brittle	Behavior of material where stress causes abrupt fracture are called the brittle.
Fracture	Fracture refers to a general term for any breaks in rock. Fractures include faults, joints, and crack,_.
Stress	Stress refers to force per unit area. May be compression, tension, or shear.
Force	Force refers to a push or pull that affects motion. The product of mass and acceleration of a material.
Depth	Depth refers to vertical distance from still-water level to the bottom.
Slope	The degree of inclination to the horizontal is the slope.
Viscosity	Resistance to flow is called viscosity.
Sand	Sand refers to an unconsolidated mixture of inorganic soil consisting of small but easily distinguishable grains ranging in size from about.062 mm to 2.0 mm.
Experiment	Experiment refers to procedure a scientist uses to study some phenomenon under known conditions. Some experiments are conducted in the laboratory, but others are conducted in nature. The resulting scientific data or facts must be verified or confirmed by repeated observation.
Rocks	An aggregate of one or more minerals rather large in area are rocks. The three classes of rocks are the following: Igneous rock - crystalline rocks formed from molten material. Sedimentary rock - A rock resulting from the consolidation of loose sediment that has accumulated in layers. Metamorphic rock - Rock that has formed from preexisting rock as a result of heat or pressure.
Bedrock	Solid rock lying beneath loose soil or unconsolidated sediment is called bedrock.
Bed	The bottom of a watercourse, or any body of water is called a bed.
Steady state	When input equals output in a system, there is no net change and the system stem is said to be in a steady state.
State	State refers to an expression of the internal form of matter. Water exists in three states: solid, liquid, and gas. A solid has a fixed volume and fixed shape; a liquid has a fixed volume but no fixed shape; and a gas has neither fixed volume nor fixed shape.
Head	A comparatively high promontory with either a CLIFF or steep face. It extends into a large body of water, such as a sea or lake. An unnamed HEAD is usually called a headland. The section of RIP CURRENT which has widened out seaward of the BREAKERS, also called head of
Equilibrium	A point of rest. A system that does not tend to undergo any change of its own accord but remains in a single, fixed condition is said to be in equilibrium. Compare with steady state.
Input	Input refers to matter, energy, or information entering a system. Compare output, throughput.
Output	Output refers to matter, energy, or information leaving a system. Compare with input, throughput.
Cross section	A two-dimensional drawing showing features in the vertical plane as in a canvon wall or road cut is referred to as cross section.
Degree	An arbitrary measure of temperature. One degree Celsius _ 1.8 degrees Fahrenheit.

Base	A substance that combines with a hydrogen ion in solution is called the base.
Tension	A state of stress that tends to pull the body apart is called tension.
Conditions	Conditions refers to physical or chemical attributes of the environment that, while not being consumed, influence biological processes and population growth. Examples are temperature, salinity, and acidity. Compare resources.
Moraine	Moraine refers to an accumulation of earth, stones, etc., deposited by a glacier, usually in the form of a mound, ridge or other prominence on the terrain.
Load	The quantity of sediment transported by a current. It includes the suspended load of small particles in the water, and the bedload of large particles that move along the bottom.
Surge	Surge refers to long-interval variations in velocity and pressure in fluid flow, not necessarily periodic, perhaps even transient in nature. The name applied to wave motion with a period intermediate between that of an ordinary wind wave and that of the tide. Changes in water level as a result of meteorological forcing causing a difference between the recorded water level and that predicted using harmonic analysis, may be positive or negative.
Well	Well refers to a hole, generally cylindrical and usually walled or lined with pipe, that is dug or drilled into the ground to penetrate an aquifer below the zone of saturation.
Point	Point refers to the extreme end of a cape, or the outer end of any land area protruding into the water, usually less prominent than a cape. A low profile shoreline promontory of more or less triangular shape, the top of which extends seaward.
Dam	Dam refers to structure built in rivers or estuaries, basically to separate water at both sides and/or to retain water at one side.
Avalanche	A large mass of snow, ice, soil, or rock that moves rapidly downslope under the pull of gravity is referred to as an avalanche.
Reach	Reach refers to an arm of the ocean extending into the land. A straight section of restricted waterway of considerable extent; may be similar to a narrows, except much longer in extent.
Radiation	Fast-moving particles or waves of energy are called radiation.
Zone of saturation	Area where all available pores in soil and rock in the earth's crust are filled by water is a zone of saturation.
Saturation	A chemical state whereby the maximum amount of solute is dissolved under the given conditions is saturation.
Water table	The upper surface of a zone of saturation, where the body of groundwater is not confined by an overlying impermeable formation is a water table. Where an overlying confining formation exists, the aquifer in question has no water table.
Concentration	Amount of a chemical in a particular volume or weight of air, water, soil, or other medium is referred to as concentration.
Spring	A place where groundwater flows out onto the surface is a spring.
Wind	Wind refers to the mass movement of air.
Fjord	Fjord refers to a long, narrow arm of the sea, usually formed by entrance of the sea into a deep glacial trough.
Cliff	A high steep face of rock is referred to as cliff.
Range	Land used for grazing is referred to as the range.
Drag	The resistance to movement of an organism induced by the fluid through which it swims is

called drag.

Mantle	Zone of the earth's interior between its core and its crust. Compare with core, crust is the mantle.
Soil	Soil refers to a layer of weathered, unconsolidated material on top of bedrock; often also defined as containing organic matter and being capable of supporting plant growth.
Map	Map refers to a representation of Earth's surface usually depicting mostly land areas.
Network	Network refers to a set consisting of stations for which geometric relationships have been determined and which are so related that removal of one station from the set will affect the relationships between the other stations; and lines connecting the stations to show this interdependence.
Ridge	Ridge refers to the volcanic mountain ranges that lie along the spreading centers on the floors of the oceans.
Climate	Physical properties of the troposphere of an area based on analysis of its weather records over a long period. The two main factors determining an area's climate are temperature, with its seasonal variations, and the amount and distri.
Lake	Large natural body of standing fresh water formed when water from precipitation, land runoff, or groundwater flow fills a depression in the earth created by glaciation, earth movement, volcanic activity, or a giant meteorit are called the lake.
Recession	Recession refers to a continuing landward movement of the shoreline. A net landward movement of the shoreline over a specified time.
Embayment	Embayment refers to an indentation in a shoreline forming an open BAY. The formation of a
Subsidence	Sinking or down warping of a part of the earth's surface is called subsidence.
Iceberg	Iceberg refers to a large fragment of drifting ice that broke off the terminus of a glacier.
Till	Poorly sorted sediment that is deposited by glaciers is referred to as till.
Mud	Mud refers to a mixture of silt and clay sized particles.
Depression	Depression refers to a general term signifying any depressed or lower area in the ocean floor.
Technology	Technology refers to the creation of new products and processes intended to improve our efficiency, chances for survival, comfort level, and quality of life. Compare with science.
Coast	A strip of land of indefinite length and width that extends from the SEASHORE inland to the first major change in terrain features is referred to as coast.
Flood	Period when tide level is rising; often taken to mean the flood current which occurs during this period. A flow above the CARRYING CAPACITY of a CHANNEL.
Nearshore	Nearshore in beach terminology refers to an indefinite zone extending seaward from the shoreline well beyond the breaker zone. The zone which extends from the swash zone to the position marking the start of the offshore zone, typically at water depths of the order of 20
Shoreline	Shoreline refers to the intersection of a specified plane of water with the shore. All of the water areas of the state, including reservoirs and their associated uplands, together with the lands underlying them.
Delta	ALLUVIAL DEPOSIT, usually triangular, at the mouth of a river of other stream. It is normally built up only where there is no tidal or CURRENT action capable of removing the sediment as fast as it is deposited, and hence the DELTA builds forward from the COASTLINE. A TIDAL DELTA is a similar deposit at the mouth of a tidal INLET, put there by TIDAL CURRENTS. A WAVE DELTA

	is a deposit made by large waves which run over the top of a SPIT or BAR beach and down the landward side.
Limestone	A sedimentary rock composed dominantly of calcium carbonate, either precipitated from seawater or deposited as shell debris are called the limestone.
Gravel	Gravel refers to loose, rounded fragments of rock, larger than sand, but smaller than cobbles. Small stones and pebbles, or a mixture of these with sand.
Harbor	A water area nearly surrounded by land, sea walls, BREAKWATERS or artificial dikes, forming a safe anchorage for ships is referred to as harbor.
Landmark	Landmark refers to a conspicious object, natural or man-made, located near or on land, which aids in fixing the position of an observer.
Bed load	Bed load refers to heavy or large sediment particles that travel near or on the bed.
Runoff	Fresh water from precipitation and melting ice that flows on the earth's surface into nearby streams, lakes, wetlands, and reservoir is referred to as runoff.
Duration	In forecasting waves, the length of time the wind blows in essentially the same direction over the FETCH is a duration.
Ore deposits	Earth materials in which metals are concentrated in high concentrations, sufficient to be mined are called ore deposits.
Flooding	The natural process whereby waters emerge from their stream channel to cover part of the floodplain. Natural flooding is not a problem until people choose to build homes and other structures on floodplains.
Observations	Information obtained through one or more of the five senses or through instruments that extend the senses are observations.
Bay	Bay refers to a recess or inlet in the shore of a sea or lake between two capes or headlands, not as large as a gulf but larger than a cove.
Peninsula	An elongated portion of land nearly surrounded by water and connected to a larger body of land, usually by a neck or an isthmus is called a peninsula.
Theory	A general explanation of a characteristic of nature consistently supported by observation or experiment is referred to as a theory.
Peat	Peat refers to an organic deposit consisting predominantly of partly decayed plant matter.
Radiometric dating	The process of determining the age of rocks by observing the ratio of unstable radioactive elements to stable decay products is radiometric dating.
Law	A large construct explaining events in nature that have been observed to occur with unvarying uniformity under the same conditions is called law.
Key	Key refers to a low, insular BANK of sand, coral, etc., as one of the islets off the southern coast of Florida.
Coastline	Coastline refers to technically, the line that forms the boundary between the COAST and the SHORE. Commonly, the line that forms the boundary between land and the water. The line where terrestrial processes give way to marine processes, TIDAL CURRENTS, wind waves, etc.
Continental shelf	Continental shelf refers to the zone bordering a continent extending from the line of permanent immersion to the DEPTH, usually about 100 m to 200 m, where there is a marked or rather steep descent toward the great depths. The area under active LITTORAL processes during the Holocene period. The region of the oceanic bottom that extends outward from the shoreline with an average slope of less than 1.

Go to **Cram101.com** for the Practice Tests for this Chapter.

Cape	Cape refers to a relatively extensive land area jutting seaward from a continent or large island which prominently marks a change in, or interrupts notably, the coastal trend; a prominent feature.
Situation	The relative geographic location of a site that makes it a good location for a city is a situation.
Gulf	A relatively large portion of sea, partly enclosed by land is called gulf.
Basin	A large submarine depression of a generally circular, elliptical or oval shape is called a basin.
Energy	Capacity to do work by performing mechanical, physical, chemical, or electrical tasks or to cause a heat transfer between two objects at different temperatures is an energy.
Resources	Resources refer to substances that can be consumed by an organism and, as a result, become unavailable to other organisms.
Escarpment	A more or less continuous line of CLIFFS or steep slopes facing in one general direction which are caused by EROSION or faulting, also called SCARP is an escarpment.
Estuary	Estuary refers to a semi-enclosed coastal body of water which has a free connection with the open sea. The seawater is usually measurably diluted with freshwater. The part of the river that is affected by tides. The zone or area of water in which freshwater and saltwater mingle and water is usually brackish due to daily mixing and layering of fresh and salt water.
Undercutting	Undercutting refers to erosion of material at the foot of a cliff or bank; a sea cliff, or river bank on the outside of a meander. Ultimately, the overhang collapses, and the process is repeated.
Sea level rise	Sea level rise refers to the long-term trend in mean sea level.
Terrestrial	Terrestrial refers to pertaining to land. Compare with aquatic.
Matter	Matter refers to anything that has mass and takes up space. On earth, where gravity is present, we weigh an object to determine its mass.
Dredging	Excavation or displacement of the bottom or SHORELINE of a water body. Dredging can be accomplished with mechanical or hydraulic machines. Most is done to maintain channel depths or berths for navigational purposes; other dredging is for shellfish harvesting or for cleanup of polluted sediments.
Strait	A relatively narrow waterway between two larger bodies of water is a strait.
Desert	Desert refers to biome in which evaporation exceeds precipitation and the average amount of precipitation is less than 25 centimeters a year. Such areas have little vegetation or have widely spaced, mostly low vegetation. Compare forest, grassland.
Chalk	A white, soft limestone consisting dominantly of the shells of foraminifera is a chalk.
Unconsolidated	Unconsolidated with regards to sediment grains refers to loose, separate, or unattached to one another.
Wave	An oscillatory movement in a body of water manifested by an alternate rise and fall of the surface is called a wave. Disturbances of the surface of a liquid body, as the ocean, in the form of a ridge, swell or hump. The term wave by itself usually refers to the term surface gravity wave.
Uplift	The rising of one part of the Earth's crust relative to another part is referred to as uplift.

285

Loess	Loess refers to extensive deposits of wind-blown fine sediment commonly winnowed from glacially dumped debris.
Dunes	Dunes refers to accumulations of windblown sand on the BACKSHORE, usually in the form of small hills or ridges, stabilized by vegetation or control structures. A type of bed form indicating significant sediment transport over a sandy seabed.
Pack ice	Numerous separate pieces of sea ice that have been packed together in dense concentrations are referred to as pack ice.
Trough	A long and broad submarine depression with gently sloping sides is referred to as a trough.
Clay	Clay refers to a fine grained sediment with a typical grain size less than 0.004 mm. Possesses electromagnetic properties which bind the grains together to give a bulk strength or cohesion.
Ripple marks	Ripple marks refer to undulations produced by fluid movement over sediments. Oscillatory currents produce symmetric ripples whereas a well-defined current direction produces asymmetrical ripples. The crest line of ripples may be straight or sinuous. The characteristic features of ripples depend upon current velocity, particle size, persistence of current direction and whether the fluid is air or water. Sand dunes may be regarded as a special kind of 'super'-ripple.
Permafrost	Permafrost refers to a perennially frozen layer of the soil that forms when the water there freezes. It is found in arctic tundra.
Basalt	Basalt refers to a dark, fine-grained igneous rock composed of minerals enriched in ferromagnesian silicates.
Red clay	Red clay refers to a descriptive term applied to pelagic or abyssal clay deposits of the deep sea; they range in color from red to brown and tend to accumulate slowly in the deepest and remotest parts of the oceans far from the influx of other types of sediments.
Calcareous	Composed of calcium carbonate is referred to as calcareous.
Magnitude	An assessment of the size of an event is a magnitude. Magnitude scales exist for earthquakes, volcanic eruptions. hurricanes, and tornadoes. For earthquakes, different magnitudes are calculated for the same earthquake when different types of seismic waves are used.
Discharge	The volume of water flowing in a stream per unit of time is the discharge.
Plants	Eukaryotic, mostly multicelled organisms such as algae, mosses, ferns, flowers, cacti, grasses, beans, wheat, rice, and trees are plants. These organisms use photosynthesis to produce organic nutrients for themselves and for other organisms is referred to as plants.
Food	General term for organic molecules capable of providing energy to heterotrophs when combined with oxygen during biochemical respiration is called food.
Sedimentary rocks	Sedimentary rocks refer to rocks that have formed by the compaction and cementation of sediment.
Storm	Local or regional atmospheric disturbance characterized by strong winds often accompanied by precipitation is referred to as a storm.
Fossils	Skeletons, bones, shells, body parts, leaves, seeds, or impressions of such items that provide recognizable evidence of organisms that lived long ago are called fossils.
Fossil	The remains or traces of organisms preserved in rocks or ancient sediment are referred to as fossil.
Climatic change	Change in mean annual temperature and other aspects of climate over periods of time ranging from decades to hundreds of years to several million years is referred to as the climatic

Go to **Cram101.com** for the Practice Tests for this Chapter.

change.

Climate change	Refers to any long-term trend in MEAN SEA LEVEL, wave height, wind speed, drift rate etc are called the climate change.
Atmosphere	Atmosphere refers to the whole mass of air surrounding the earth.
Orbit	Orbit in ocean waves refers to the circular pattern of water particle movement at the air-sea interface.
Heat	Total kinetic energy of all the randomly moving atoms, ions, or molecules within a given substance, excluding the overall motion of the whole object. This form of kinetic energy flows from one body to another when there is a temperature difference betwe is referred to as heat.
Threshold	Threshold refers to a point in the operation of a system at which a change occurs. With respect to toxicology, it is a level below which effects are not observable and above which effects become apparent.
Forcing	With respect to global change, processes capable of changing global temperature, such as changes in solar energy emitted from the sun, or volcanic activity, we have forcing.
Carbon dioxide	Molecule of carbon and oxygen present in the atmosphere at approximately 350 ppm. Emissions of carbon dioxide resulting from burning of fossil fuels are thought to be co:.itributing to potential global warming through an enhanced greenhouse effect.
Current	Current refers to the flowing of water, or other liquid or gas. That portion of a stream of water which is moving with a velocity much greater than the average or in which the progress of the water is principally concentrated. Ocean currents can be classified in a number of different ways.
Isotopes	Two or more forms of a chemical element that have the same number of protons but different mass numbers because of different numbers of neutrons in their nuclei is referred to as isotopes.
Glass	Glass refers to matter created when magma cools too quickly for atoms to arrange themselves into the ordered atomic structures of minerals. Most glasses are supercooled liquids.
Ion	Ion refers to atom or group of atoms with one or more positive or negative electrical charges. Compare atone, molecule.
Magnetic field	A region where magnetic forces affect any magnetized bodies or electric currents is the magnetic field. Earth is surrounded by a magnetic field.
Isotope	An atom with a specified number of protons and a specified number of neutrons is the isotope.
Plates	Various-sized areas of earth's lithosphere that move slowly around with the mantle's flowing asthenosphere are plates. Most earthquakes and volcanoes occur around the boundaries of these plates.
Landlocked	Enclosed by land, or nearly enclosed, as a HARBOR is referred to as landlocked.
Weathering	Physical and chemical processes in which solid rock exposed at earth's surface is changed to separate solid particles and dissolved material, which can then be moved to another place as sediment is referred to as weathering.
Greenhouse gases	Greenhouse gases refers to gases in the earth's lower atmosphere that cause the greenhouse effect. Examples are carbon dioxide, chloro.
Weather	Weather refers to short-term changes in the temperature, barometric pressure, humidity, precipitation, sunshine, cloud cover, wind direction and speed, and other conditions in the troposphere at a given place and time. Compare with climate.

Rings	Rings refers to large whirl-like eddies created by meander cutoffs of strong geostropic currents such as the Gulf Stream. They either have warm-water centers or cold-water centers.
Forest	Forest refers to biome with enough average annual precipitation to support growth of various species of trees and smaller forms of vegetation. Compare desert, grassland.
Ash	The loose debris that is ejected from an erupting volcano is called ash.
Quaternary	The youngest geologic period, includes the present time is a quaternary. The latest period of time in the stratigraphic column, 0 - 2 million years, represented by local accumulations of glacial and post-glacial deposits, which continue, without change of fauna, from the top of the Pliocene. The quaternary appears to be an artificial division of time to separate pre-human from post-human sedimentation. As thus defined, the quaternary is increasing in duration as man's ancestry becomes longer.
Geology	Geology refers to the science which treats of the origin, history and structure of the Earth, as recorded in rocks; together with the forces and processes now operating to modify rocks.
Feedback	A kind of system response that occurs when output of the system also serves as input leading to changes in the system is called feedback.
Slide	In mass wasting, movement of a descending mass along a plane approximately parallel to the slope of the surface is referred to as slide.

Shoreline	Shoreline refers to the intersection of a specified plane of water with the shore. All of the water areas of the state, including reservoirs and their associated uplands, together with the lands underlying them.
Tides	The periodic rise and fall of the Earth's water surface as a consequence of the gravitational attraction of the Moon and the Sun, which are called tides.
Seismic	Referring to vibrations in the Earth produced by earthquakes is referred to as seismic.
Sea	Sea refers to the ocean. A large body of salt water, second in rank to an ocean, more or less landlocked and generally part of, or connected with, an ocean or a larger sea. State of the ocean or lake surface, in regard to waves.
Density	Density refers to the ratio of a mass to a unit volume specified as grams per cubic centimeter.
Energy	Capacity to do work by performing mechanical, physical, chemical, or electrical tasks or to cause a heat transfer between two objects at different temperatures is an energy.
Gravity	The attraction between bodies of matter is a gravity.
Nearshore	Nearshore in beach terminology refers to an indefinite zone extending seaward from the shoreline well beyond the breaker zone. The zone which extends from the swash zone to the position marking the start of the offshore zone, typically at water depths of the order of 20
System	A set of components that function and interact in some regular and theoretically predictable manner is called a system.
Shore	That strip of ground bordering any body of water which is alternately exposed, or covered by tides and/or waves is a shore. A shore of unconsolidated material is usually called a beach.
Sand	Sand refers to an unconsolidated mixture of inorganic soil consisting of small but easily distinguishable grains ranging in size from about.062 mm to 2.0 mm.
Sediment	Loose, fragments of rocks, minerals or organic material which are transported from their source for varying distances and deposited by air, wind, ice and water are sediment. Other sediments are precipitated from the overlying water or form chemically, in place. Sediment includes all the unconsolidated materials on the sea floor. The fine-grained material deposited by water or wind.
Pleistocene	An epoch of the Quaternary Period characterized by several glacial ages is referred to as the Pleistocene.
Stream	Stream refers to any flow of water; a current. A course of water flowing along a bed in the earth.
Erosion	Erosion refers to wearing away of the land by natural forces. On a beach, the carrying away of beach material by wave action, tidal currents or by DEFLATION. The wearing away of land by the action of natural forces.
Plants	Eukaryotic, mostly multicelled organisms such as algae, mosses, ferns, flowers, cacti, grasses, beans, wheat, rice, and trees are plants. These organisms use photosynthesis to produce organic nutrients for themselves and for other organisms is referred to as plants.
Coast	A strip of land of indefinite length and width that extends from the SEASHORE inland to the first major change in terrain features is referred to as coast.
Terrace	A horizontal or nearly horizontal natural or artificial topographic feature interrupting a steeper slop, sometimes occurring in a series is called a terrace.
Wave	An oscillatory movement in a body of water manifested by an alternate rise and fall of the surface is called a wave. Disturbances of the surface of a liquid body, as the ocean, in the

Go to Cram101.com for the Practice Tests for this Chapter.

form of a ridge, swell or hump. The term wave by itself usually refers to the term surface gravity wave.

Uplift	The rising of one part of the Earth's crust relative to another part is referred to as uplift.
Base	A substance that combines with a hydrogen ion in solution is called the base.
Concentration	Amount of a chemical in a particular volume or weight of air, water, soil, or other medium is referred to as concentration.
Population	Group of individual organisms of the same species living within a particular area is referred to as a population.
Refraction	The process by which the direction of a wave moving in shallow water at an angle to the bottom contours is changed is refraction. The part of the wave moving shoreward in shallower water travels more slowly than that portion in deeper water, causing the wave to turn or bend to become parallel to the contour.
Longshore drift	Movement of sediments approximately parallel to the COASTLINE is called longshore drift.
Undercutting	Undercutting refers to erosion of material at the foot of a cliff or bank; a sea cliff, or river bank on the outside of a meander. Ultimately, the overhang collapses, and the process is repeated.
Cliff	A high steep face of rock is referred to as cliff.
Equilibrium	A point of rest. A system that does not tend to undergo any change of its own accord but remains in a single, fixed condition is said to be in equilibrium. Compare with steady state.
Current	Current refers to the flowing of water, or other liquid or gas. That portion of a stream of water which is moving with a velocity much greater than the average or in which the progress of the water is principally concentrated. Ocean currents can be classified in a number of different ways.
Barrier islands	Long, thin, low offshore islands of sediment that generally run parallel to the shore along some coasts are referred to as barrier islands.
Ocean	The great body of salt water which occupies two-thirds of the surface of the Earth, or one of its major subdivisions is called an ocean.
Wind	Wind refers to the mass movement of air.
Wavelength	The horizontal distance between corresponding points on successive waves, such as from crest to crest or trough to trough is called a wavelength.
Wave height	The vertical distance between the crest and the trough is referred to as wave height.
Trough	A long and broad submarine depression with gently sloping sides is referred to as a trough.
Morphology	Morphology refers to river/estuary/lake/seabed form and its change with time.
Wave crest	Wave crest refers to the highest part of the wave. That part of the wave above still water level.
Wave trough	The lowest part of the waveform between crests is a wave trough. Also that part of a wave below still water level.
Wave period	Wave period refers to the time required for two successive wave crests to pass a fixed point. The time, in seconds, required for a wave crest to traverse a distance equal to one wavelength.
Orbit	Orbit in ocean waves refers to the circular pattern of water particle movement at the air-sea

interface.

Depth	Depth refers to vertical distance from still-water level to the bottom.
Wave base	Wave base refers to the plane or depths to which waves may erode the bottom in shallow water.
Shallow water	Shallow water refers to water of such depths that surface waves are noticeably affected by bottom topography. Typically this implies water depths equivalent to less than half the wave length.
Friction	The resistance to motion of two bodies in contact is a friction.
Point	Point refers to the extreme end of a cape, or the outer end of any land area protruding into the water, usually less prominent than a cape. A low profile shoreline promontory of more or less triangular shape, the top of which extends seaward.
Range	Land used for grazing is referred to as the range.
Surf	Collective term for breakers is surf. The wave activity in the area between the shoreline and the outermost limit of breakers. The term surf in literature usually refers to the breaking waves on shore and on reefs when accompanied by a roaring noise caused by the larger waves breaking.
Breaker	Breaker refers to a wave that has become so steep that the crest of the wave topples forward, moving faster than the main body of the wave. Breakers may be roughly classified into four kinds, although there is much overlap.
Swash	Swash refers to same as uprush. A body of dashing, splashing water. A bar over which the ocean washes.
Slope	The degree of inclination to the horizontal is the slope.
Surge	Surge refers to long-interval variations in velocity and pressure in fluid flow, not necessarily periodic, perhaps even transient in nature. The name applied to wave motion with a period intermediate between that of an ordinary wind wave and that of the tide. Changes in water level as a result of meteorological forcing causing a difference between the recorded water level and that predicted using harmonic analysis, may be positive or negative.
Gravel	Gravel refers to loose, rounded fragments of rock, larger than sand, but smaller than cobbles. Small stones and pebbles, or a mixture of these with sand.
Force	Force refers to a push or pull that affects motion. The product of mass and acceleration of a material.
Backwash	Backwash refers to the seaward return of the water following the uprush of the waves. Also called backrush or run down. Water of waves thrown back by an obstruction such as a ship, breakwater, cliff, etc.
Storm	Local or regional atmospheric disturbance characterized by strong winds often accompanied by precipitation is referred to as a storm.
Key	Key refers to a low, insular BANK of sand, coral, etc., as one of the islets off the southern coast of Florida.
Well	Well refers to a hole, generally cylindrical and usually walled or lined with pipe, that is dug or drilled into the ground to penetrate an aquifer below the zone of saturation.
Drag	The resistance to movement of an organism induced by the fluid through which it swims is called drag.
Frictional drag	Frictional drag refers to the retarding force associated with the surface of a body moving through a fluid or gas, or with a fluid or gas moving across a body surface. Sometimes called

Go to **Cram101.com** for the Practice Tests for this Chapter.

Go to **Cram101.com** for the Practice Tests for this Chapter.
And, **NEVER** highlight a book again!

surface drag.

Dispersion	Dispersion refers to act of dispersing, or state of being dispersed. The separation of waves by virtue of their differing rates of travel.
Front	The boundary between two air masses with different temperatures and densitie is referred to as front.
Headland	A land mass having a considerable ELEVATIO is referred to as headland.
Beach face	Beach face refers to the section of the beach normally exposed to the action of wave uprush. The foreshore of the beach.
Breaker zone	Breaker zone refers to the zone within which waves approaching the COASTLINE commence breaking, typically in water DEPTHS of between 5 m and 10 m.
Zone	Division or province of the ocean with homogeneous characteristics is referred to as a zone.
Saltation	Saltation refers to a term used to describe the movement of a particle being transported by wind or water, which is too heavy to remain in suspension. The particle is rolled forward by the current, generates lift and rises, loses the forward momentum supplying the lift and settles to the floor, where the process is repeated. The size of the particles, which can be saltated, depends upon the velocity of the current and its density.
River	A natural stream of water larger than a brook or creek is a river.
Wave direction	Wave direction refers to the direction from which the waves are coming.
Pile	A long substantial pole of wood, concrete or metal, driven into the earth or sea bed to serve as a support or protection is called pile.
Rip current	Rip current refers to a strong surface current of short duration flowing seaward from the shore. It usually appears as a visible band of agitated water and is the return movement of water piled up on the shore by incoming waves and wind.
Offshore	Offshore in beach terminology refers to the comparatively flat zone of variable width, extending from the shoreface to the edge of the continental shelf. It is continually submerged. The direction seaward from the shore. The zone beyond the nearshore zone where sediment motion induced by waves alone effectively ceases and where the influence of the sea bed on wave action is small in comparison with the effect of wind. The breaker zone directly seaward of the low tide line.
Surf zone	The nearshore zone along which the waves become breakers as they approach the shore is referred to as the surf zone.
Harbor	A water area nearly surrounded by land, sea walls, BREAKWATERS or artificial dikes, forming a safe anchorage for ships is referred to as harbor.
Site	A factor considering the summation of all environmental features of a location that influences the placement of a city is a site.
Head	A comparatively high promontory with either a CLIFF or steep face. It extends into a large body of water, such as a sea or lake. An unnamed HEAD is usually called a headland. The section of RIP CURRENT which has widened out seaward of the BREAKERS, also called head of
Submarine canyon	V-shaped valleys that run across the continental shelf and down the continental slope are referred to as submarine canyon.
Breakwater	Breakwater refers to a structure protecting a HARBOR, anchorage, or BASIN from waves. Offshore structure aligned parallel to the SHORE, sometimes shore-connected, that provides protection from waves.

Dredge	A metal collar and collecting bag that is dragged along the bottom to sample rock, sediment, or bottom organisms is referred to as the dredge.
Dredging	Excavation or displacement of the bottom or SHORELINE of a water body. Dredging can be accomplished with mechanical or hydraulic machines. Most is done to maintain channel depths or berths for navigational purposes; other dredging is for shellfish harvesting or for cleanup of polluted sediments.
Sea cliff	Cliff situated at the seaward edge of the coast is called a sea cliff.
Development	Development refers to change from a society that is largely rural, agricultural, illiterate, and poor, with a rapidly growing population, to one that is mostly urban, industrial, educated, and wealthy, with a slowly growing or stationary population.
Tidal flats	Tidal flats refer to marshy or muddy areas covered and uncovered by the rise and fall of the tide. A tidal marsh. Marshy or muddy areas of the seabed, which are covered and uncovered by the rise and fall of tidal water.
Topography	Topography refers to the form of the features of the actual surface of the Earth in a particular region considered collectively.
Subsidence	Sinking or down warping of a part of the earth's surface is called subsidence.
Power	Power refers to the time rate of doing work.
Rocks	An aggregate of one or more minerals rather large in area are rocks. The three classes of rocks are the following: Igneous rock - crystalline rocks formed from molten material. Sedimentary rock - A rock resulting from the consolidation of loose sediment that has accumulated in layers. Metamorphic rock - Rock that has formed from preexisting rock as a result of heat or pressure.
Limestone	A sedimentary rock composed dominantly of calcium carbonate, either precipitated from seawater or deposited as shell debris are called the limestone.
Chemical	One of the millions of different elements and compounds found naturally or synthesized by human is referred to as chemical.
Bed load	Bed load refers to heavy or large sediment particles that travel near or on the bed.
Load	The quantity of sediment transported by a current. It includes the suspended load of small particles in the water, and the bedload of large particles that move along the bottom.
Bedrock	Solid rock lying beneath loose soil or unconsolidated sediment is called bedrock.
Debris	Debris refers to any accumulation of rock fragments; detritus.
Tide	The periodic rising and falling of the water that results from gravitational attraction of the moon and sun acting upon the rotating earth is the tide. Although the accompanying horizontal movement of the water resulting from the same cause is also sometimes called the tide, it is preferable to designate the latter as tidal current, reserving the name tide for the vertical movement.
Weathering	Physical and chemical processes in which solid rock exposed at earth's surface is changed to separate solid particles and dissolved material, which can then be moved to another place as sediment is referred to as weathering.
Mass	The amount of material in an object is the mass.
Rock	Any material that makes up a large, natural, continuous part of earth's crust is rock.
Degree	An arbitrary measure of temperature. One degree Celsius _ 1.8 degrees Fahrenheit.
Fault	A fracture in rock along which there has been an observable amount of displacement. Faults

are rarely single planar units; normally they occur as parallel to sub-parallel sets of planes along which movement has taken place to a greater or lesser extent. Such sets are called fault or fracture-zones.

Fall	Fall refers to a mass moving nearly vertical and downward under the influence of gravity.
Coastline	Coastline refers to technically, the line that forms the boundary between the COAST and the SHORE. Commonly, the line that forms the boundary between land and the water. The line where terrestrial processes give way to marine processes, TIDAL CURRENTS, wind waves, etc.
Recent	A synonym of Holocene is called recent.
Dunes	Dunes refers to accumulations of windblown sand on the BACKSHORE, usually in the form of small hills or ridges, stabilized by vegetation or control structures. A type of bed form indicating significant sediment transport over a sandy seabed.
Ash	The loose debris that is ejected from an erupting volcano is called ash.
Matter	Matter refers to anything that has mass and takes up space. On earth, where gravity is present, we weigh an object to determine its mass.
Seashore	Seashore refers to all ground between the ordinary high-water and low-water mark. The shore of the sea or ocean.
Bay	Bay refers to a recess or inlet in the shore of a sea or lake between two capes or headlands, not as large as a gulf but larger than a cove.
Cape	Cape refers to a relatively extensive land area jutting seaward from a continent or large island which prominently marks a change in, or interrupts notably, the coastal trend; a prominent feature.
Jetty	Jetty refers to on open seacoasts, a structure extending into a body of water to direct and confine the stream or tidal flow to a selected CHANNEL, or to prevent shoaling. Jetties are built at the mouth of a river or ENTRANCE to a BAY to help deepen and stabilize a CHANNEL and facilitate navigation. A structure usually projecting out into the SEA at the mouth of a river for the purpose of protecting a navigational channel, a harbor or to influence water currents.
Zoning	Regulating how various parcels of land can be used is zoning.
Input	Input refers to matter, energy, or information entering a system. Compare output, throughput.
Coral	Coral refers to any of more than 6,000 species of small cnidarians, many of which are capable of generating hard calcareous skeletons.
Migration	A term that refers to the habit of some animals is a migration.
Slumping	The sliding of large, cohering blocks of sediment or rock downslope under the influence of gravity is called slumping.
Recession	Recession refers to a continuing landward movement of the shoreline. A net landward movement of the shoreline over a specified time.
Unconsolidated	Unconsolidated with regards to sediment grains refers to loose, separate, or unattached to one another.
Clay	Clay refers to a fine grained sediment with a typical grain size less than 0.004 mm. Possesses electromagnetic properties which bind the grains together to give a bulk strength or cohesion.
Spit	A long narrow accumulation of sand or shingle, lying generally in line with the coast, with one end attached to the land the other projecting into the sea or across the mouth of an

estuary is called a spit.

Baymouth bar	Bar extending partly or entirely across the mouth of a bay is called a baymouth bar.
Bar	An offshore ridge or mound of sand, gravel, or other unconsolidated material which is submerged, especially at the mouth of a river or estuary, or lying parallel to, and a short distance from, the beach is referred to as a bar.
Tombolo	Coastal formation of beach material developed by refraction, diffraction and longshore drift to form a 'neck' connecting a coast to an offshore island or breakwater is a tombolo.
Output	Output refers to matter, energy, or information leaving a system. Compare with input, throughput.
Backshore	Backshore refers to the upper part of the active beach above the normal reach of the tides, but affected by large waves occurring during a high. The accretion or erosion zone, located landward of ordinary high tide, which is normally wetted only by storm tides.
Deep water	Deep water in regard to waves, where DEPTH is greater than one-half the WAVE LENGTH. Deep-water conditions are said to exist when the surf waves are not affected by conditions on the bottom.
Continent	Continent refers to lower-density masses of rock, exposed as about 40 percent of the Earth's surface: 29 percent as land and I 1 percent as the floor of shallow seas.
Lagoon	A shallow body of water, as a pond or lake, which usually has a shallow restricted INLET from the se is referred to as lagoon.
Dike	Dike refers to sometimes written as dyke; earth structure along a sea or river in order to protect LITTORAL lands from flooding by high water; DIKES along rivers are sometimes called levees.
Fire	The rapid combination of oxygen with organic material to produce flame, heat, and light is referred to as the fire.
Barrier beach	Barrier beach refers to a bar essentially parallel to the shore, which has been built up so that its crest rises above the normal high water level. Also called barrier island and offshore barrier.
High tide	The high-water position corresponding to a tidal crest are called the high tide.
Barrier island	A detached portion of a barrier beach between two inlets is a barrier island.
Sediment transport	Sediment transport refers to the main agencies by which sedimentary materials are moved. Running water and wind are the most widespread transporting agents. In both cases, three mechanisms operate, although the particle size of the transported material involved is very different, owing to the differences in density and viscosity of air and water.
Sand waves	Longshore sand waves are large-scale features that maintain form while migrating along the shore with speeds on the order of kilometers per year. Large-scale asymmetrical bedforms in sandy river beds having high length to height ratios and continuous crestlines.
Delta	ALLUVIAL DEPOSIT, usually triangular, at the mouth of a river of other stream. It is normally built up only where there is no tidal or CURRENT action capable of removing the sediment as fast as it is deposited, and hence the DELTA builds forward from the COASTLINE. A TIDAL DELTA is a similar deposit at the mouth of a tidal INLET, put there by TIDAL CURRENTS. A WAVE DELTA is a deposit made by large waves which run over the top of a SPIT or BAR beach and down the landward side.
Tidal inlet	Tidal inlet refers to a channel or opening through a barrier island that admits the tidal flow of water.

Go to **Cram101.com** for the Practice Tests for this Chapter.

Inlet	Inlet refers to a narrow strip of water running into the land or between islands. An arm of the sea that is long compared to its width, and that may extend a considerable distance inland.
Embayed	Formed into a BAY or bays; as an embayed shore.
Conditions	Conditions refers to physical or chemical attributes of the environment that, while not being consumed, influence biological processes and population growth. Examples are temperature, salinity, and acidity. Compare resources.
Tectonics	The study of the major structural features of the Earth's crust or the broad structure of a region is referred to as tectonics.
Atmosphere	Atmosphere refers to the whole mass of air surrounding the earth.
Flooding	The natural process whereby waters emerge from their stream channel to cover part of the floodplain. Natural flooding is not a problem until people choose to build homes and other structures on floodplains.
Washover fan	The fan-shaped accumulation of sand on the landward side of a barrier island that is deposited by storm waves, which overtop the island, is referred to as a washover fan.
Storm surge	A rise or piling-up of water against shore, produced by strong winds blowing onshore is a storm surge. A storm surge is most severe when it occurs in conjunction with a high tide.
Hurricane	A cyclonic storm, usually of tropic origin, covering an extensive area, and containing winds in excess of 75 miles per hour is the hurricane.
Community	Community refers to populations of all species living and interacting in an area at a particular time.
Algae	Algae refers to simple marine and freshwater plants, unicellular and multicellular, that lack roots, stems, and leaves.
Invertebrates	Invertebrates refers to animals that have no backbones. Compare vertebrates.
Invertebrate	Animal lacking a backbone is the invertebrate.
Reef	A ridge of rock or other material lying just below the surface of the sea is called a reef.
Barrier reef	A coral reef growing around the periphery of an island, but separated from it by a lagoon is referred to as a barrier reef.
Beach erosion	The carrying away of beach materials by wave action, tidal currents, littoral currents or wind is referred to as beach erosion.
Ecology	Ecology refers to study of the interactions of living organisms with one another and with their nonliving environment of matter and energy; study of the structure and functions of nature.
Temperature	Temperature refers to a measure of the average speed of motion of the atoms, ions, or molecules in a substance or combination of substances at a given moment. Compare with heat.
Salinity	Salinity refers to number of grams of salt per thousand grams of seawater, usually expressed in parts per thousand.
Water depth	Distance between the seabed and the still water level is referred to as water depth.
Latitude	Latitude refers to distance from the equator. Compare altitude.
Mud	Mud refers to a mixture of silt and clay sized particles.
Parts per thousand	Parts per thousand refer to a unit of salinity; 35% indicates that 35 grams of salt are contained in 1,000 grams of seawater. In other words, salt comprises 3.5% by weight of a

volume of seawater.

Flood	Period when tide level is rising; often taken to mean the flood current which occurs during this period. A flow above the CARRYING CAPACITY of a CHANNEL.
Food	General term for organic molecules capable of providing energy to heterotrophs when combined with oxygen during biochemical respiration is called food.
Runoff	Fresh water from precipitation and melting ice that flows on the earth's surface into nearby streams, lakes, wetlands, and reservoir is referred to as runoff.
Continental shelf	Continental shelf refers to the zone bordering a continent extending from the line of permanent immersion to the DEPTH, usually about 100 m to 200 m, where there is a marked or rather steep descent toward the great depths. The area under active LITTORAL processes during the Holocene period. The region of the oceanic bottom that extends outward from the shoreline with an average slope of less than 1.
Atoll	A ring-shaped coral reef that surrounds a lagoon is called an atoll.
Calcareous	Composed of calcium carbonate is referred to as calcareous.
Coral reef	Formation produced by massive colonies containing billions of tiny coral animals, called polyps, that secrete a stony substance around themselves for protection. When the corals die, their empty outer skeletons form layers that caus is referred to as coral reef.
Theory	A general explanation of a characteristic of nature consistently supported by observation or experiment is referred to as a theory.
Basalt	Basalt refers to a dark, fine-grained igneous rock composed of minerals enriched in ferromagnesian silicates.
Volcano	Vent or fissure in the earth's surface through which magma, liquid lava, and gases are released into the environment is a volcano.
Polynesia	A large group of Pacific islands lying east of Melanesia and Micronesia and extending from the Hawaiian Islands south to New Zealand and east to Easter Island is called Polynesia.
Ice age	Ice age refers to one of several periods of low temperature during the last million years.
Coastal processes	Collective term covering the action of natural forces on the shoreline, and the nearshore seabed are called coastal processes.
Fringing reef	A reef that is growing at the edge of a landmass without an intervening lagoon is referred to as fringing reef.
Plate tectonics	Plate tectonics refer to the theory of geophysical processes that explains the movements of lithospheric plates and the processes that occur at their boundaries.
Convergence	Resemblance among species belonging to different taxonomic groups as the result from adaptation to similar environments is the convergence.
Map	Map refers to a representation of Earth's surface usually depicting mostly land areas.
Subduction zone	Elongate region in which the sea floor slides beneath a continent or island arc is referred to as the subduction zone.
Seismicity	The distribution, frequency, and magnitude of earthquakes in an area are called seismicity.
Elevation	Elevation refers to the distance of a point above a specified surface of constant potential; the distance is measured along the direction of gravity between the point and the surface.
Accumulation	Buildup of matter, energy, or information in a system is referred to as accumulation.
Ridge	Ridge refers to the volcanic mountain ranges that lie along the spreading centers on the

309

floors of the oceans.

Relief	The difference in elevation between the highest and lowest points in an area is called relief.
Lithosphere	Lithosphere refers to outer shell of the earth, composed of the crust and the rigid, outermost part of the mantle outside of the asthenosphere; material found in the earth's plates.
Rift	Rift refers to the valley created at a pull-apart zone.
Stability	Ability of a living system to withstand or recover from externally imposed changes or stresses is called stability.
Passive continental margin	A subsiding continental margin situated in a nontectonic setting away from a lithospheric plate boundary is a passive continental margin.
Continental margin	The drowned edges of continents consisting of the continental shelf, the continental slope, and the continental rise is called continental margin.
Cross section	A two-dimensional drawing showing features in the vertical plane as in a canyon wall or road cut is referred to as cross section.
Passive margin	Passive margin refers to a continental margin near an area of lithospheric plate divergence.
Seacoast	Seacoast refers to the coast adjacent to the sea or ocean.
Gulf	A relatively large portion of sea, partly enclosed by land is called gulf.
Island arc	A curved group of volcanic islands associated with a deep oceanic trench and subduction zone is called island arc.
Climate	Physical properties of the troposphere of an area based on analysis of its weather records over a long period. The two main factors determining an area's climate are temperature, with its seasonal variations, and the amount and distri.
Terrestrial	Terrestrial refers to pertaining to land. Compare with aquatic.
Valley	An elongated depression, usually with an outlet, between bluffs or between ranges of hills or mountains is a valley.
Moraine	Moraine refers to an accumulation of earth, stones, etc., deposited by a glacier, usually in the form of a mound, ridge or other prominence on the terrain.
Marsh	Marsh refers to a tract of soft, wet land, usually vegetated by reeds, grasses and occasionally small shrubs. Soft, wet area periodically or continuously flooded to a shallow depth, usually characterized by a particular subclass of grasses, cattails and other low plants.
Mangrove	Large flowering shrub or tree that grows in dense thickets or forests along muddy or silty tropical coasts is called the mangrove.
Centrifugal force	Centrifugal force refers to an apparent force exerted outward from a rotating object. The faster the rate of rotation and the longer the radius of rotation, the stronger the apparent force.
Water level	Elevation of a particular point or small patch on the surface of a body of water above a specific point or surface, averaged over a period of time sufficiently long to remove the effects of short period disturbances is referred to as water level.
Tsunami	Tsunami refers to a large, high-velocity wave generated by displacement of the sea floor; also called seismic sea wave. Commonly misnamed tidal wave.

Meteorite	A stony or metallic body from space that passed through the atmosphere and landed on the surface of the Earth is referred to as a meteorite.
Magnitude	An assessment of the size of an event is a magnitude. Magnitude scales exist for earthquakes, volcanic eruptions. hurricanes, and tornadoes. For earthquakes, different magnitudes are calculated for the same earthquake when different types of seismic waves are used.
Earthquake	Shaking of the ground resulting either from the fracturing and displacement of rock, producing a fault, or from subsequent movement along the fault is the earthquake.
Depression	Depression refers to a general term signifying any depressed or lower area in the ocean floor.
Swell	Waves that have traveled a long distance from their generating area and have been sorted out by travel into long waves of the same approximate period are referred to as a swell.
Pebble	Pebble refers to a rock fragment between 4 and 64 mm in diameter.
Trench	A long narrow submarine depression with relatively steep sides is called a trench.
Observations	Information obtained through one or more of the five senses or through instruments that extend the senses are observations.
Estuary	Estuary refers to a semi-enclosed coastal body of water which has a free connection with the open sea. The seawater is usually measurably diluted with freshwater. The part of the river that is affected by tides. The zone or area of water in which freshwater and saltwater mingle and water is usually brackish due to daily mixing and layering of fresh and salt water.
Fjord	Fjord refers to a long, narrow arm of the sea, usually formed by entrance of the sea into a deep glacial trough.
Geomorphology	Geomorphology refers to that branch of physical geography which deals with the form of the Earth, the general configuration of its surface, the distribution of the land, water, etc. The investigation of the history of geologic changes through the interpretation of topographic forms.
Resources	Resources refer to substances that can be consumed by an organism and, as a result, become unavailable to other organisms.
Geology	Geology refers to the science which treats of the origin, history and structure of the Earth, as recorded in rocks; together with the forces and processes now operating to modify rocks.
Feedback	A kind of system response that occurs when output of the system also serves as input leading to changes in the system is called feedback.
Slide	In mass wasting, movement of a descending mass along a plane approximately parallel to the slope of the surface is referred to as slide.

313

Eolian	Describes sediment that was deposited by wind are called eolian.
Wind	Wind refers to the mass movement of air.
Erosion	Erosion refers to wearing away of the land by natural forces. On a beach, the carrying away of beach material by wave action, tidal currents or by DEFLATION. The wearing away of land by the action of natural forces.
Abrasion	Frictional erosion by material transported by wind and waves is referred to as abrasion.
Sand	Sand refers to an unconsolidated mixture of inorganic soil consisting of small but easily distinguishable grains ranging in size from about.062 mm to 2.0 mm.
Dunes	Dunes refers to accumulations of windblown sand on the BACKSHORE, usually in the form of small hills or ridges, stabilized by vegetation or control structures. A type of bed form indicating significant sediment transport over a sandy seabed.
Precipitation	Water in the form of rain, sleet, hail, and snow that falls from the atmosphere onto the land and bodies of water is called precipitation.
Evaporation	Evaporation refers to conversion of a liquid into a gas.
Desert	Desert refers to biome in which evaporation exceeds precipitation and the average amount of precipitation is less than 25 centimeters a year. Such areas have little vegetation or have widely spaced, mostly low vegetation. Compare forest, grassland.
Valley	An elongated depression, usually with an outlet, between bluffs or between ranges of hills or mountains is a valley.
Loess	Loess refers to extensive deposits of wind-blown fine sediment commonly winnowed from glacially dumped debris.
Food	General term for organic molecules capable of providing energy to heterotrophs when combined with oxygen during biochemical respiration is called food.
Ocean	The great body of salt water which occupies two-thirds of the surface of the Earth, or one of its major subdivisions is called an ocean.
Sediment	Loose, fragments of rocks, minerals or organic material which are transported from their source for varying distances and deposited by air, wind, ice and water are sediment. Other sediments are precipitated from the overlying water or form chemically, in place. Sediment includes all the unconsolidated materials on the sea floor. The fine-grained material deposited by water or wind.
System	A set of components that function and interact in some regular and theoretically predictable manner is called a system.
Energy	Capacity to do work by performing mechanical, physical, chemical, or electrical tasks or to cause a heat transfer between two objects at different temperatures is an energy.
Planet	A smaller, usually nonluminous body orbiting a star is a planet.
Groundwater	Groundwater refers to water that sinks into the soil and is stored in slowly flowing and slowly renewed underground reservoirs called aquifers; underground water in the zone of saturation, below the water table. Compare runoff, surface water.
Deflation	Deflation refers to the removal of loose material from a beach or other land surface by wind action.
Well	Well refers to a hole, generally cylindrical and usually walled or lined with pipe, that is dug or drilled into the ground to penetrate an aquifer below the zone of saturation.
Rocks	An aggregate of one or more minerals rather large in area are rocks. The three classes of

Go to **Cram101.com** for the Practice Tests for this Chapter.

315

rocks are the following: Igneous rock - crystalline rocks formed from molten material. Sedimentary rock - A rock resulting from the consolidation of loose sediment that has accumulated in layers. Metamorphic rock - Rock that has formed from preexisting rock as a result of heat or pressure.

Unconsolidated	Unconsolidated with regards to sediment grains refers to loose, separate, or unattached to one another.
Saltation	Saltation refers to a term used to describe the movement of a particle being transported by wind or water, which is too heavy to remain in suspension. The particle is rolled forward by the current, generates lift and rises, loses the forward momentum supplying the lift and settles to the floor, where the process is repeated. The size of the particles, which can be saltated, depends upon the velocity of the current and its density.
Creep	Creep refers to very slow, continuous downslope movement of soil or debris.
Atmosphere	Atmosphere refers to the whole mass of air surrounding the earth.
Windward	The direction from which the wind is blowing is windward.
Slope	The degree of inclination to the horizontal is the slope.
Constancy	Ability of a living system, such as a population, to maintain a certain size. Compare inertia, resilienc is called constancy.
Rock	Any material that makes up a large, natural, continuous part of earth's crust is rock.
Debris	Debris refers to any accumulation of rock fragments; detritus.
Desertification	Desertification refers to conversion of rangeland, rainfed cropland, or irrigated cropland to desertlike land, with a drop in agricultural productivity of 10% or more. It is usually caused by a combination of overgrazing, soil erosion, prolonged drought, and climate change.
Open system	Open system refers to a system, such as a living organism, in which both matter and energy are exchanged between the system and the environment. Compare with closed system.
Heat	Total kinetic energy of all the randomly moving atoms, ions, or molecules within a given substance, excluding the overall motion of the whole object. This form of kinetic energy flows from one body to another when there is a temperature difference betwe is referred to as heat.
Kinetic energy	Kinetic energy refers to energy that matter has because of its mass and speed or velocity. Compare potential energy.
Radiation	Fast-moving particles or waves of energy are called radiation.
Latitude	Latitude refers to distance from the equator. Compare altitude.
Coriolis effect	Coriolis effect, force due to the Earth's rotation, capable of generating currents. It causes moving bodies to be deflected to the right in the Northern Hemisphere and to the left in the Southern Hemisphere. The 'force' is proportional to the speed and latitude of the moving object. It is zero at the equator and maximum at the poles.
Runoff	Fresh water from precipitation and melting ice that flows on the earth's surface into nearby streams, lakes, wetlands, and reservoir is referred to as runoff.
Convection	Convection refers to the vertical transport of a fluid or the transfer of heat in fluids.
Trade winds	Surface winds within the Hadley cells, centered at about 15° latitude, which approach from the northeast in the Northern Hemisphere and from the southeast in the Southern Hemisphere are referred to as trade winds.
Temperature	Temperature refers to a measure of the average speed of motion of the atoms, ions, or

molecules in a substance or combination of substances at a given moment. Compare with heat.

Range	Land used for grazing is referred to as the range.
Pleistocene	An epoch of the Quaternary Period characterized by several glacial ages is referred to as the Pleistocene.
Weathering	Physical and chemical processes in which solid rock exposed at earth's surface is changed to separate solid particles and dissolved material, which can then be moved to another place as sediment is referred to as weathering.
Gravel	Gravel refers to loose, rounded fragments of rock, larger than sand, but smaller than cobbles. Small stones and pebbles, or a mixture of these with sand.
Power	Power refers to the time rate of doing work.
Chemical	One of the millions of different elements and compounds found naturally or synthesized by human is referred to as chemical.
Climate	Physical properties of the troposphere of an area based on analysis of its weather records over a long period. The two main factors determining an area's climate are temperature, with its seasonal variations, and the amount and distri.
Cement	Minerals such as silica and carbonate that are chemically precipitated in the pores of sediments, binding the grain are referred to as cement.
Delta	ALLUVIAL DEPOSIT, usually triangular, at the mouth of a river of other stream. It is normally built up only where there is no tidal or CURRENT action capable of removing the sediment as fast as it is deposited, and hence the DELTA builds forward from the COASTLINE. A TIDAL DELTA is a similar deposit at the mouth of a tidal INLET, put there by TIDAL CURRENTS. A WAVE DELTA is a deposit made by large waves which run over the top of a SPIT or BAR beach and down the landward side.
Subsidence	Sinking or down warping of a part of the earth's surface is called subsidence.
Development	Development refers to change from a society that is largely rural, agricultural, illiterate, and poor, with a rapidly growing population, to one that is mostly urban, industrial, educated, and wealthy, with a slowly growing or stationary population.
Bedrock	Solid rock lying beneath loose soil or unconsolidated sediment is called bedrock.
Calcareous	Composed of calcium carbonate is referred to as calcareous.
Basin	A large submarine depression of a generally circular, elliptical or oval shape is called a basin.
Mud	Mud refers to a mixture of silt and clay sized particles.
Conditions	Conditions refers to physical or chemical attributes of the environment that, while not being consumed, influence biological processes and population growth. Examples are temperature, salinity, and acidity. Compare resources.
Stone	Stone refers to quarried or artificially broken rock for use in construction.
Pebbles	Beach material usually well-rounded and between about 4 mm to 64 mm diameter are pebbles.
Ash	The loose debris that is ejected from an erupting volcano is called ash.
Soil	Soil refers to a layer of weathered, unconsolidated material on top of bedrock; often also defined as containing organic matter and being capable of supporting plant growth.
Stream	Stream refers to any flow of water; a current. A course of water flowing along a bed in the earth.

Go to **Cram101.com** for the Practice Tests for this Chapter.

Pebble	Pebble refers to a rock fragment between 4 and 64 mm in diameter.
Viscosity	Resistance to flow is called viscosity.
Cohesion	Attachment of water molecules to each other by hydrogen bonds is a cohesion.
Drag	The resistance to movement of an organism induced by the fluid through which it swims is called drag.
Zone	Division or province of the ocean with homogeneous characteristics is referred to as a zone.
Sedimentary rocks	Sedimentary rocks refer to rocks that have formed by the compaction and cementation of sediment.
Shield volcano	Shield volcano refers to a very wide volcano built of low-viscosity lavas.
Volcano	Vent or fissure in the earth's surface through which magma, liquid lava, and gases are released into the environment is a volcano.
Elastic	Elastic refers to behavior of material where stress causes deformation that is recoverable; when stress stops, the material returns to its original state.
Gravity	The attraction between bodies of matter is a gravity.
Chain reaction	Multiple nuclear fissions, taking place within a certain mass of a fissionable isotope, that release an enormous amount of energy in a short time is a chain reaction.
Bed	The bottom of a watercourse, or any body of water is called a bed.
Storm	Local or regional atmospheric disturbance characterized by strong winds often accompanied by precipitation is referred to as a storm.
Soil erosion	Soil erosion refers to movement of soil components, especially topsoil, from one place to another, usually by exposure to wind, flowing water, or both. This natural process can be greatly accelerated by human activities that remove vegetation from soil.
Point	Point refers to the extreme end of a cape, or the outer end of any land area protruding into the water, usually less prominent than a cape. A low profile shoreline promontory of more or less triangular shape, the top of which extends seaward.
Fall	Fall refers to a mass moving nearly vertical and downward under the influence of gravity.
Density current	A current powered by gravity such as a turbidity current is a density current.
Current	Current refers to the flowing of water, or other liquid or gas. That portion of a stream of water which is moving with a velocity much greater than the average or in which the progress of the water is principally concentrated. Ocean currents can be classified in a number of different ways.
Turbulent flow	Turbulent flow refers to any flow, which is not laminar, i.e., the stream lines of the fluid, instead of remaining parallel, become confused and intermingled.
Slide	In mass wasting, movement of a descending mass along a plane approximately parallel to the slope of the surface is referred to as slide.
Clay	Clay refers to a fine grained sediment with a typical grain size less than 0.004 mm. Possesses electromagnetic properties which bind the grains together to give a bulk strength or cohesion.
Reach	Reach refers to an arm of the ocean extending into the land. A straight section of restricted waterway of considerable extent; may be similar to a narrows, except much longer in extent.
River	A natural stream of water larger than a brook or creek is a river.

Go to **Cram101.com** for the Practice Tests for this Chapter.

Sea	Sea refers to the ocean. A large body of salt water, second in rank to an ocean, more or less landlocked and generally part of, or connected with, an ocean or a larger sea. State of the ocean or lake surface, in regard to waves.
Coast	A strip of land of indefinite length and width that extends from the SEASHORE inland to the first major change in terrain features is referred to as coast.
Coral	Coral refers to any of more than 6,000 species of small cnidarians, many of which are capable of generating hard calcareous skeletons.
Star	A massive sphere of incandescent gases powered by the conversion of hydrogen to helium and other heavier elements is a star.
Ripple	Ripple refers to the light fretting or ruffling on the surface of the water caused by a breeze. The smallest class of waves and one in which the force of restoration is, to a significant degree, both surface tension and gravity.
Migration	A term that refers to the habit of some animals is a migration.
Ripple marks	Ripple marks refer to undulations produced by fluid movement over sediments. Oscillatory currents produce symmetric ripples whereas a well-defined current direction produces asymmetrical ripples. The crest line of ripples may be straight or sinuous. The characteristic features of ripples depend upon current velocity, particle size, persistence of current direction and whether the fluid is air or water. Sand dunes may be regarded as a special kind of 'super'-ripple.
Sand waves	Longshore sand waves are large-scale features that maintain form while migrating along the shore with speeds on the order of kilometers per year. Large-scale asymmetrical bedforms in sandy river beds having high length to height ratios and continuous crestlines.
Forest	Forest refers to biome with enough average annual precipitation to support growth of various species of trees and smaller forms of vegetation. Compare desert, grassland.
Sand dune	Dune formed of sand is called a sand dune.
Accumulation	Buildup of matter, energy, or information in a system is referred to as accumulation.
Slip face	Slip face refers to the steep, downwind slope of a dune; formed from loose, cascading sand that generally keeps the slope at the angle of repose.
Angle of repose	The maximum slope at which soils and loose materials on the banks of canals, rivers or embankments stay stable is called the angle of repose.
Map	Map refers to a representation of Earth's surface usually depicting mostly land areas.
Mineral	A naturally occurring, inorganic, crystalline solid that has a definite chemical composition and possesses characteristic physical properties is a mineral.
Gypsum	An evaporite deposit composed of hydrous calcium sulfat is a gypsum.
Onshore	A direction landward from the sea is called onshore.
Ridge	Ridge refers to the volcanic mountain ranges that lie along the spreading centers on the floors of the oceans.
Peninsula	An elongated portion of land nearly surrounded by water and connected to a larger body of land, usually by a neck or an isthmus is called a peninsula.
Blowout	DEPRESSION on the land surface caused by wind EROSION is called blowout.
Continent	Continent refers to lower-density masses of rock, exposed as about 40 percent of the Earth's surface: 29 percent as land and I 1 percent as the floor of shallow seas.

Climatic change	Change in mean annual temperature and other aspects of climate over periods of time ranging from decades to hundreds of years to several million years is referred to as the climatic change.
Plates	Various-sized areas of earth's lithosphere that move slowly around with the mantle's flowing asthenosphere are plates. Most earthquakes and volcanoes occur around the boundaries of these plates.
Cliff	A high steep face of rock is referred to as cliff.
Ice age	Ice age refers to one of several periods of low temperature during the last million years.
Climate change	Refers to any long-term trend in MEAN SEA LEVEL, wave height, wind speed, drift rate etc are called the climate change.
Reef	A ridge of rock or other material lying just below the surface of the sea is called a reef.
Dip	The angle of inclination measured in degrees from the horizontal is called dip.
Dispersion	Dispersion refers to act of dispersing, or state of being dispersed. The separation of waves by virtue of their differing rates of travel.
Ness	Ness refers to roughly triangular promontory of land jutting into the sea, often consisting of mobile material.
Thermal	The energy of the random motion of atoms and molecules is referred to as thermal.
Glacier	Large masses of moving ice on land derived by the recrystallization of snow into ice under pressure are referred to as glacier.
Network	Network refers to a set consisting of stations for which geometric relationships have been determined and which are so related that removal of one station from the set will affect the relationships between the other stations; and lines connecting the stations to show this interdependence.
Nutrients	Nutrients refer to chemicals such as phosphorus and nitrogen that, when released into water sources, may cause pollution events such as eutrophication.
Fertility	Fertility refers to the ability to produce offspring; the proportion of births to population.
Load	The quantity of sediment transported by a current. It includes the suspended load of small particles in the water, and the bedload of large particles that move along the bottom.
Degradation	The geologic process by means of which various parts of the surface of the earth are worn away and their general level lowered, by the action of wind and water is referred to as degradation.
Overgrazing	Destruction of vegetation when too many grazing animals feed too long and exceed the carrying capacity of a rangeland area is referred to as overgrazing.
Response	The amount of health damage caused by exposure to a certain dose of a harmful substance or form of radiation is a response.
Deforestation	Removal of trees from a forested area without adequate replanting is a deforestation.
Compaction	The decrease in volume and porosity of a sediment via burial is referred to as compaction.
Reduce	With respect to waste management, reduce refers to practices that will reduce the amount of waste we produce.
Lake	Large natural body of standing fresh water formed when water from precipitation, land runoff, or groundwater flow fills a depression in the earth created by glaciation, earth movement, volcanic activity, or a giant meteorit are called the lake.

Go to **Cram101.com** for the Practice Tests for this Chapter.

Drought	Condition in which an area does not get enough water because of lower-than-normal precipitation, higher-than-normal temperatures that increase evaporation, or both is called drought.
Shore	That strip of ground bordering any body of water which is alternately exposed, or covered by tides and/or waves is a shore. A shore of unconsolidated material is usually called a beach.
Weather	Weather refers to short-term changes in the temperature, barometric pressure, humidity, precipitation, sunshine, cloud cover, wind direction and speed, and other conditions in the troposphere at a given place and time. Compare with climate.
Shoreline	Shoreline refers to the intersection of a specified plane of water with the shore. All of the water areas of the state, including reservoirs and their associated uplands, together with the lands underlying them.
Situation	The relative geographic location of a site that makes it a good location for a city is a situation.
Annual	Annual refers to plant that grows, sets seed, and dies in one growing season. Compare to perennial.
Population	Group of individual organisms of the same species living within a particular area is referred to as a population.
Growth rate	The net increase in some factor per unit time. In ecology, the growth rate of a population is sometimes measured as the increase in numbers of individuals or biomass per unit time and sometimes as a percentage increase in numbers or biomass per unit time.
Environment	Environment refers to all external conditions and factors, living and nonliving, that affect an organism or other specified system during its lifetime; the earth's life-support systems for us and for all other forms of life-another term for solar capita.
Science	Attempts to discover order in nature and use that knowledge to make predictions about what should happen in nature is called science.
Observations	Information obtained through one or more of the five senses or through instruments that extend the senses are observations.
Hurricane	A cyclonic storm, usually of tropic origin, covering an extensive area, and containing winds in excess of 75 miles per hour is the hurricane.
Mantle	Zone of the earth's interior between its core and its crust. Compare with core, crust is the mantle.
Key	Key refers to a low, insular BANK of sand, coral, etc., as one of the islets off the southern coast of Florida.
Sorting	Sorting refers to a process of selection and separation of sediment grains according to their grain size.
State	State refers to an expression of the internal form of matter. Water exists in three states: solid, liquid, and gas. A solid has a fixed volume and fixed shape; a liquid has a fixed volume but no fixed shape; and a gas has neither fixed volume nor fixed shape.
Geomorphology	Geomorphology refers to that branch of physical geography which deals with the form of the Earth, the general configuration of its surface, the distribution of the land, water, etc. The investigation of the history of geologic changes through the interpretation of topographic forms.
Resources	Resources refer to substances that can be consumed by an organism and, as a result, become unavailable to other organisms.

Geology	Geology refers to the science which treats of the origin, history and structure of the Earth, as recorded in rocks; together with the forces and processes now operating to modify rocks.
Feedback	A kind of system response that occurs when output of the system also serves as input leading to changes in the system is called feedback.
Plate tectonics	Plate tectonics refer to the theory of geophysical processes that explains the movements of lithospheric plates and the processes that occur at their boundaries.
Tectonics	The study of the major structural features of the Earth's crust or the broad structure of a region is referred to as tectonics.
Seismicity	The distribution, frequency, and magnitude of earthquakes in an area are called seismicity.
Radioactivity	Nuclear change in which unstable nuclei of atoms spontaneously shoot out 'chunks' of mass, energy, or both, at a fixed rate is called radioactivity. The three principal types of radioactivity are gamma rays and fast-moving alpha particles and beta particles.
Core	Core refers to a cylindrical sample extracted from a beach or seabed to investigate the types and DEPTHS of sediment layers. An inner, often much less permeable portion of a BREAKWATER, or BARRIER beach.
Lithosphere	Lithosphere refers to outer shell of the earth, composed of the crust and the rigid, outermost part of the mantle outside of the asthenosphere; material found in the earth's plates.
Subduction	Subduction refers to a process in which one lithospheric plate descends beneath another.
Oceanic crust	Oceanic crust refers to the outermost rock shell of the earth, some 5 kilometers thick, that underlies ocean basins; it is composed of basalt and sedimentary layers.
Crust	Solid outer zone of the earth. It consists of oceanic crust and continental crust. Compare core, mantle.

Plate tectonics	Plate tectonics refer to the theory of geophysical processes that explains the movements of lithospheric plates and the processes that occur at their boundaries.
Tectonics	The study of the major structural features of the Earth's crust or the broad structure of a region is referred to as tectonics.
Relief	The difference in elevation between the highest and lowest points in an area is called relief.
Map	Map refers to a representation of Earth's surface usually depicting mostly land areas.
Trench	A long narrow submarine depression with relatively steep sides is called a trench.
Shore	That strip of ground bordering any body of water which is alternately exposed, or covered by tides and/or waves is a shore. A shore of unconsolidated material is usually called a beach.
Zone	Division or province of the ocean with homogeneous characteristics is referred to as a zone.
Mantle	Zone of the earth's interior between its core and its crust. Compare with core, crust is the mantle.
Sediment	Loose, fragments of rocks, minerals or organic material which are transported from their source for varying distances and deposited by air, wind, ice and water are sediment. Other sediments are precipitated from the overlying water or form chemically, in place. Sediment includes all the unconsolidated materials on the sea floor. The fine-grained material deposited by water or wind.
System	A set of components that function and interact in some regular and theoretically predictable manner is called a system.
Subduction	Subduction refers to a process in which one lithospheric plate descends beneath another.
Fold	Fold refers to wavy geologic structures formed by the compression and bending of sedimentary layers.
Compression	A state of stress that causes a pushing together or contraction is called compression.
Transform plate boundary	Transform plate boundary refers to a place where crustal plates shear laterally past one another. Crust is neither produced nor destroyed at this type of junction.
Plates	Various-sized areas of earth's lithosphere that move slowly around with the mantle's flowing asthenosphere are plates. Most earthquakes and volcanoes occur around the boundaries of these plates.
Andesite	Andesite refers to a common volcanic rock found in the volcanic arcs of subduction zones; it is intermediate in composition between the granitic crust of the continents and the basaltic crust of the oceans.
Magma	Molten rock below the earth's surface is referred to as magma.
Depth	Depth refers to vertical distance from still-water level to the bottom.
Crust	Solid outer zone of the earth. It consists of oceanic crust and continental crust. Compare core, mantle.
Wave	An oscillatory movement in a body of water manifested by an alternate rise and fall of the surface is called a wave. Disturbances of the surface of a liquid body, as the ocean, in the form of a ridge, swell or hump. The term wave by itself usually refers to the term surface gravity wave.
Shoreline	Shoreline refers to the intersection of a specified plane of water with the shore. All of the water areas of the state, including reservoirs and their associated uplands, together with the lands underlying them.

Go to **Cram101.com** for the Practice Tests for this Chapter.

Continental shelf	Continental shelf refers to the zone bordering a continent extending from the line of permanent immersion to the DEPTH, usually about 100 m to 200 m, where there is a marked or rather steep descent toward the great depths. The area under active LITTORAL processes during the Holocene period. The region of the oceanic bottom that extends outward from the shoreline with an average slope of less than 1.
Coral	Coral refers to any of more than 6,000 species of small cnidarians, many of which are capable of generating hard calcareous skeletons.
Coastline	Coastline refers to technically, the line that forms the boundary between the COAST and the SHORE. Commonly, the line that forms the boundary between land and the water. The line where terrestrial processes give way to marine processes, TIDAL CURRENTS, wind waves, etc.
Fault	A fracture in rock along which there has been an observable amount of displacement. Faults are rarely single planar units; normally they occur as parallel to sub-parallel sets of planes along which movement has taken place to a greater or lesser extent. Such sets are called fault or fracture-zones.
Gulf	A relatively large portion of sea, partly enclosed by land is called gulf.
Ridge	Ridge refers to the volcanic mountain ranges that lie along the spreading centers on the floors of the oceans.
Climate change	Refers to any long-term trend in MEAN SEA LEVEL, wave height, wind speed, drift rate etc are called the climate change.
Theory	A general explanation of a characteristic of nature consistently supported by observation or experiment is referred to as a theory.
Planet	A smaller, usually nonluminous body orbiting a star is a planet.
Ocean	The great body of salt water which occupies two-thirds of the surface of the Earth, or one of its major subdivisions is called an ocean.
Continental drift	Continental drift refers to the process whereby continents are and have been in motion relative to one another across the Earth's surface.
Science	Attempts to discover order in nature and use that knowledge to make predictions about what should happen in nature is called science.
Geology	Geology refers to the science which treats of the origin, history and structure of the Earth, as recorded in rocks; together with the forces and processes now operating to modify rocks.
Oceanic crust	Oceanic crust refers to the outermost rock shell of the earth, some 5 kilometers thick, that underlies ocean basins; it is composed of basalt and sedimentary layers.
Development	Development refers to change from a society that is largely rural, agricultural, illiterate, and poor, with a rapidly growing population, to one that is mostly urban, industrial, educated, and wealthy, with a slowly growing or stationary population.
Topography	Topography refers to the form of the features of the actual surface of the Earth in a particular region considered collectively.
Seismic	Referring to vibrations in the Earth produced by earthquakes is referred to as seismic.
Transform fault	Transform fault refers to area where earth's lithospheric plates move in opposite but parallel directions along a fracture in the lithosphere. Compare with convergent plate boundary, divergent plate boundary.
Slide	In mass wasting, movement of a descending mass along a plane approximately parallel to the slope of the surface is referred to as slide.

Heat	Total kinetic energy of all the randomly moving atoms, ions, or molecules within a given substance, excluding the overall motion of the whole object. This form of kinetic energy flows from one body to another when there is a temperature difference betwe is referred to as heat.
Radioactivity	Nuclear change in which unstable nuclei of atoms spontaneously shoot out 'chunks' of mass, energy, or both, at a fixed rate is called radioactivity. The three principal types of radioactivity are gamma rays and fast-moving alpha particles and beta particles.
Core	Core refers to a cylindrical sample extracted from a beach or seabed to investigate the types and DEPTHS of sediment layers. An inner, often much less permeable portion of a BREAKWATER, or BARRIER beach.
Convection	Convection refers to the vertical transport of a fluid or the transfer of heat in fluids.
Drag	The resistance to movement of an organism induced by the fluid through which it swims is called drag.
Friction	The resistance to motion of two bodies in contact is a friction.
Fossil	The remains or traces of organisms preserved in rocks or ancient sediment are referred to as fossil.
Catastrophe	Catastrophe refers to a situation or event that causes significant damage to people and property, such that recovery and/or rehabilitation is a long and involved process. Examples of natural catastrophes include hurricanes, volcanic eruptions, large wildfires, and floods.
Flood	Period when tide level is rising; often taken to mean the flood current which occurs during this period. A flow above the CARRYING CAPACITY of a CHANNEL.
Alfred Wegener	German scientist who proposed the theory of continental drift in 1912 is called Alfred Wegener.
Fossils	Skeletons, bones, shells, body parts, leaves, seeds, or impressions of such items that provide recognizable evidence of organisms that lived long ago are called fossils.
Mass	The amount of material in an object is the mass.
Pangaea	Pangaea refers to the megacontinent of the Mesozoic Era that consisted of all of the present-day continents joined together into a single unit.
Rock	Any material that makes up a large, natural, continuous part of earth's crust is rock.
Rocks	An aggregate of one or more minerals rather large in area are rocks. The three classes of rocks are the following: Igneous rock - crystalline rocks formed from molten material. Sedimentary rock - A rock resulting from the consolidation of loose sediment that has accumulated in layers. Metamorphic rock - Rock that has formed from preexisting rock as a result of heat or pressure.
Observations	Information obtained through one or more of the five senses or through instruments that extend the senses are observations.
Physical geology	A large division of geology concerned with earth materials, changes of the surface and interior of the earth, and the forces that cause those changes is referred to as physical geology.
Species	Group of organisms that resemble one another in appearance, behavior, chemical makeup and processes, and genetic structure is a species. Organisms that reproduce sexually are classified as members of the same species only if they can breed with one another and produce offspring.
Point	Point refers to the extreme end of a cape, or the outer end of any land area protruding into

the water, usually less prominent than a cape. A low profile shoreline promontory of more or less triangular shape, the top of which extends seaward.

Continent	Continent refers to lower-density masses of rock, exposed as about 40 percent of the Earth's surface: 29 percent as land and I 1 percent as the floor of shallow seas.
Migration	A term that refers to the habit of some animals is a migration.
Plants	Eukaryotic, mostly multicelled organisms such as algae, mosses, ferns, flowers, cacti, grasses, beans, wheat, rice, and trees are plants. These organisms use photosynthesis to produce organic nutrients for themselves and for other organisms is referred to as plants.
Coast	A strip of land of indefinite length and width that extends from the SEASHORE inland to the first major change in terrain features is referred to as coast.
Cape	Cape refers to a relatively extensive land area jutting seaward from a continent or large island which prominently marks a change in, or interrupts notably, the coastal trend; a prominent feature.
Igneous rocks	Igneous rocks refers to rocks formed from the solidification of magma. They are extrusive if they crystallize on the surface of the Earth and intrusive if they crystallize beneath the surface.
Latitude	Latitude refers to distance from the equator. Compare altitude.
Sea	Sea refers to the ocean. A large body of salt water, second in rank to an ocean, more or less landlocked and generally part of, or connected with, an ocean or a larger sea. State of the ocean or lake surface, in regard to waves.
Accumulation	Buildup of matter, energy, or information in a system is referred to as accumulation.
Glacier	Large masses of moving ice on land derived by the recrystallization of snow into ice under pressure are referred to as glacier.
Coal	Solid, combustible mixture of organic compounds with 30-98% carbon by weight, mixed with various amounts of water and small amounts of sulfur and nitrogen compounds. It is formed in several stages as the remains of plants are subjected to heat and press is a coal.
Granite	A light-colored, coarsegrained, intrusive igneous rock composed mainly of quartz and feldspar and that typifies the continental crust are called the granite.
Lithosphere	Lithosphere refers to outer shell of the earth, composed of the crust and the rigid, outermost part of the mantle outside of the asthenosphere; material found in the earth's plates.
Hypothesis	In science, an explanation set forth in a manner that can be tested and is capable of being disproved. A tested hypothesis is accepted until and unless it has been disproved.
Evaporite	A type of sediment precipitated from an aqueous solution, usually by the evaporation of water from a basin with restricted circulation is the evaporite.
Desert	Desert refers to biome in which evaporation exceeds precipitation and the average amount of precipitation is less than 25 centimeters a year. Such areas have little vegetation or have widely spaced, mostly low vegetation. Compare forest, grassland.
Coral reef	Formation produced by massive colonies containing billions of tiny coral animals, called polyps, that secrete a stony substance around themselves for protection. When the corals die, their empty outer skeletons form layers that caus is referred to as coral reef.
Reef	A ridge of rock or other material lying just below the surface of the sea is called a reef.
Sand	Sand refers to an unconsolidated mixture of inorganic soil consisting of small but easily

distinguishable grains ranging in size from about.062 mm to 2.0 mm.

Gypsum	An evaporite deposit composed of hydrous calcium sulfat is a gypsum.
Technology	Technology refers to the creation of new products and processes intended to improve our efficiency, chances for survival, comfort level, and quality of life. Compare with science.
Paleomagnetism	The 'fossil,' or remanent, magnetic field of a rock is called paleomagnetism.
Valley	An elongated depression, usually with an outlet, between bluffs or between ranges of hills or mountains is a valley.
Rift valley	Rift valley refers to the fault-bounded valley found along the crest of many ocean ridges; it is created by tensional stresses that accompany the process of sea-floor spreading.
Tension	A state of stress that tends to pull the body apart is called tension.
Range	Land used for grazing is referred to as the range.
Continental crust	The light, buoyant granitic rock that underlies continental masses and averages about 35 kilometers in thickness is called continental crust.
Dredging	Excavation or displacement of the bottom or SHORELINE of a water body. Dredging can be accomplished with mechanical or hydraulic machines. Most is done to maintain channel depths or berths for navigational purposes; other dredging is for shellfish harvesting or for cleanup of polluted sediments.
Basalt	Basalt refers to a dark, fine-grained igneous rock composed of minerals enriched in ferromagnesian silicates.
Seafloor spreading	Seafloor spreading refers to the theory that new ocean crust forms at spreading centers, most of which are on the ocean floor, and pushes the continents aside. Power is thought to be provided by convection currents in Earth's upper mantle.
Convection current	A single closedflow circuit of rising warm material and falling cool material is a convection current.
Current	Current refers to the flowing of water, or other liquid or gas. That portion of a stream of water which is moving with a velocity much greater than the average or in which the progress of the water is principally concentrated. Ocean currents can be classified in a number of different ways.
Magnetic field	A region where magnetic forces affect any magnetized bodies or electric currents is the magnetic field. Earth is surrounded by a magnetic field.
Bar	An offshore ridge or mound of sand, gravel, or other unconsolidated material which is submerged, especially at the mouth of a river or estuary, or lying parallel to, and a short distance from, the beach is referred to as a bar.
Magnetic pole	Magnetic pole refers to the point where the Earth's magnetic field flows back into the ground. Currently, this point is near the North Pole.
Magnetism	Magnetism refers to a group of physical phenomena associated with moving electricity.
Mineral	A naturally occurring, inorganic, crystalline solid that has a definite chemical composition and possesses characteristic physical properties is a mineral.
Sedimentary rocks	Sedimentary rocks refer to rocks that have formed by the compaction and cementation of sediment.
Temperature	Temperature refers to a measure of the average speed of motion of the atoms, ions, or molecules in a substance or combination of substances at a given moment. Compare with heat.

Go to **Cram101.com** for the Practice Tests for this Chapter.

Force	Force refers to a push or pull that affects motion. The product of mass and acceleration of a material.
Volcanic rock	Rock formed by solidification of magma at the Earth's surface is a volcanic rock.
Paleomagnetic time scale	A detailed chronology of the history of the polarity reversals of the earth's magnetic field is called the paleomagnetic time scale.
Pleistocene	An epoch of the Quaternary Period characterized by several glacial ages is referred to as the Pleistocene.
Recent	A synonym of Holocene is called recent.
Ocean basin	Ocean basin refers to deep-ocean floor made of basaltic crust. Compare with continental margin.
Basin	A large submarine depression of a generally circular, elliptical or oval shape is called a basin.
Offshore	Offshore in beach terminology refers to the comparatively flat zone of variable width, extending from the shoreface to the edge of the continental shelf. It is continually submerged. The direction seaward from the shore. The zone beyond the nearshore zone where sediment motion induced by waves alone effectively ceases and where the influence of the sea bed on wave action is small in comparison with the effect of wind. The breaker zone directly seaward of the low tide line.
Bore	A steep wave that moves upriver during the flooding tide is the bore.
Oceanography	That science treating of the oceans, their forms, physical features and phenomena is referred to as oceanography.
Red clay	Red clay refers to a descriptive term applied to pelagic or abyssal clay deposits of the deep sea; they range in color from red to brown and tend to accumulate slowly in the deepest and remotest parts of the oceans far from the influx of other types of sediments.
Clay	Clay refers to a fine grained sediment with a typical grain size less than 0.004 mm. Possesses electromagnetic properties which bind the grains together to give a bulk strength or cohesion.
Plankton	Plankton refers to small plant organisms and animal organisms that float in aquatic ecosystems. Compare with benthos, nekton.
Upwelling	The process by which water rises from a deeper to a shallower depths, usually as a result of offshore surface water flow is called upwelling. It is most prominent where persistent wind blows parallel to a coastline so that the resultant Ekman transport moves surface water away from the coast.
Chalk	A white, soft limestone consisting dominantly of the shells of foraminifera is a chalk.
Load	The quantity of sediment transported by a current. It includes the suspended load of small particles in the water, and the bedload of large particles that move along the bottom.
Earthquake	Shaking of the ground resulting either from the fracturing and displacement of rock, producing a fault, or from subsequent movement along the fault is the earthquake.
Subduction zone	Elongate region in which the sea floor slides beneath a continent or island arc is referred to as the subduction zone.
Accretion	The accumulation of sediment, deposited by natural fluid flow processes is accretion.
Passive margin	Passive margin refers to a continental margin near an area of lithospheric plate divergence.
Well	Well refers to a hole, generally cylindrical and usually walled or lined with pipe, that is

Go to **Cram101.com** for the Practice Tests for this Chapter.

dug or drilled into the ground to penetrate an aquifer below the zone of saturation.

Divergent plate boundary

Divergent plate boundary refers to area where earth's lithospheric plates move apart in opposite directions. Compare convergent plate boundary, transform fault.

Site

A factor considering the summation of all environmental features of a location that influences the placement of a city is a site.

Fissure

A narrow parting or crack in rock are called the fissure.

Continental margin

The drowned edges of continents consisting of the continental shelf, the continental slope, and the continental rise is called continental margin.

Metamorphism

The changes in minerals and rock textures that occur with the elevated temperatures and pressures below the Earth's surface are referred to as metamorphism.

Convergent plate boundary

Area where earth's lithospheric plates are pushed togethe is called convergent plate boundary.

Asthenosphere

Earth materials located at a depth of 100 km to 700 km below the Earth's surface, underneath the tectonic plates is called the asthenosphere.

Rhyolite

Rhyolite refers to a volcanic rock typical of continents. Typically forms from high viscosity magma.

Density

Density refers to the ratio of a mass to a unit volume specified as grams per cubic centimeter.

Convergence

Resemblance among species belonging to different taxonomic groups as the result from adaptation to similar environments is the convergence.

Topographic map

A map on which elevations are shown by means of contour lines is referred to as a topographic

Fracture zone

A linear zone of highly irregular, faulted topography that is oriented perpendicular to ocean-spreading ridges is referred to as the fracture zone.

Longitude

The angular distance to the east or west of the prime meridian that runs through Greenwich, England are called the longitude.

Situation

The relative geographic location of a site that makes it a good location for a city is a situation.

Oscillation

Oscillation refers to a periodic motion backward and forward. To vibrate or vary above and below a mean value.

Cell

Smallest living unit of an organism. Each cell is encased in an outer membrane or wall and contains genetic material and other parts to perform its life function. Organisms such as bacteria consist of only one cell, but most of the organisms we are.

Energy

Capacity to do work by performing mechanical, physical, chemical, or electrical tasks or to cause a heat transfer between two objects at different temperatures is an energy.

Reach

Reach refers to an arm of the ocean extending into the land. A straight section of restricted waterway of considerable extent; may be similar to a narrows, except much longer in extent.

Thermal equilibrium

The condition in which the total heat coming into a system is balanced by the total heat leaving the system is referred to as thermal equilibrium.

Equilibrium

A point of rest. A system that does not tend to undergo any change of its own accord but remains in a single, fixed condition is said to be in equilibrium. Compare with steady state.

Magnetometer

Magnetometer refers to a device that measures the amount and direction of residual magnetism in a rock sample.

343

Degree	An arbitrary measure of temperature. One degree Celsius _ 1.8 degrees Fahrenheit.
Gravity	The attraction between bodies of matter is a gravity.
Shear	The failure of a body where the mass on one side slides past the portion on the other side is a shear.
Base	A substance that combines with a hydrogen ion in solution is called the base.
Mesosphere	Mesosphere refers to the third layer of the atmosphere; found above the stratosphere. Compare with stratosphere, thermosphere, troposphere.
Front	The boundary between two air masses with different temperatures and densitie is referred to as front.
Lava	Molten rock that is extruded out of volcanoes is called lava.
Lake	Large natural body of standing fresh water formed when water from precipitation, land runoff, or groundwater flow fills a depression in the earth created by glaciation, earth movement, volcanic activity, or a giant meteorit are called the lake.
Crater	Crater refers to an abrupt basin commonly rimmed by ejected material. In volcanoes, craters form by outwarcc explosion, are commonly less than 2 km diameter, and occur at the summit of a volcanic cone. Similar rimmed basins form by impacts with meteorites, asteroids, and.
Elevation	Elevation refers to the distance of a point above a specified surface of constant potential; the distance is measured along the direction of gravity between the point and the surface.
Key	Key refers to a low, insular BANK of sand, coral, etc., as one of the islets off the southern coast of Florida.
Magnetic anomaly	A disturbance of the Earth's magnetic field created by magnetized rock in the Earth's crust is referred to as magnetic anomaly.
Community	Community refers to populations of all species living and interacting in an area at a particular time.
Cross section	A two-dimensional drawing showing features in the vertical plane as in a canvon wall or road cut is referred to as cross section.
Boulder	A rounded rock on a beach, greater than 256 mm in diameter, larger than a cobbl is a boulder.
Resources	Resources refer to substances that can be consumed by an organism and, as a result, become unavailable to other organisms.
Feedback	A kind of system response that occurs when output of the system also serves as input leading to changes in the system is called feedback.

Seismicity	The distribution, frequency, and magnitude of earthquakes in an area are called seismicity.
Plate tectonics	Plate tectonics refer to the theory of geophysical processes that explains the movements of lithospheric plates and the processes that occur at their boundaries.
Tectonics	The study of the major structural features of the Earth's crust or the broad structure of a region is referred to as tectonics.
Earthquake	Shaking of the ground resulting either from the fracturing and displacement of rock, producing a fault, or from subsequent movement along the fault is the earthquake.
Fault	A fracture in rock along which there has been an observable amount of displacement. Faults are rarely single planar units; normally they occur as parallel to sub-parallel sets of planes along which movement has taken place to a greater or lesser extent. Such sets are called fault or fracture-zones.
Lithosphere	Lithosphere refers to outer shell of the earth, composed of the crust and the rigid, outermost part of the mantle outside of the asthenosphere; material found in the earth's plates.
Planet	A smaller, usually nonluminous body orbiting a star is a planet.
Tectonic forces	Forces generated from within the earth that result in uplift, movement, or deformations of part of the earth's crust are tectonic forces.
Plates	Various-sized areas of earth's lithosphere that move slowly around with the mantle's flowing asthenosphere are plates. Most earthquakes and volcanoes occur around the boundaries of these plates.
Subduction	Subduction refers to a process in which one lithospheric plate descends beneath another.
Network	Network refers to a set consisting of stations for which geometric relationships have been determined and which are so related that removal of one station from the set will affect the relationships between the other stations; and lines connecting the stations to show this interdependence.
Seismic	Referring to vibrations in the Earth produced by earthquakes is referred to as seismic.
Depth	Depth refers to vertical distance from still-water level to the bottom.
Magnitude	An assessment of the size of an event is a magnitude. Magnitude scales exist for earthquakes, volcanic eruptions. hurricanes, and tornadoes. For earthquakes, different magnitudes are calculated for the same earthquake when different types of seismic waves are used.
Seismic waves	A long-period wave caused by an underwater seismic disturbance or volcanic eruption is a seismic waves.
Base	A substance that combines with a hydrogen ion in solution is called the base.
Power	Power refers to the time rate of doing work.
Event	Event refers to an occurrence meeting specified conditions, e.g. damage, a threshold wave height or a threshold water level.
Energy	Capacity to do work by performing mechanical, physical, chemical, or electrical tasks or to cause a heat transfer between two objects at different temperatures is an energy.
Rocks	An aggregate of one or more minerals rather large in area are rocks. The three classes of rocks are the following: Igneous rock - crystalline rocks formed from molten material. Sedimentary rock - A rock resulting from the consolidation of loose sediment that has accumulated in layers. Metamorphic rock - Rock that has formed from preexisting rock as a result of heat or pressure.

Rock	Any material that makes up a large, natural, continuous part of earth's crust is rock.
Uplift	The rising of one part of the Earth's crust relative to another part is referred to as uplift.
Subsidence	Sinking or down warping of a part of the earth's surface is called subsidence.
Risk	Risk refers to the probability that something undesirable will happen from deliberate or accidental exposure.
Transform fault	Transform fault refers to area where earth's lithospheric plates move in opposite but parallel directions along a fracture in the lithosphere. Compare with convergent plate boundary, divergent plate boundary.
Zone	Division or province of the ocean with homogeneous characteristics is referred to as a zone.
Mantle	Zone of the earth's interior between its core and its crust. Compare with core, crust is the mantle.
Core	Core refers to a cylindrical sample extracted from a beach or seabed to investigate the types and DEPTHS of sediment layers. An inner, often much less permeable portion of a BREAKWATER, or BARRIER beach.
Asthenosphere	Earth materials located at a depth of 100 km to 700 km below the Earth's surface, underneath the tectonic plates is called the asthenosphere.
Upwelling	The process by which water rises from a deeper to a shallower depths, usually as a result of offshore surface water flow is called upwelling. It is most prominent where persistent wind blows parallel to a coastline so that the resultant Ekman transport moves surface water away from the coast.
Downwelling	A downward movement of surface water caused by onshore Ekman transport, converging CURRENTS or when a water mass becomes more dense than the surrounding water is called downwelling.
Convection	Convection refers to the vertical transport of a fluid or the transfer of heat in fluids.
Magnetic field	A region where magnetic forces affect any magnetized bodies or electric currents is the magnetic field. Earth is surrounded by a magnetic field.
Elastic	Elastic refers to behavior of material where stress causes deformation that is recoverable; when stress stops, the material returns to its original state.
Theory	A general explanation of a characteristic of nature consistently supported by observation or experiment is referred to as a theory.
Experiment	Experiment refers to procedure a scientist uses to study some phenomenon under known conditions. Some experiments are conducted in the laboratory, but others are conducted in nature. The resulting scientific data or facts must be verified or confirmed by repeated observation.
Precision	Precision refers to a measure of reproducibility, or how closely a series of measurements of the same quantity agree with one another. Compare with accuracy.
Strain	A change in torn or size of a body due to external forces is called strain.
Point	Point refers to the extreme end of a cape, or the outer end of any land area protruding into the water, usually less prominent than a cape. A low profile shoreline promontory of more or less triangular shape, the top of which extends seaward.
Epicenter	The point on the earth's surface directly above the spot in the earth where the earthquake originated is referred to as epicenter.
Seismograph	Seismograph refers to an instrument that detects and records earth movement associated with

	earthquakes and other disturbances.
Wave	An oscillatory movement in a body of water manifested by an alternate rise and fall of the surface is called a wave. Disturbances of the surface of a liquid body, as the ocean, in the form of a ridge, swell or hump. The term wave by itself usually refers to the term surface gravity wave.
Shear	The failure of a body where the mass on one side slides past the portion on the other side is a shear.
Slide	In mass wasting, movement of a descending mass along a plane approximately parallel to the slope of the surface is referred to as slide.
Range	Land used for grazing is referred to as the range.
Seismic wave	Seismic wave refers to a low-frequency wave generated by the forces that cause earthquakes. Some kinds of seismic waves can pass through Earth.
P wave	P wave refers to first seismic wave to reach a seismometer. Movement is by alternating push-pull pulses that travel through solid, liquid, and gas.
Disturbance	Disturbance refers to a discrete event in time that disrupts an ecosystem or community. Examples of natural disturbances include fires, hurricanes, tornadoes, droughts, and floods. Examples of humancaused disturbances include deforestation, overgrazing, and plowing.
S wave	S wave refers to the second seismic wave to arrive at the seismometer. S waves move through solids only.
Ocean	The great body of salt water which occupies two-thirds of the surface of the Earth, or one of its major subdivisions is called an ocean.
Crust	Solid outer zone of the earth. It consists of oceanic crust and continental crust. Compare core, mantle.
Degree	An arbitrary measure of temperature. One degree Celsius _ 1.8 degrees Fahrenheit.
Amplitude	Half of the peak-to-trough range of a wave is the amplitude.
Unconsolidated	Unconsolidated with regards to sediment grains refers to loose, separate, or unattached to one another.
Sediment	Loose, fragments of rocks, minerals or organic material which are transported from their source for varying distances and deposited by air, wind, ice and water are sediment. Other sediments are precipitated from the overlying water or form chemically, in place. Sediment includes all the unconsolidated materials on the sea floor. The fine-grained material deposited by water or wind.
Bedrock	Solid rock lying beneath loose soil or unconsolidated sediment is called bedrock.
Observations	Information obtained through one or more of the five senses or through instruments that extend the senses are observations.
Richter scale	A logarithmic measure of earthquake magnitude is called a Richter scale. A great earthquake measures above 8 on the Richter scale.
Recent	A synonym of Holocene is called recent.
Dam	Dam refers to structure built in rivers or estuaries, basically to separate water at both sides and/or to retain water at one side.
Population	Group of individual organisms of the same species living within a particular area is referred to as a population.

Go to **Cram101.com** for the Practice Tests for this Chapter.

Probability	A mathematical statement about how likely it is that something will happen is probability.
Well	Well refers to a hole, generally cylindrical and usually walled or lined with pipe, that is dug or drilled into the ground to penetrate an aquifer below the zone of saturation.
Soil	Soil refers to a layer of weathered, unconsolidated material on top of bedrock; often also defined as containing organic matter and being capable of supporting plant growth.
Transform plate boundary	Transform plate boundary refers to a place where crustal plates shear laterally past one another. Crust is neither produced nor destroyed at this type of junction.
Fire	The rapid combination of oxygen with organic material to produce flame, heat, and light is referred to as the fire.
Map	Map refers to a representation of Earth's surface usually depicting mostly land areas.
Sand	Sand refers to an unconsolidated mixture of inorganic soil consisting of small but easily distinguishable grains ranging in size from about.062 mm to 2.0 mm.
Compaction	The decrease in volume and porosity of a sediment via burial is referred to as compaction.
Duration	In forecasting waves, the length of time the wind blows in essentially the same direction over the FETCH is a duration.
Terrestrial	Terrestrial refers to pertaining to land. Compare with aquatic.
Coast	A strip of land of indefinite length and width that extends from the SEASHORE inland to the first major change in terrain features is referred to as coast.
Slumping	The sliding of large, cohering blocks of sediment or rock downslope under the influence of gravity is called slumping.
Shoreline	Shoreline refers to the intersection of a specified plane of water with the shore. All of the water areas of the state, including reservoirs and their associated uplands, together with the lands underlying them.
Oil	The liquid form of petroleum consisting of a complex mixture of large hydrocarbon molecules is referred to as oil.
Harbor	A water area nearly surrounded by land, sea walls, BREAKWATERS or artificial dikes, forming a safe anchorage for ships is referred to as harbor.
Reverse fault	A dip-slip fault where the upper fault block has moved upward in response to compressional stresses is referred to as a reverse fault.
Reach	Reach refers to an arm of the ocean extending into the land. A straight section of restricted waterway of considerable extent; may be similar to a narrows, except much longer in extent.
Valley	An elongated depression, usually with an outlet, between bluffs or between ranges of hills or mountains is a valley.
Bay	Bay refers to a recess or inlet in the shore of a sea or lake between two capes or headlands, not as large as a gulf but larger than a cove.
Dock	The slip or waterway between two piers, or cut into the land, for the reception of ships is called dock.
Natural gas	Underground deposits of gases consisting of 50-90% by weight methane gas and small amounts of heavier gaseous hydrocarbon compounds such as propane and butane are called natural gas.
Port	Port refers to a place where vessels may discharge or receive cargo.
Strand	Strand refers to shore or beach of the ocean or a large lake. The land bordering any large body of water, especially a sea or an arm of the ocean.

Go to **Cram101.com** for the Practice Tests for this Chapter.

Ridge	Ridge refers to the volcanic mountain ranges that lie along the spreading centers on the floors of the oceans.
Subduction zone	Elongate region in which the sea floor slides beneath a continent or island arc is referred to as the subduction zone.
Trench	A long narrow submarine depression with relatively steep sides is called a trench.
Granite	A light-colored, coarsegrained, intrusive igneous rock composed mainly of quartz and feldspar and that typifies the continental crust are called the granite.
Shallow water	Shallow water refers to water of such depths that surface waves are noticeably affected by bottom topography. Typically this implies water depths equivalent to less than half the wave length.
Heave	Heave refers to the vertical rise or fall of the waves or the sea. The translational movement of a craft parallel to its vertical axis. The net transport of a floating body resulting from wave action.
Tsunami	Tsunami refers to a large, high-velocity wave generated by displacement of the sea floor; also called seismic sea wave. Commonly misnamed tidal wave.
System	A set of components that function and interact in some regular and theoretically predictable manner is called a system.
Desert	Desert refers to biome in which evaporation exceeds precipitation and the average amount of precipitation is less than 25 centimeters a year. Such areas have little vegetation or have widely spaced, mostly low vegetation. Compare forest, grassland.
Radar	An instrument for determining the distance and direction to an object by measuring the time needed for radio signals to travel from the instrument to the object and back, and by measuring the angle through which the instrument's antenna has traveled is referred to as radar.
Oscillation	Oscillation refers to a periodic motion backward and forward. To vibrate or vary above and below a mean value.
Magma	Molten rock below the earth's surface is referred to as magma.
Sea ice	Frozen seawater as opposed to glacial ice on land is referred to as sea ice.
Inlet	Inlet refers to a narrow strip of water running into the land or between islands. An arm of the sea that is long compared to its width, and that may extend a considerable distance inland.
Chalk	A white, soft limestone consisting dominantly of the shells of foraminifera is a chalk.
Matter	Matter refers to anything that has mass and takes up space. On earth, where gravity is present, we weigh an object to determine its mass.
Debris	Debris refers to any accumulation of rock fragments; detritus.
Avalanche	A large mass of snow, ice, soil, or rock that moves rapidly downslope under the pull of gravity is referred to as an avalanche.
Flooding	The natural process whereby waters emerge from their stream channel to cover part of the floodplain. Natural flooding is not a problem until people choose to build homes and other structures on floodplains.
Tides	The periodic rise and fall of the Earth's water surface as a consequence of the gravitational attraction of the Moon and the Sun, which are called tides.
Weather	Weather refers to short-term changes in the temperature, barometric pressure, humidity,

Go to **Cram101.com** for the Practice Tests for this Chapter.

	precipitation, sunshine, cloud cover, wind direction and speed, and other conditions in the troposphere at a given place and time. Compare with climate.
Water table	The upper surface of a zone of saturation, where the body of groundwater is not confined by an overlying impermeable formation is a water table. Where an overlying confining formation exists, the aquifer in question has no water table.
Site	A factor considering the summation of all environmental features of a location that influences the placement of a city is a site.
Host	Plant or animal on which a parasite feeds is referred to as host.
Elevation	Elevation refers to the distance of a point above a specified surface of constant potential; the distance is measured along the direction of gravity between the point and the surface.
Hypothesis	In science, an explanation set forth in a manner that can be tested and is capable of being disproved. A tested hypothesis is accepted until and unless it has been disproved.
Stress	Stress refers to force per unit area. May be compression, tension, or shear.
Threshold	Threshold refers to a point in the operation of a system at which a change occurs. With respect to toxicology, it is a level below which effects are not observable and above which effects become apparent.
Accumulation	Buildup of matter, energy, or information in a system is referred to as accumulation.
Relief	The difference in elevation between the highest and lowest points in an area is called relief.
Monitoring	Monitoring refers to the process of collecting data on a regular basis at specific sites to provide a database from which to evaluate change.
Response	The amount of health damage caused by exposure to a certain dose of a harmful substance or form of radiation is a response.
Community	Community refers to populations of all species living and interacting in an area at a particular time.
Zoning	Regulating how various parcels of land can be used is zoning.
Development	Development refers to change from a society that is largely rural, agricultural, illiterate, and poor, with a rapidly growing population, to one that is mostly urban, industrial, educated, and wealthy, with a slowly growing or stationary population.
Mud	Mud refers to a mixture of silt and clay sized particles.
Pebbles	Beach material usually well-rounded and between about 4 mm to 64 mm diameter are pebbles.
Continent	Continent refers to lower-density masses of rock, exposed as about 40 percent of the Earth's surface: 29 percent as land and I 1 percent as the floor of shallow seas.
Island arc	A curved group of volcanic islands associated with a deep oceanic trench and subduction zone is called island arc.
Current	Current refers to the flowing of water, or other liquid or gas. That portion of a stream of water which is moving with a velocity much greater than the average or in which the progress of the water is principally concentrated. Ocean currents can be classified in a number of different ways.
Rift valley	Rift valley refers to the fault-bounded valley found along the crest of many ocean ridges; it is created by tensional stresses that accompany the process of sea-floor spreading.
Fracture	Fracture refers to a general term for any breaks in rock. Fractures include faults, joints,

and crack,_.

Concentration	Amount of a chemical in a particular volume or weight of air, water, soil, or other medium is referred to as concentration.
Convergence	Resemblance among species belonging to different taxonomic groups as the result from adaptation to similar environments is the convergence.
Topography	Topography refers to the form of the features of the actual surface of the Earth in a particular region considered collectively.
Front	The boundary between two air masses with different temperatures and densitie is referred to as front.
Mass	The amount of material in an object is the mass.
Density	Density refers to the ratio of a mass to a unit volume specified as grams per cubic centimeter.
Refraction	The process by which the direction of a wave moving in shallow water at an angle to the bottom contours is changed is refraction. The part of the wave moving shoreward in shallower water travels more slowly than that portion in deeper water, causing the wave to turn or bend to become parallel to the contour.
Wave velocity	Speed at which the individual waveform advances, defined as the wavelength divided by the wave period is called wave velocity.
Chemical	One of the millions of different elements and compounds found naturally or synthesized by human is referred to as chemical.
Accuracy	Accuracy refers to the extent to which a measurement agrees with the accepted or correct value for that quantity, based on careful measurements by many people over a long time. Compare to precision.
Discontinuity	Discontinuity refers to a marked or abrupt change in the property of a substance, such as water temperature, salinity, or density. Or a contact between different substances, such as the air-sea interface or the Moho, which separates the crust from the mantle.
Mohorovicic discontinuity	A compositional and density discontinuity marking the interface between the rocks of the crust and the mantle is a Mohorovicic discontinuity.
Continental crust	The light, buoyant granitic rock that underlies continental masses and averages about 35 kilometers in thickness is called continental crust.
Oceanic crust	Oceanic crust refers to the outermost rock shell of the earth, some 5 kilometers thick, that underlies ocean basins; it is composed of basalt and sedimentary layers.
Mineral	A naturally occurring, inorganic, crystalline solid that has a definite chemical composition and possesses characteristic physical properties is a mineral.
Ductile	Ductile refers to behavior of material where stress causes permanent flow or strain.
Key	Key refers to a low, insular BANK of sand, coral, etc., as one of the islets off the southern coast of Florida.
Lower mantle	The rigid portion of Earth's mantle below the asthenosphere are called the lower mantle.
State	State refers to an expression of the internal form of matter. Water exists in three states: solid, liquid, and gas. A solid has a fixed volume and fixed shape; a liquid has a fixed volume but no fixed shape; and a gas has neither fixed volume nor fixed shape.
Temperature	Temperature refers to a measure of the average speed of motion of the atoms, ions, or molecules in a substance or combination of substances at a given moment. Compare with heat.

Go to **Cram101.com** for the Practice Tests for this Chapter.

Heat	Total kinetic energy of all the randomly moving atoms, ions, or molecules within a given substance, excluding the overall motion of the whole object. This form of kinetic energy flows from one body to another when there is a temperature difference betwe is referred to as heat.
Chart	A map that depicts mostly water and the adjoining land areas is referred to as chart.
Thermal	The energy of the random motion of atoms and molecules is referred to as thermal.
Fall	Fall refers to a mass moving nearly vertical and downward under the influence of gravity.
Plume	Plume refers to an arm of magna rising upward from the mantle.
Cross section	A two-dimensional drawing showing features in the vertical plane as in a canvon wall or road cut is referred to as cross section.
Brittle	Behavior of material where stress causes abrupt fracture are called the brittle.
Dip	The angle of inclination measured in degrees from the horizontal is called dip.
Convergent plate boundary	Area where earth's lithospheric plates are pushed togethe is called convergent plate boundary.
Science	Attempts to discover order in nature and use that knowledge to make predictions about what should happen in nature is called science.
River	A natural stream of water larger than a brook or creek is a river.
Geophysics	The study of the physical characteristics and properties of the Earth is a geophysics.
Geology	Geology refers to the science which treats of the origin, history and structure of the Earth, as recorded in rocks; together with the forces and processes now operating to modify rocks.
Resources	Resources refer to substances that can be consumed by an organism and, as a result, become unavailable to other organisms.
Feedback	A kind of system response that occurs when output of the system also serves as input leading to changes in the system is called feedback.

Plate	One of about a dozen rigid segments of Earth's lithosphere that move independently is a plate. The plate consists of continental or oceanic crust and the cool, rigid upper mantle directly below the crust.
Ocean	The great body of salt water which occupies two-thirds of the surface of the Earth, or one of its major subdivisions is called an ocean.
Resources	Resources refer to substances that can be consumed by an organism and, as a result, become unavailable to other organisms.
Radar	An instrument for determining the distance and direction to an object by measuring the time needed for radio signals to travel from the instrument to the object and back, and by measuring the angle through which the instrument's antenna has traveled is referred to as radar.
Ridge	Ridge refers to the volcanic mountain ranges that lie along the spreading centers on the floors of the oceans.
System	A set of components that function and interact in some regular and theoretically predictable manner is called a system.
Planet	A smaller, usually nonluminous body orbiting a star is a planet.
Rift	Rift refers to the valley created at a pull-apart zone.
Basalt	Basalt refers to a dark, fine-grained igneous rock composed of minerals enriched in ferromagnesian silicates.
Mantle	Zone of the earth's interior between its core and its crust. Compare with core, crust is the mantle.
Rocks	An aggregate of one or more minerals rather large in area are rocks. The three classes of rocks are the following: Igneous rock - crystalline rocks formed from molten material. Sedimentary rock - A rock resulting from the consolidation of loose sediment that has accumulated in layers. Metamorphic rock - Rock that has formed from preexisting rock as a result of heat or pressure.
Crust	Solid outer zone of the earth. It consists of oceanic crust and continental crust. Compare core, mantle.
Sea	Sea refers to the ocean. A large body of salt water, second in rank to an ocean, more or less landlocked and generally part of, or connected with, an ocean or a larger sea. State of the ocean or lake surface, in regard to waves.
Rift valley	Rift valley refers to the fault-bounded valley found along the crest of many ocean ridges; it is created by tensional stresses that accompany the process of sea-floor spreading.
Valley	An elongated depression, usually with an outlet, between bluffs or between ranges of hills or mountains is a valley.
Lake	Large natural body of standing fresh water formed when water from precipitation, land runoff, or groundwater flow fills a depression in the earth created by glaciation, earth movement, volcanic activity, or a giant meteorit are called the lake.
Fracture	Fracture refers to a general term for any breaks in rock. Fractures include faults, joints, and crack,_.
Natural gas	Underground deposits of gases consisting of 50-90% by weight methane gas and small amounts of heavier gaseous hydrocarbon compounds such as propane and butane are called natural gas.
Oceanic crust	Oceanic crust refers to the outermost rock shell of the earth, some 5 kilometers thick, that underlies ocean basins; it is composed of basalt and sedimentary layers.

Go to **Cram101.com** for the Practice Tests for this Chapter.

Plates	Various-sized areas of earth's lithosphere that move slowly around with the mantle's flowing asthenosphere are plates. Most earthquakes and volcanoes occur around the boundaries of these plates.
Swell	Waves that have traveled a long distance from their generating area and have been sorted out by travel into long waves of the same approximate period are referred to as a swell.
Zone	Division or province of the ocean with homogeneous characteristics is referred to as a zone.
Lithosphere	Lithosphere refers to outer shell of the earth, composed of the crust and the rigid, outermost part of the mantle outside of the asthenosphere; material found in the earth's plates.
Sediment	Loose, fragments of rocks, minerals or organic material which are transported from their source for varying distances and deposited by air, wind, ice and water are sediment. Other sediments are precipitated from the overlying water or form chemically, in place. Sediment includes all the unconsolidated materials on the sea floor. The fine-grained material deposited by water or wind.
Decompression melting	The most common process creating magma is by reducing pressure on hot rock, not by adding more heat is a decompression melting.
Magma	Molten rock below the earth's surface is referred to as magma.
Metamorphism	The changes in minerals and rock textures that occur with the elevated temperatures and pressures below the Earth's surface are referred to as metamorphism.
Sedimentary rocks	Sedimentary rocks refer to rocks that have formed by the compaction and cementation of sediment.
Continental margin	The drowned edges of continents consisting of the continental shelf, the continental slope, and the continental rise is called continental margin.
Divergent plate boundary	Divergent plate boundary refers to area where earth's lithospheric plates move apart in opposite directions. Compare convergent plate boundary, transform fault.
Ocean basin	Ocean basin refers to deep-ocean floor made of basaltic crust. Compare with continental margin.
Basin	A large submarine depression of a generally circular, elliptical or oval shape is called a basin.
Rock	Any material that makes up a large, natural, continuous part of earth's crust is rock.
Observations	Information obtained through one or more of the five senses or through instruments that extend the senses are observations.
Sonar	Sonar refers to an acronym for sound navigation and ranging; an instrument used to locate objects underwater by reflecting sound waves.
Relief	The difference in elevation between the highest and lowest points in an area is called relief.
Dredge	A metal collar and collecting bag that is dragged along the bottom to sample rock, sediment, or bottom organisms is referred to as the dredge.
Topography	Topography refers to the form of the features of the actual surface of the Earth in a particular region considered collectively.
Core	Core refers to a cylindrical sample extracted from a beach or seabed to investigate the types and DEPTHS of sediment layers. An inner, often much less permeable portion of a BREAKWATER, or BARRIER beach.

Go to **Cram101.com** for the Practice Tests for this Chapter.

Seismic reflection	Seismic reflection refers to the return of part of the energy of seismic waves to the earth's surface after the waves bounce off a rock boundary.
Reflection	The process by which the energy of the wave is returned seaward is a reflection.
Paleomagnetism	The 'fossil,' or remanent, magnetic field of a rock is called paleomagnetism.
Seismicity	The distribution, frequency, and magnitude of earthquakes in an area are called seismicity.
Gravity	The attraction between bodies of matter is a gravity.
Heat	Total kinetic energy of all the randomly moving atoms, ions, or molecules within a given substance, excluding the overall motion of the whole object. This form of kinetic energy flows from one body to another when there is a temperature difference betwe is referred to as heat.
Gulf	A relatively large portion of sea, partly enclosed by land is called gulf.
Subsidence	Sinking or down warping of a part of the earth's surface is called subsidence.
Elevation	Elevation refers to the distance of a point above a specified surface of constant potential; the distance is measured along the direction of gravity between the point and the surface.
Temperature	Temperature refers to a measure of the average speed of motion of the atoms, ions, or molecules in a substance or combination of substances at a given moment. Compare with heat.
Topographic map	A map on which elevations are shown by means of contour lines is referred to as a topographic
Map	Map refers to a representation of Earth's surface usually depicting mostly land areas.
Vertical exaggeration	Vertical exaggeration refers to the exaggeration of the vertical scale relative to the horizontal scale in a topographic profile or section.
Depth	Depth refers to vertical distance from still-water level to the bottom.
Water depth	Distance between the seabed and the still water level is referred to as water depth.
Seafloor spreading	Seafloor spreading refers to the theory that new ocean crust forms at spreading centers, most of which are on the ocean floor, and pushes the continents aside. Power is thought to be provided by convection currents in Earth's upper mantle.
Coast	A strip of land of indefinite length and width that extends from the SEASHORE inland to the first major change in terrain features is referred to as coast.
Shield volcano	Shield volcano refers to a very wide volcano built of low-viscosity lavas.
Volcano	Vent or fissure in the earth's surface through which magma, liquid lava, and gases are released into the environment is a volcano.
Crater	Crater refers to an abrupt basin commonly rimmed by ejected material. In volcanoes, craters form by outwarcc explosion, are commonly less than 2 km diameter, and occur at the summit of a volcanic cone. Similar rimmed basins form by impacts with meteorites, asteroids, and.
Thinning	The timber harvesting practice of selectively removing only smaller or poorly formed trees is called thinning.
Fault	A fracture in rock along which there has been an observable amount of displacement. Faults are rarely single planar units; normally they occur as parallel to sub-parallel sets of planes along which movement has taken place to a greater or lesser extent. Such sets are called fault or fracture-zones.
Transform fault	Transform fault refers to area where earth's lithospheric plates move in opposite but parallel directions along a fracture in the lithosphere. Compare with convergent plate boundary, divergent plate boundary.

Go to **Cram101.com** for the Practice Tests for this Chapter.

Trough	A long and broad submarine depression with gently sloping sides is referred to as a trough.
NOAA	NOAA refers to National Oceanic and Atmospheric Administration, founded within the U.S. Department of Commerce in 1970 to facilitate commercial uses of the ocean.
Climate	Physical properties of the troposphere of an area based on analysis of its weather records over a long period. The two main factors determining an area's climate are temperature, with its seasonal variations, and the amount and distri.
Environment	Environment refers to all external conditions and factors, living and nonliving, that affect an organism or other specified system during its lifetime; the earth's life-support systems for us and for all other forms of life-another term for solar capita.
Terrestrial	Terrestrial refers to pertaining to land. Compare with aquatic.
Extinction	Extinction refers to complete disappearance of a species from the earth. This happens when a species cannot adapt and successfully reproduce under new environmental conditions or when it evolves into one or more new species. Compare speciation.
Species	Group of organisms that resemble one another in appearance, behavior, chemical makeup and processes, and genetic structure is a species. Organisms that reproduce sexually are classified as members of the same species only if they can breed with one another and produce offspring.
Lava	Molten rock that is extruded out of volcanoes is called lava.
Well	Well refers to a hole, generally cylindrical and usually walled or lined with pipe, that is dug or drilled into the ground to penetrate an aquifer below the zone of saturation.
Range	Land used for grazing is referred to as the range.
Community	Community refers to populations of all species living and interacting in an area at a particular time.
Nutrients	Nutrients refer to chemicals such as phosphorus and nitrogen that, when released into water sources, may cause pollution events such as eutrophication.
Energy	Capacity to do work by performing mechanical, physical, chemical, or electrical tasks or to cause a heat transfer between two objects at different temperatures is an energy.
Matter	Matter refers to anything that has mass and takes up space. On earth, where gravity is present, we weigh an object to determine its mass.
Bacteria	Bacteria refer to prokaryotic, one-celled organisms. Some transmit diseases. Most act as decomposers and get the nutrients they need by breaking down complex organic compounds in the tissues of living or dead organisms into simpler inorganic nutrient compounds.
Habitat	The place where an organism lives is called habitat.
Plume	Plume refers to an arm of magna rising upward from the mantle.
Site	A factor considering the summation of all environmental features of a location that influences the placement of a city is a site.
Monitoring	Monitoring refers to the process of collecting data on a regular basis at specific sites to provide a database from which to evaluate change.
Magnitude	An assessment of the size of an event is a magnitude. Magnitude scales exist for earthquakes, volcanic eruptions. hurricanes, and tornadoes. For earthquakes, different magnitudes are calculated for the same earthquake when different types of seismic waves are used.
Normal fault	Normal fault refers to high-angle faults with one block dropping down relative to another block; they denote tension and are found at the axial rift valleys of ocean spreading ridges.

Go to **Cram101.com** for the Practice Tests for this Chapter.

369

Subduction zone	Elongate region in which the sea floor slides beneath a continent or island arc is referred to as the subduction zone.
Brittle	Behavior of material where stress causes abrupt fracture are called the brittle.
Magnetic field	A region where magnetic forces affect any magnetized bodies or electric currents is the magnetic field. Earth is surrounded by a magnetic field.
Earthquake	Shaking of the ground resulting either from the fracturing and displacement of rock, producing a fault, or from subsequent movement along the fault is the earthquake.
Bar	An offshore ridge or mound of sand, gravel, or other unconsolidated material which is submerged, especially at the mouth of a river or estuary, or lying parallel to, and a short distance from, the beach is referred to as a bar.
Pore	Pore refers to an opening or void space in; oil or rock.
Isostatic equilibrium	Isostatic equilibrium refers to balanced support of lighter material in a heavier, displaced supporting matrix. Analogous to buoyancy in a liquid.
Equilibrium	A point of rest. A system that does not tend to undergo any change of its own accord but remains in a single, fixed condition is said to be in equilibrium. Compare with steady state.
Cross section	A two-dimensional drawing showing features in the vertical plane as in a canyon wall or road cut is referred to as cross section.
Magnetism	Magnetism refers to a group of physical phenomena associated with moving electricity.
Seismic wave	Seismic wave refers to a low-frequency wave generated by the forces that cause earthquakes. Some kinds of seismic waves can pass through Earth.
Wave	An oscillatory movement in a body of water manifested by an alternate rise and fall of the surface is called a wave. Disturbances of the surface of a liquid body, as the ocean, in the form of a ridge, swell or hump. The term wave by itself usually refers to the term surface gravity wave.
Density	Density refers to the ratio of a mass to a unit volume specified as grams per cubic centimeter.
Asthenosphere	Earth materials located at a depth of 100 km to 700 km below the Earth's surface, underneath the tectonic plates is called the asthenosphere.
Continental crust	The light, buoyant granitic rock that underlies continental masses and averages about 35 kilometers in thickness is called continental crust.
Dredging	Excavation or displacement of the bottom or SHORELINE of a water body. Dredging can be accomplished with mechanical or hydraulic machines. Most is done to maintain channel depths or berths for navigational purposes; other dredging is for shellfish harvesting or for cleanup of polluted sediments.
Mud	Mud refers to a mixture of silt and clay sized particles.
P wave	P wave refers to first seismic wave to reach a seismometer. Movement is by alternating push-pull pulses that travel through solid, liquid, and gas.
Wave velocity	Speed at which the individual waveform advances, defined as the wavelength divided by the wave period is called wave velocity.
Continent	Continent refers to lower-density masses of rock, exposed as about 40 percent of the Earth's surface: 29 percent as land and I 1 percent as the floor of shallow seas.
Peninsula	An elongated portion of land nearly surrounded by water and connected to a larger body of land, usually by a neck or an isthmus is called a peninsula.

Go to **Cram101.com** for the Practice Tests for this Chapter.

Shore	That strip of ground bordering any body of water which is alternately exposed, or covered by tides and/or waves is a shore. A shore of unconsolidated material is usually called a beach.
Soil	Soil refers to a layer of weathered, unconsolidated material on top of bedrock; often also defined as containing organic matter and being capable of supporting plant growth.
Desert	Desert refers to biome in which evaporation exceeds precipitation and the average amount of precipitation is less than 25 centimeters a year. Such areas have little vegetation or have widely spaced, mostly low vegetation. Compare forest, grassland.
Stream	Stream refers to any flow of water; a current. A course of water flowing along a bed in the earth.
Erosion	Erosion refers to wearing away of the land by natural forces. On a beach, the carrying away of beach material by wave action, tidal currents or by DEFLATION. The wearing away of land by the action of natural forces.
Lead	Lead refers to a heavy metal that is an important constituent of automobile batteries and other industrial products. A toxic metal capable of causing environmental disruption and producing a health problem to people and other living organisms.
Clay	Clay refers to a fine grained sediment with a typical grain size less than 0.004 mm. Possesses electromagnetic properties which bind the grains together to give a bulk strength or cohesion.
Calcareous	Composed of calcium carbonate is referred to as calcareous.
Siliceous	Material whose composition is silicate is referred to as siliceous.
Foraminifera	Planktonic and benthonic protozoans that have a test composed of calcium carbonate is a foraminifera.
Diatoms	Diatoms refers to microscopic, unicellular phytoplankton possessing silica valves; they are responsible for much of the ocean's primary production.
Chert	Chert refers to a hard siliceous rock composed of opaline silica derived from the hard parts of microscopic plants and animals.
Sand	Sand refers to an unconsolidated mixture of inorganic soil consisting of small but easily distinguishable grains ranging in size from about .062 mm to 2.0 mm.
Turbidity	Turbidity refers to a condition of a liquid due to fine visible material in suspension, which may not be of sufficient size to be seen as individual particles by the naked eye but which prevents the passage of light through the liquid. A measure of fine suspended matter in liquids.
Mass	The amount of material in an object is the mass.
Igneous rocks	Igneous rocks refers to rocks formed from the solidification of magma. They are extrusive if they crystallize on the surface of the Earth and intrusive if they crystallize beneath the surface.
Debris	Debris refers to any accumulation of rock fragments; detritus.
Dike	Dike refers to sometimes written as dyke; earth structure along a sea or river in order to protect LITTORAL lands from flooding by high water; DIKES along rivers are sometimes called levees.
Geology	Geology refers to the science which treats of the origin, history and structure of the Earth, as recorded in rocks; together with the forces and processes now operating to modify rocks.
Base	A substance that combines with a hydrogen ion in solution is called the base.

Go to **Cram101.com** for the Practice Tests for this Chapter.

Development	Development refers to change from a society that is largely rural, agricultural, illiterate, and poor, with a rapidly growing population, to one that is mostly urban, industrial, educated, and wealthy, with a slowly growing or stationary population.
Ductile	Ductile refers to behavior of material where stress causes permanent flow or strain.
Mineral	A naturally occurring, inorganic, crystalline solid that has a definite chemical composition and possesses characteristic physical properties is a mineral.
Concentration	Amount of a chemical in a particular volume or weight of air, water, soil, or other medium is referred to as concentration.
Fracture zone	A linear zone of highly irregular, faulted topography that is oriented perpendicular to ocean-spreading ridges is referred to as the fracture zone.
Cliff	A high steep face of rock is referred to as cliff.
Submersible	Submersible refers to a vessel that can submerge.
Metamorphic rock	Metamorphic rock refers to rock produced when a preexisting rock is subjected to high temperatures, high pressures, chemically active fluids, or a combination of these agents. Compare with igneous rock, sedimentary rock.
Mantle plume	The upwelling of hot material into and through the lithosphere, with magma spilling out onto the earth's surface and building thick volcanic piles is called a mantle plume. If the lithosphere is moving relative to the plume or 'hot spot,' a linear trail of volcanoes is produced.
Rhyolite	Rhyolite refers to a volcanic rock typical of continents. Typically forms from high viscosity magma.
Groundwater	Groundwater refers to water that sinks into the soil and is stored in slowly flowing and slowly renewed underground reservoirs called aquifers; underground water in the zone of saturation, below the water table. Compare runoff, surface water.
Fissure	A narrow parting or crack in rock are called the fissure.
Weathering	Physical and chemical processes in which solid rock exposed at earth's surface is changed to separate solid particles and dissolved material, which can then be moved to another place as sediment is referred to as weathering.
Soil profile	Cross-sectional view of the horizons in a soil is referred to as the soil profile.
Slide	In mass wasting, movement of a descending mass along a plane approximately parallel to the slope of the surface is referred to as slide.
River	A natural stream of water larger than a brook or creek is a river.
Front	The boundary between two air masses with different temperatures and densitie is referred to as front.
Catastrophe	Catastrophe refers to a situation or event that causes significant damage to people and property, such that recovery and/or rehabilitation is a long and involved process. Examples of natural catastrophes include hurricanes, volcanic eruptions, large wildfires, and floods.
Ash	The loose debris that is ejected from an erupting volcano is called ash.
Population	Group of individual organisms of the same species living within a particular area is referred to as a population.
Famine	Famine refers to widespread malnutrition and starvation in a particular area because of a shortage of food, usually caused by drought, war, flood, earthquake, or other catastrophic events that disrupts food production and distribution.

Go to **Cram101.com** for the Practice Tests for this Chapter.

Radiation	Fast-moving particles or waves of energy are called radiation.
Weather	Weather refers to short-term changes in the temperature, barometric pressure, humidity, precipitation, sunshine, cloud cover, wind direction and speed, and other conditions in the troposphere at a given place and time. Compare with climate.
Glacier	Large masses of moving ice on land derived by the recrystallization of snow into ice under pressure are referred to as glacier.
Flood	Period when tide level is rising; often taken to mean the flood current which occurs during this period. A flow above the CARRYING CAPACITY of a CHANNEL.
Bedrock	Solid rock lying beneath loose soil or unconsolidated sediment is called bedrock.
Accretion	The accumulation of sediment, deposited by natural fluid flow processes is accretion.
Igneous rock	Rock formed when molten rock material wells up from earth's interior, cools, and solidifies into rock masses. Compare metamorphic rock, sedimentary roc is an igneous rock.
Silicate minerals	The most important group of rock-forming minerals is called silicate minerals.
Convection	Convection refers to the vertical transport of a fluid or the transfer of heat in fluids.
Mixture	Mixture refers to combination of two or more elements and compounds.
Divergence	Divergence refers to to move apart from a common source.
Crystallization	The growth of minerals in a fluid such as magma is a crystallization.
Residual	The components of water level not attributable to astronomical effects are called residual.
Conditions	Conditions refers to physical or chemical attributes of the environment that, while not being consumed, influence biological processes and population growth. Examples are temperature, salinity, and acidity. Compare resources.
Upwelling	The process by which water rises from a deeper to a shallower depths, usually as a result of offshore surface water flow is called upwelling. It is most prominent where persistent wind blows parallel to a coastline so that the resultant Ekman transport moves surface water away from the coast.
Conduction	The transfer of heat usually in solids whereby energy is passed from particle to particle by thermal agitation is referred to as conduction.
Hydrothermal vent	Hydrothermal vent refers to spring of hot, mineral- and gas-rich seawater found on some oceanic ridges in zones of active seafloor spreading.
Offshore	Offshore in beach terminology refers to the comparatively flat zone of variable width, extending from the shoreface to the edge of the continental shelf. It is continually submerged. The direction seaward from the shore. The zone beyond the nearshore zone where sediment motion induced by waves alone effectively ceases and where the influence of the sea bed on wave action is small in comparison with the effect of wind. The breaker zone directly seaward of the low tide line.
State	State refers to an expression of the internal form of matter. Water exists in three states: solid, liquid, and gas. A solid has a fixed volume and fixed shape; a liquid has a fixed volume but no fixed shape; and a gas has neither fixed volume nor fixed shape.
Compression	A state of stress that causes a pushing together or contraction is called compression.
Chemical	One of the millions of different elements and compounds found naturally or synthesized by human is referred to as chemical.

Facies	Facies refers to the sum total of features such as sedimentary rock type, MINERAL content, SEDIMENTARY STRUCTURES, BEDDING characteristics, fossil content, etc. which characterise a sediment as having been deposited in a given environment.
Reach	Reach refers to an arm of the ocean extending into the land. A straight section of restricted waterway of considerable extent; may be similar to a narrows, except much longer in extent.
Ore deposits	Earth materials in which metals are concentrated in high concentrations, sufficient to be mined are called ore deposits.
Annual	Annual refers to plant that grows, sets seed, and dies in one growing season. Compare to perennial.
Discharge	The volume of water flowing in a stream per unit of time is the discharge.
Recycle	Recycle is integral part of waste management that attempts to identify resources in the waste stream that may be collected and reused.
Atmosphere	Atmosphere refers to the whole mass of air surrounding the earth.
Abyssal hill	A relatively small hill, typically of volcanic origin, rising no more than 1,000 meters above the sea floor is an abyssal hill.
Silica	A compound with a composition, such as quartz in granite and opal in the shells of radiolaria is referred to as silica.
Depression	Depression refers to a general term signifying any depressed or lower area in the ocean floor.
Oil	The liquid form of petroleum consisting of a complex mixture of large hydrocarbon molecules is referred to as oil.
Reserves	Reserves refer to resources that have been identified from which a usable mineral can be extracted profitably at present prices with current mining.
Caldera	Caldera refers to a large, basinshaped volcanic depression. Roughly circular in map view, that forms by a piston-like collapse of a cylinder of overlying rock into an underlying, partially evacuated magma chamber.
Volcanic rock	Rock formed by solidification of magma at the Earth's surface is a volcanic rock.
Seismic refraction	The bending of seismic waves as they pass from one material to another is called seismic refraction.
Refraction	The process by which the direction of a wave moving in shallow water at an angle to the bottom contours is changed is refraction. The part of the wave moving shoreward in shallower water travels more slowly than that portion in deeper water, causing the wave to turn or bend to become parallel to the contour.
Chalk	A white, soft limestone consisting dominantly of the shells of foraminifera is a chalk.
Gravel	Gravel refers to loose, rounded fragments of rock, larger than sand, but smaller than cobbles. Small stones and pebbles, or a mixture of these with sand.
Avalanche	A large mass of snow, ice, soil, or rock that moves rapidly downslope under the pull of gravity is referred to as an avalanche.
Gypsum	An evaporite deposit composed of hydrous calcium sulfat is a gypsum.
Evaporite	A type of sediment precipitated from an aqueous solution, usually by the evaporation of water from a basin with restricted circulation is the evaporite.
Evaporation	Evaporation refers to conversion of a liquid into a gas.

379

Recent	A synonym of Holocene is called recent.
Alluvial fan	A cone-shaped deposit of sediment that forms at the mouth of steep mountain streams in desert regions is an alluvial fan.
Clastic sediments	Clastic sediments refers to deposits of fragments of preexisting rocks that have been transported from their point of origin.
Continental shelf	Continental shelf refers to the zone bordering a continent extending from the line of permanent immersion to the DEPTH, usually about 100 m to 200 m, where there is a marked or rather steep descent toward the great depths. The area under active LITTORAL processes during the Holocene period. The region of the oceanic bottom that extends outward from the shoreline with an average slope of less than 1.
Passive margin	Passive margin refers to a continental margin near an area of lithospheric plate divergence.
Event	Event refers to an occurrence meeting specified conditions, e.g. damage, a threshold wave height or a threshold water level.
Flood basalt	Tremendous outpourings of basaltic lava that form thick, extensive plateaus are called flood basalt.
Ion	Ion refers to atom or group of atoms with one or more positive or negative electrical charges. Compare atone, molecule.
Slope	The degree of inclination to the horizontal is the slope.
Tensional stress	A force that pulls apart rocks or parts of a structure is tensional stress.
Stress	Stress refers to force per unit area. May be compression, tension, or shear.
Reef	A ridge of rock or other material lying just below the surface of the sea is called a reef.
Slump	In mass wasting, movement along a curved surface in which the upper part moves vertically downward while the lower part moves outward is a slump.
Lagoon	A shallow body of water, as a pond or lake, which usually has a shallow restricted INLET from the se is referred to as lagoon.
Passive continental margin	A subsiding continental margin situated in a nontectonic setting away from a lithospheric plate boundary is a passive continental margin.
Conglomerate	A sedimentary rock dominated by gravel is the conglomerate.
Accumulation	Buildup of matter, energy, or information in a system is referred to as accumulation.
Deep water	Deep water in regard to waves, where DEPTH is greater than one-half the WAVE LENGTH. Deep-water conditions are said to exist when the surf waves are not affected by conditions on the bottom.
Till	Poorly sorted sediment that is deposited by glaciers is referred to as till.
Sedimentary rock	Rock that forms from the accumulated products of erosion and in some cases from the compacted shells, skeletons, and other remains of dead organisms is sedimentary rock. Compare with igneous rock, metamorphic roc is a sedimentary rock.
Spring	A place where groundwater flows out onto the surface is a spring.
Acceleration	Acceleration refers to cause to move faster. The rate of change of motion.
Limestone	A sedimentary rock composed dominantly of calcium carbonate, either precipitated from seawater or deposited as shell debris are called the limestone.

Go to **Cram101.com** for the Practice Tests for this Chapter.

Pangaea	Pangaea refers to the megacontinent of the Mesozoic Era that consisted of all of the present-day continents joined together into a single unit.
Laurasia	The aggregation of North America, Europe, and Asia into a large continental mass that comprised the northern half of the megacontinent Pangaea during the Mesozoic Era is a laurasia.
Gondwanaland	Gondwanaland is the large aggregation of continents-South America, Africa, India, Australia, Antarctica-that formed the southern half of the megacontinent Pangaea during the Mesozoic
Tethys Sea	An immense seaway that separated Gondwanaland from Laurasia during the Mesozoic Era is called Tethys Sea.
Outcrop	A surface exposure of bare rock, not covered by soil or vegetation is called outcrop.
Key	Key refers to a low, insular BANK of sand, coral, etc., as one of the islets off the southern coast of Florida.
Abyssal plain	A flat area on the deep-sea floor having a very gentle slope of less than one meter per kilometer, and consisting chiefly of graded terrigenous sediments known as turbidites is an abyssal plain.
Magnetic anomaly	A disturbance of the Earth's magnetic field created by magnetized rock in the Earth's crust is referred to as magnetic anomaly.
Tectonics	The study of the major structural features of the Earth's crust or the broad structure of a region is referred to as tectonics.
Feedback	A kind of system response that occurs when output of the system also serves as input leading to changes in the system is called feedback.
Plate tectonics	Plate tectonics refer to the theory of geophysical processes that explains the movements of lithospheric plates and the processes that occur at their boundaries.

Ocean	The great body of salt water which occupies two-thirds of the surface of the Earth, or one of its major subdivisions is called an ocean.
Planet	A smaller, usually nonluminous body orbiting a star is a planet.
Fracture	Fracture refers to a general term for any breaks in rock. Fractures include faults, joints, and crack,_.
System	A set of components that function and interact in some regular and theoretically predictable manner is called a system.
Transform fault	Transform fault refers to area where earth's lithospheric plates move in opposite but parallel directions along a fracture in the lithosphere. Compare with convergent plate boundary, divergent plate boundary.
Fault	A fracture in rock along which there has been an observable amount of displacement. Faults are rarely single planar units; normally they occur as parallel to sub-parallel sets of planes along which movement has taken place to a greater or lesser extent. Such sets are called fault or fracture-zones.
Plate	One of about a dozen rigid segments of Earth's lithosphere that move independently is a plate. The plate consists of continental or oceanic crust and the cool, rigid upper mantle directly below the crust.
Ridge	Ridge refers to the volcanic mountain ranges that lie along the spreading centers on the floors of the oceans.
Tectonics	The study of the major structural features of the Earth's crust or the broad structure of a region is referred to as tectonics.
Plates	Various-sized areas of earth's lithosphere that move slowly around with the mantle's flowing asthenosphere are plates. Most earthquakes and volcanoes occur around the boundaries of these plates.
Lithosphere	Lithosphere refers to outer shell of the earth, composed of the crust and the rigid, outermost part of the mantle outside of the asthenosphere; material found in the earth's plates.
Shear	The failure of a body where the mass on one side slides past the portion on the other side is a shear.
Compression	A state of stress that causes a pushing together or contraction is called compression.
Topography	Topography refers to the form of the features of the actual surface of the Earth in a particular region considered collectively.
Temperature	Temperature refers to a measure of the average speed of motion of the atoms, ions, or molecules in a substance or combination of substances at a given moment. Compare with heat.
Fracture zone	A linear zone of highly irregular, faulted topography that is oriented perpendicular to ocean-spreading ridges is referred to as the fracture zone.
Zone	Division or province of the ocean with homogeneous characteristics is referred to as a zone.
Basalt	Basalt refers to a dark, fine-grained igneous rock composed of minerals enriched in ferromagnesian silicates.
Metamorphism	The changes in minerals and rock textures that occur with the elevated temperatures and pressures below the Earth's surface are referred to as metamorphism.
Rocks	An aggregate of one or more minerals rather large in area are rocks. The three classes of rocks are the following: Igneous rock - crystalline rocks formed from molten material.

	Sedimentary rock - A rock resulting from the consolidation of loose sediment that has accumulated in layers. Metamorphic rock - Rock that has formed from preexisting rock as a result of heat or pressure.
Well	Well refers to a hole, generally cylindrical and usually walled or lined with pipe, that is dug or drilled into the ground to penetrate an aquifer below the zone of saturation.
Mantle	Zone of the earth's interior between its core and its crust. Compare with core, crust is the mantle.
Slide	In mass wasting, movement of a descending mass along a plane approximately parallel to the slope of the surface is referred to as slide.
Range	Land used for grazing is referred to as the range.
Relief	The difference in elevation between the highest and lowest points in an area is called relief.
Transform plate boundary	Transform plate boundary refers to a place where crustal plates shear laterally past one another. Crust is neither produced nor destroyed at this type of junction.
Oceanic crust	Oceanic crust refers to the outermost rock shell of the earth, some 5 kilometers thick, that underlies ocean basins; it is composed of basalt and sedimentary layers.
Crust	Solid outer zone of the earth. It consists of oceanic crust and continental crust. Compare core, mantle.
Sea	Sea refers to the ocean. A large body of salt water, second in rank to an ocean, more or less landlocked and generally part of, or connected with, an ocean or a larger sea. State of the ocean or lake surface, in regard to waves.
Map	Map refers to a representation of Earth's surface usually depicting mostly land areas.
Valley	An elongated depression, usually with an outlet, between bluffs or between ranges of hills or mountains is a valley.
Front	The boundary between two air masses with different temperatures and densitie is referred to as front.
Erosion	Erosion refers to wearing away of the land by natural forces. On a beach, the carrying away of beach material by wave action, tidal currents or by DEFLATION. The wearing away of land by the action of natural forces.
River	A natural stream of water larger than a brook or creek is a river.
Stream	Stream refers to any flow of water; a current. A course of water flowing along a bed in the earth.
Weathering	Physical and chemical processes in which solid rock exposed at earth's surface is changed to separate solid particles and dissolved material, which can then be moved to another place as sediment is referred to as weathering.
Mass	The amount of material in an object is the mass.
Escarpment	A more or less continuous line of CLIFFS or steep slopes facing in one general direction which are caused by EROSION or faulting, also called SCARP is an escarpment.
Subduction zone	Elongate region in which the sea floor slides beneath a continent or island arc is referred to as the subduction zone.
Divergent plate boundary	Divergent plate boundary refers to area where earth's lithospheric plates move apart in opposite directions. Compare convergent plate boundary, transform fault.

Go to **Cram101.com** for the Practice Tests for this Chapter.

Trench	A long narrow submarine depression with relatively steep sides is called a trench.
Brittle	Behavior of material where stress causes abrupt fracture are called the brittle.
Ductile	Ductile refers to behavior of material where stress causes permanent flow or strain.
Asthenosphere	Earth materials located at a depth of 100 km to 700 km below the Earth's surface, underneath the tectonic plates is called the asthenosphere.
Seismicity	The distribution, frequency, and magnitude of earthquakes in an area are called seismicity.
Coast	A strip of land of indefinite length and width that extends from the SEASHORE inland to the first major change in terrain features is referred to as coast.
Shore	That strip of ground bordering any body of water which is alternately exposed, or covered by tides and/or waves is a shore. A shore of unconsolidated material is usually called a beach.
Depth	Depth refers to vertical distance from still-water level to the bottom.
Reach	Reach refers to an arm of the ocean extending into the land. A straight section of restricted waterway of considerable extent; may be similar to a narrows, except much longer in extent.
Cliff	A high steep face of rock is referred to as cliff.
Rock	Any material that makes up a large, natural, continuous part of earth's crust is rock.
Elevation	Elevation refers to the distance of a point above a specified surface of constant potential; the distance is measured along the direction of gravity between the point and the surface.
Trough	A long and broad submarine depression with gently sloping sides is referred to as a trough.
Oceanography	That science treating of the oceans, their forms, physical features and phenomena is referred to as oceanography.
Bottom water	A general term applied to dense water masses that sink to the 'bottom' of ocean basins is a bottom water.
Igneous rocks	Igneous rocks refers to rocks formed from the solidification of magma. They are extrusive if they crystallize on the surface of the Earth and intrusive if they crystallize beneath the surface.
Slumping	The sliding of large, cohering blocks of sediment or rock downslope under the influence of gravity is called slumping.
Seismic	Referring to vibrations in the Earth produced by earthquakes is referred to as seismic.
Magma	Molten rock below the earth's surface is referred to as magma.
Lava	Molten rock that is extruded out of volcanoes is called lava.
Sediment	Loose, fragments of rocks, minerals or organic material which are transported from their source for varying distances and deposited by air, wind, ice and water are sediment. Other sediments are precipitated from the overlying water or form chemically, in place. Sediment includes all the unconsolidated materials on the sea floor. The fine-grained material deposited by water or wind.
Dike	Dike refers to sometimes written as dyke; earth structure along a sea or river in order to protect LITTORAL lands from flooding by high water; DIKES along rivers are sometimes called levees.
Network	Network refers to a set consisting of stations for which geometric relationships have been determined and which are so related that removal of one station from the set will affect the relationships between the other stations; and lines connecting the stations to show this interdependence.

Basin	A large submarine depression of a generally circular, elliptical or oval shape is called a basin.
Fold	Fold refers to wavy geologic structures formed by the compression and bending of sedimentary layers.
Uplift	The rising of one part of the Earth's crust relative to another part is referred to as uplift.
Subsidence	Sinking or down warping of a part of the earth's surface is called subsidence.
Thermal	The energy of the random motion of atoms and molecules is referred to as thermal.
Cross section	A two-dimensional drawing showing features in the vertical plane as in a canvon wall or road cut is referred to as cross section.
Development	Development refers to change from a society that is largely rural, agricultural, illiterate, and poor, with a rapidly growing population, to one that is mostly urban, industrial, educated, and wealthy, with a slowly growing or stationary population.
Metamorphic rock	Metamorphic rock refers to rock produced when a preexisting rock is subjected to high temperatures, high pressures, chemically active fluids, or a combination of these agents. Compare with igneous rock, sedimentary rock.
Density	Density refers to the ratio of a mass to a unit volume specified as grams per cubic centimeter.
Continental crust	The light, buoyant granitic rock that underlies continental masses and averages about 35 kilometers in thickness is called continental crust.
Gulf	A relatively large portion of sea, partly enclosed by land is called gulf.
Head	A comparatively high promontory with either a CLIFF or steep face. It extends into a large body of water, such as a sea or lake. An unnamed HEAD is usually called a headland. The section of RIP CURRENT which has widened out seaward of the BREAKERS, also called head of
Stress	Stress refers to force per unit area. May be compression, tension, or shear.
Triple junction	A place where three plate edges meet is called a triple junction.
Earthquake	Shaking of the ground resulting either from the fracturing and displacement of rock, producing a fault, or from subsequent movement along the fault is the earthquake.
Continent	Continent refers to lower-density masses of rock, exposed as about 40 percent of the Earth's surface: 29 percent as land and I 1 percent as the floor of shallow seas.
Peninsula	An elongated portion of land nearly surrounded by water and connected to a larger body of land, usually by a neck or an isthmus is called a peninsula.
Climate	Physical properties of the troposphere of an area based on analysis of its weather records over a long period. The two main factors determining an area's climate are temperature, with its seasonal variations, and the amount and distri.
Lake	Large natural body of standing fresh water formed when water from precipitation, land runoff, or groundwater flow fills a depression in the earth created by glaciation, earth movement, volcanic activity, or a giant meteorit are called the lake.
Chalk	A white, soft limestone consisting dominantly of the shells of foraminifera is a chalk.
Pleistocene	An epoch of the Quaternary Period characterized by several glacial ages is referred to as the Pleistocene.
Theory	A general explanation of a characteristic of nature consistently supported by observation or

experiment is referred to as a theory.

Magnitude	An assessment of the size of an event is a magnitude. Magnitude scales exist for earthquakes, volcanic eruptions. hurricanes, and tornadoes. For earthquakes, different magnitudes are calculated for the same earthquake when different types of seismic waves are used.
Energy	Capacity to do work by performing mechanical, physical, chemical, or electrical tasks or to cause a heat transfer between two objects at different temperatures is an energy.
Event	Event refers to an occurrence meeting specified conditions, e.g. damage, a threshold wave height or a threshold water level.
Fire	The rapid combination of oxygen with organic material to produce flame, heat, and light is referred to as the fire.
Oil	The liquid form of petroleum consisting of a complex mixture of large hydrocarbon molecules is referred to as oil.
Outcrop	A surface exposure of bare rock, not covered by soil or vegetation is called outcrop.
Topographic map	A map on which elevations are shown by means of contour lines is referred to as a topographic
Global positioning system	Global positioning system refers to a satellite-based navigation system that provides very accurate location at low consumer cost.
Orbit	Orbit in ocean waves refers to the circular pattern of water particle movement at the air-sea interface.
Accuracy	Accuracy refers to the extent to which a measurement agrees with the accepted or correct value for that quantity, based on careful measurements by many people over a long time. Compare to precision.
Site	A factor considering the summation of all environmental features of a location that influences the placement of a city is a site.
Power	Power refers to the time rate of doing work.
Till	Poorly sorted sediment that is deposited by glaciers is referred to as till.
Strain	A change in torn or size of a body due to external forces is called strain.
Strand	Strand refers to shore or beach of the ocean or a large lake. The land bordering any large body of water, especially a sea or an arm of the ocean.
Migration	A term that refers to the habit of some animals is a migration.
Upwelling	The process by which water rises from a deeper to a shallower depths, usually as a result of offshore surface water flow is called upwelling. It is most prominent where persistent wind blows parallel to a coastline so that the resultant Ekman transport moves surface water away from the coast.
Observations	Information obtained through one or more of the five senses or through instruments that extend the senses are observations.
Key	Key refers to a low, insular BANK of sand, coral, etc., as one of the islets off the southern coast of Florida.
Resources	Resources refer to substances that can be consumed by an organism and, as a result, become unavailable to other organisms.
Geology	Geology refers to the science which treats of the origin, history and structure of the Earth, as recorded in rocks; together with the forces and processes now operating to modify rocks.

Feedback	A kind of system response that occurs when output of the system also serves as input leading to changes in the system is called feedback.

Plate	One of about a dozen rigid segments of Earth's lithosphere that move independently is a plate. The plate consists of continental or oceanic crust and the cool, rigid upper mantle directly below the crust.
Convergence	Resemblance among species belonging to different taxonomic groups as the result from adaptation to similar environments is the convergence.
Plates	Various-sized areas of earth's lithosphere that move slowly around with the mantle's flowing asthenosphere are plates. Most earthquakes and volcanoes occur around the boundaries of these plates.
Planet	A smaller, usually nonluminous body orbiting a star is a planet.
Lithosphere	Lithosphere refers to outer shell of the earth, composed of the crust and the rigid, outermost part of the mantle outside of the asthenosphere; material found in the earth's plates.
Slide	In mass wasting, movement of a descending mass along a plane approximately parallel to the slope of the surface is referred to as slide.
Mantle	Zone of the earth's interior between its core and its crust. Compare with core, crust is the mantle.
Reach	Reach refers to an arm of the ocean extending into the land. A straight section of restricted waterway of considerable extent; may be similar to a narrows, except much longer in extent.
Chemical equilibrium	In seawater, the condition in which the proportion and amounts of dissolved salts per unit volume of ocean are nearly constant is a chemical equilibrium.
Equilibrium	A point of rest. A system that does not tend to undergo any change of its own accord but remains in a single, fixed condition is said to be in equilibrium. Compare with steady state.
Compression	A state of stress that causes a pushing together or contraction is called compression.
Erosion	Erosion refers to wearing away of the land by natural forces. On a beach, the carrying away of beach material by wave action, tidal currents or by DEFLATION. The wearing away of land by the action of natural forces.
Fold	Fold refers to wavy geologic structures formed by the compression and bending of sedimentary layers.
Sea	Sea refers to the ocean. A large body of salt water, second in rank to an ocean, more or less landlocked and generally part of, or connected with, an ocean or a larger sea. State of the ocean or lake surface, in regard to waves.
Rocks	An aggregate of one or more minerals rather large in area are rocks. The three classes of rocks are the following: Igneous rock - crystalline rocks formed from molten material. Sedimentary rock - A rock resulting from the consolidation of loose sediment that has accumulated in layers. Metamorphic rock - Rock that has formed from preexisting rock as a result of heat or pressure.
Rock	Any material that makes up a large, natural, continuous part of earth's crust is rock.
Continental crust	The light, buoyant granitic rock that underlies continental masses and averages about 35 kilometers in thickness is called continental crust.
Crust	Solid outer zone of the earth. It consists of oceanic crust and continental crust. Compare core, mantle.
Oceanic crust	Oceanic crust refers to the outermost rock shell of the earth, some 5 kilometers thick, that underlies ocean basins; it is composed of basalt and sedimentary layers.

Go to **Cram101.com** for the Practice Tests for this Chapter.

Ocean	The great body of salt water which occupies two-thirds of the surface of the Earth, or one of its major subdivisions is called an ocean.
Subduction	Subduction refers to a process in which one lithospheric plate descends beneath another.
Convergent plate boundary	Area where earth's lithospheric plates are pushed togethe is called convergent plate boundary.
Swell	Waves that have traveled a long distance from their generating area and have been sorted out by travel into long waves of the same approximate period are referred to as a swell.
Trench	A long narrow submarine depression with relatively steep sides is called a trench.
Basin	A large submarine depression of a generally circular, elliptical or oval shape is called a basin.
Continent	Continent refers to lower-density masses of rock, exposed as about 40 percent of the Earth's surface: 29 percent as land and I 1 percent as the floor of shallow seas.
Zone	Division or province of the ocean with homogeneous characteristics is referred to as a zone.
Depth	Depth refers to vertical distance from still-water level to the bottom.
Magma	Molten rock below the earth's surface is referred to as magma.
Andesite	Andesite refers to a common volcanic rock found in the volcanic arcs of subduction zones; it is intermediate in composition between the granitic crust of the continents and the basaltic crust of the oceans.
Granite	A light-colored, coarsegrained, intrusive igneous rock composed mainly of quartz and feldspar and that typifies the continental crust are called the granite.
Metamorphism	The changes in minerals and rock textures that occur with the elevated temperatures and pressures below the Earth's surface are referred to as metamorphism.
Facies	Facies refers to the sum total of features such as sedimentary rock type, MINERAL content, SEDIMENTARY STRUCTURES, BEDDING characteristics, fossil content, etc. which characterise a sediment as having been deposited in a given environment.
Terrane	Terrane refers to an isolated segment of seafloor, island arc, plateau, continental crust, or sediment transported by seafloor spreading to a position adjacent to a larger continental mass. Usually different in composition from the larger mass.
Accretion	The accumulation of sediment, deposited by natural fluid flow processes is accretion.
Gravity	The attraction between bodies of matter is a gravity.
Heat	Total kinetic energy of all the randomly moving atoms, ions, or molecules within a given substance, excluding the overall motion of the whole object. This form of kinetic energy flows from one body to another when there is a temperature difference betwe is referred to as heat.
Igneous rocks	Igneous rocks refers to rocks formed from the solidification of magma. They are extrusive if they crystallize on the surface of the Earth and intrusive if they crystallize beneath the surface.
Ridge	Ridge refers to the volcanic mountain ranges that lie along the spreading centers on the floors of the oceans.
Subduction zone	Elongate region in which the sea floor slides beneath a continent or island arc is referred to as the subduction zone.
Volcanic rock	Rock formed by solidification of magma at the Earth's surface is a volcanic rock.

399

Island arc	A curved group of volcanic islands associated with a deep oceanic trench and subduction zone is called island arc.
Metamorphic rock	Metamorphic rock refers to rock produced when a preexisting rock is subjected to high temperatures, high pressures, chemically active fluids, or a combination of these agents. Compare with igneous rock, sedimentary rock.
Topography	Topography refers to the form of the features of the actual surface of the Earth in a particular region considered collectively.
Continental margin	The drowned edges of continents consisting of the continental shelf, the continental slope, and the continental rise is called continental margin.
Range	Land used for grazing is referred to as the range.
Mass	The amount of material in an object is the mass.
Sedimentary rocks	Sedimentary rocks refer to rocks that have formed by the compaction and cementation of sediment.
Sediment	Loose, fragments of rocks, minerals or organic material which are transported from their source for varying distances and deposited by air, wind, ice and water are sediment. Other sediments are precipitated from the overlying water or form chemically, in place. Sediment includes all the unconsolidated materials on the sea floor. The fine-grained material deposited by water or wind.
Rift	Rift refers to the valley created at a pull-apart zone.
Buoyancy	The resultant upward forces, exerted by the water on a submerged or floating body, equal to the weight of the water displaced by this body is referred to as buoyancy.
Thermal	The energy of the random motion of atoms and molecules is referred to as thermal.
Development	Development refers to change from a society that is largely rural, agricultural, illiterate, and poor, with a rapidly growing population, to one that is mostly urban, industrial, educated, and wealthy, with a slowly growing or stationary population.
Density	Density refers to the ratio of a mass to a unit volume specified as grams per cubic centimeter.
Basalt	Basalt refers to a dark, fine-grained igneous rock composed of minerals enriched in ferromagnesian silicates.
Suture zone	A relatively narrow zone along collisional plate boundaries where two large continental masses are welded into one unit is the suture zone.
River	A natural stream of water larger than a brook or creek is a river.
Glacier	Large masses of moving ice on land derived by the recrystallization of snow into ice under pressure are referred to as glacier.
Lava	Molten rock that is extruded out of volcanoes is called lava.
Seamount	Conical mountain rising 1000 m or more above the sea floor is a seamount.
Temperature	Temperature refers to a measure of the average speed of motion of the atoms, ions, or molecules in a substance or combination of substances at a given moment. Compare with heat.
Dip	The angle of inclination measured in degrees from the horizontal is called dip.
Brittle	Behavior of material where stress causes abrupt fracture are called the brittle.
Point	Point refers to the extreme end of a cape, or the outer end of any land area protruding into the water, usually less prominent than a cape. A low profile shoreline promontory of more or

Go to **Cram101.com** for the Practice Tests for this Chapter.

less triangular shape, the top of which extends seaward.

Situation	The relative geographic location of a site that makes it a good location for a city is a situation.
Asthenosphere	Earth materials located at a depth of 100 km to 700 km below the Earth's surface, underneath the tectonic plates is called the asthenosphere.
Gradient	A measure of slope in meters of rise or fall per meter of horizontal distance. More general, a change of a value per unit of distance, e.g. the GRADIENT in longshore transport causes EROSION or ACCRETION. With reference to winds or currents, the rate of increase or decrease in speed, usually in the vertical; or the curve that represents this rate.
Convection	Convection refers to the vertical transport of a fluid or the transfer of heat in fluids.
Ductile	Ductile refers to behavior of material where stress causes permanent flow or strain.
Recent	A synonym of Holocene is called recent.
Seismic	Referring to vibrations in the Earth produced by earthquakes is referred to as seismic.
Elevation	Elevation refers to the distance of a point above a specified surface of constant potential; the distance is measured along the direction of gravity between the point and the surface.
Energy	Capacity to do work by performing mechanical, physical, chemical, or electrical tasks or to cause a heat transfer between two objects at different temperatures is an energy.
Seismic wave	Seismic wave refers to a low-frequency wave generated by the forces that cause earthquakes. Some kinds of seismic waves can pass through Earth.
Wave	An oscillatory movement in a body of water manifested by an alternate rise and fall of the surface is called a wave. Disturbances of the surface of a liquid body, as the ocean, in the form of a ridge, swell or hump. The term wave by itself usually refers to the term surface gravity wave.
Seismicity	The distribution, frequency, and magnitude of earthquakes in an area are called seismicity.
Earthquake	Shaking of the ground resulting either from the fracturing and displacement of rock, producing a fault, or from subsequent movement along the fault is the earthquake.
Map	Map refers to a representation of Earth's surface usually depicting mostly land areas.
Concentration	Amount of a chemical in a particular volume or weight of air, water, soil, or other medium is referred to as concentration.
Volcano	Vent or fissure in the earth's surface through which magma, liquid lava, and gases are released into the environment is a volcano.
Key	Key refers to a low, insular BANK of sand, coral, etc., as one of the islets off the southern coast of Florida.
Cross section	A two-dimensional drawing showing features in the vertical plane as in a canyon wall or road cut is referred to as cross section.
Front	The boundary between two air masses with different temperatures and densitie is referred to as front.
Fracture	Fracture refers to a general term for any breaks in rock. Fractures include faults, joints, and crack,_.
Stress	Stress refers to force per unit area. May be compression, tension, or shear.
Elastic	Elastic refers to behavior of material where stress causes deformation that is recoverable; when stress stops, the material returns to its original state.

Mineral	A naturally occurring, inorganic, crystalline solid that has a definite chemical composition and possesses characteristic physical properties is a mineral.
Seismic waves	A long-period wave caused by an underwater seismic disturbance or volcanic eruption is a seismic waves.
Fossils	Skeletons, bones, shells, body parts, leaves, seeds, or impressions of such items that provide recognizable evidence of organisms that lived long ago are called fossils.
Unconsolidated	Unconsolidated with regards to sediment grains refers to loose, separate, or unattached to one another.
Pile	A long substantial pole of wood, concrete or metal, driven into the earth or sea bed to serve as a support or protection is called pile.
Blade	Blade refers to algal equivalent of a vascular plant's leaf. Also called a frond.
Thinning	The timber harvesting practice of selectively removing only smaller or poorly formed trees is called thinning.
Uplift	The rising of one part of the Earth's crust relative to another part is referred to as uplift.
Mixture	Mixture refers to combination of two or more elements and compounds.
Thrust fault	Thrust fault refers to a low-angle fault formed by compression, whereby the upper block is forced over the lower block causing the crust to shorten in length.
Fault	A fracture in rock along which there has been an observable amount of displacement. Faults are rarely single planar units; normally they occur as parallel to sub-parallel sets of planes along which movement has taken place to a greater or lesser extent. Such sets are called fault or fracture-zones.
Igneous rock	Rock formed when molten rock material wells up from earth's interior, cools, and solidifies into rock masses. Compare metamorphic rock, sedimentary roc is an igneous rock.
Sonar	Sonar refers to an acronym for sound navigation and ranging; an instrument used to locate objects underwater by reflecting sound waves.
Coast	A strip of land of indefinite length and width that extends from the SEASHORE inland to the first major change in terrain features is referred to as coast.
Seismic reflection	Seismic reflection refers to the return of part of the energy of seismic waves to the earth's surface after the waves bounce off a rock boundary.
Reflection	The process by which the energy of the wave is returned seaward is a reflection.
Base	A substance that combines with a hydrogen ion in solution is called the base.
Desert	Desert refers to biome in which evaporation exceeds precipitation and the average amount of precipitation is less than 25 centimeters a year. Such areas have little vegetation or have widely spaced, mostly low vegetation. Compare forest, grassland.
Well	Well refers to a hole, generally cylindrical and usually walled or lined with pipe, that is dug or drilled into the ground to penetrate an aquifer below the zone of saturation.
Load	The quantity of sediment transported by a current. It includes the suspended load of small particles in the water, and the bedload of large particles that move along the bottom.
Subsidence	Sinking or down warping of a part of the earth's surface is called subsidence.
Clastic sediments	Clastic sediments refers to deposits of fragments of preexisting rocks that have been transported from their point of origin.

Ash	The loose debris that is ejected from an erupting volcano is called ash.
Natural resources	Natural resources refer to nutrients and minerals in the soil and deeper layers of the earth's crust, water, wild and domesticated plants and animals, air, and other resources produced by the earth's natural processes. Compare with human capital, manufactured capital, and solar capital.
Resources	Resources refer to substances that can be consumed by an organism and, as a result, become unavailable to other organisms.
Coal	Solid, combustible mixture of organic compounds with 30-98% carbon by weight, mixed with various amounts of water and small amounts of sulfur and nitrogen compounds. It is formed in several stages as the remains of plants are subjected to heat and press is a coal.
Oil	The liquid form of petroleum consisting of a complex mixture of large hydrocarbon molecules is referred to as oil.
Natural gas	Underground deposits of gases consisting of 50-90% by weight methane gas and small amounts of heavier gaseous hydrocarbon compounds such as propane and butane are called natural gas.
Pebbles	Beach material usually well-rounded and between about 4 mm to 64 mm diameter are pebbles.
Pebble	Pebble refers to a rock fragment between 4 and 64 mm in diameter.
Valley	An elongated depression, usually with an outlet, between bluffs or between ranges of hills or mountains is a valley.
Shore	That strip of ground bordering any body of water which is alternately exposed, or covered by tides and/or waves is a shore. A shore of unconsolidated material is usually called a beach.
Accretionary prism	The region of a subduction zone located between the deep-sea trench and volcanic arc, where thick sedimentary units and volcanic rocks are squeezed and wrinkled by compression to form thrust faults and tight folds is referred to as an accretionary prism.
Passive margin	Passive margin refers to a continental margin near an area of lithospheric plate divergence.
Passive continental margin	A subsiding continental margin situated in a nontectonic setting away from a lithospheric plate boundary is a passive continental margin.
Lead	Lead refers to a heavy metal that is an important constituent of automobile batteries and other industrial products. A toxic metal capable of causing environmental disruption and producing a health problem to people and other living organisms.
Tectonics	The study of the major structural features of the Earth's crust or the broad structure of a region is referred to as tectonics.
Lake	Large natural body of standing fresh water formed when water from precipitation, land runoff, or groundwater flow fills a depression in the earth created by glaciation, earth movement, volcanic activity, or a giant meteorit are called the lake.
Relief	The difference in elevation between the highest and lowest points in an area is called relief.
Shear	The failure of a body where the mass on one side slides past the portion on the other side is a shear.
Chalk	A white, soft limestone consisting dominantly of the shells of foraminifera is a chalk.
Drag	The resistance to movement of an organism induced by the fluid through which it swims is called drag.
Viscous	Ease of flow is viscous. The more viscous a substance, the less readily it flows.

Go to **Cram101.com** for the Practice Tests for this Chapter.

Rhyolite	Rhyolite refers to a volcanic rock typical of continents. Typically forms from high viscosity magma.
Crystallization	The growth of minerals in a fluid such as magma is a crystallization.
Fissure	A narrow parting or crack in rock are called the fissure.
Silica	A compound with a composition, such as quartz in granite and opal in the shells of radiolaria is referred to as silica.
High water	High water refers to maximum height reached by a rising tide. The height may be solely due to the periodic tidal forces or it may have superimposed upon it the effects of prevailing meteorological conditions. Nontechnically, also called the HIGH TIDE.
Pore	Pore refers to an opening or void space in; oil or rock.
Divergent plate boundary	Divergent plate boundary refers to area where earth's lithospheric plates move apart in opposite directions. Compare convergent plate boundary, transform fault.
Ore	Part of a metal-yielding material that can be economically extracted at a given time is ore. An ore typically contains two parts: the ore mineral, which contains the desired metal, and waste mineral material.
Volatile	Substances that readily become gases when pressure is decreased, or temperature increased are referred to as volatile.
Debris	Debris refers to any accumulation of rock fragments; detritus.
Peninsula	An elongated portion of land nearly surrounded by water and connected to a larger body of land, usually by a neck or an isthmus is called a peninsula.
Force	Force refers to a push or pull that affects motion. The product of mass and acceleration of a material.
Composite volcano	Composite volcano refers to a volcano constructed of alternating layers of pyroclastic debris and lava flows. Syn. stratovolcano.
Pyroclastic	Pertaining to magma and volcanic rock blasted up into the air is referred to as pyroclastic.
Fall	Fall refers to a mass moving nearly vertical and downward under the influence of gravity.
Caldera	Caldera refers to a large, basinshaped volcanic depression. Roughly circular in map view, that forms by a piston-like collapse of a cylinder of overlying rock into an underlying, partially evacuated magma chamber.
Atmosphere	Atmosphere refers to the whole mass of air surrounding the earth.
Tsunami	Tsunami refers to a large, high-velocity wave generated by displacement of the sea floor; also called seismic sea wave. Commonly misnamed tidal wave.
Avalanche	A large mass of snow, ice, soil, or rock that moves rapidly downslope under the pull of gravity is referred to as an avalanche.
Population	Group of individual organisms of the same species living within a particular area is referred to as a population.
Glass	Glass refers to matter created when magma cools too quickly for atoms to arrange themselves into the ordered atomic structures of minerals. Most glasses are supercooled liquids.
Pumice	Pumice refers to volcanic glass so full of holes that it commonly floats on water.
Crater	Crater refers to an abrupt basin commonly rimmed by ejected material. In volcanoes, craters form by outwarcc explosion, are commonly less than 2 km diameter, and occur at the summit of a volcanic cone. Similar rimmed basins form by impacts with meteorites, asteroids, and.

Lava dome	Lava dome refers to a mountain or hill made from highly viscous lava, which plugs the central conduit of volcanoes.
State	State refers to an expression of the internal form of matter. Water exists in three states: solid, liquid, and gas. A solid has a fixed volume and fixed shape; a liquid has a fixed volume but no fixed shape; and a gas has neither fixed volume nor fixed shape.
Forest	Forest refers to biome with enough average annual precipitation to support growth of various species of trees and smaller forms of vegetation. Compare desert, grassland.
Monitoring	Monitoring refers to the process of collecting data on a regular basis at specific sites to provide a database from which to evaluate change.
Magnitude	An assessment of the size of an event is a magnitude. Magnitude scales exist for earthquakes, volcanic eruptions. hurricanes, and tornadoes. For earthquakes, different magnitudes are calculated for the same earthquake when different types of seismic waves are used.
Slope	The degree of inclination to the horizontal is the slope.
Altitude	Altitude refers to the height above sea level. Compare to latitude.
Conditions	Conditions refers to physical or chemical attributes of the environment that, while not being consumed, influence biological processes and population growth. Examples are temperature, salinity, and acidity. Compare resources.
Environment	Environment refers to all external conditions and factors, living and nonliving, that affect an organism or other specified system during its lifetime; the earth's life-support systems for us and for all other forms of life-another term for solar capita.
Soil	Soil refers to a layer of weathered, unconsolidated material on top of bedrock; often also defined as containing organic matter and being capable of supporting plant growth.
Magnetic field	A region where magnetic forces affect any magnetized bodies or electric currents is the magnetic field. Earth is surrounded by a magnetic field.
Core	Core refers to a cylindrical sample extracted from a beach or seabed to investigate the types and DEPTHS of sediment layers. An inner, often much less permeable portion of a BREAKWATER, or BARRIER beach.
Wind	Wind refers to the mass movement of air.
Magnetism	Magnetism refers to a group of physical phenomena associated with moving electricity.
Magnetometer	Magnetometer refers to a device that measures the amount and direction of residual magnetism in a rock sample.
Proton	Positively charged particle in the nuclei of all atoms is a proton. Each proton has a relative mass of 1 and a single positive charge.
Current	Current refers to the flowing of water, or other liquid or gas. That portion of a stream of water which is moving with a velocity much greater than the average or in which the progress of the water is principally concentrated. Ocean currents can be classified in a number of different ways.
Frequency	Number of events in a given time interval. For earthquakes, it is the number of cycles of seismic waves that pass in a second; frequency = l/period.
Power	Power refers to the time rate of doing work.
Offshore	Offshore in beach terminology refers to the comparatively flat zone of variable width, extending from the shoreface to the edge of the continental shelf. It is continually submerged. The direction seaward from the shore. The zone beyond the nearshore zone where

Go to **Cram101.com** for the Practice Tests for this Chapter.

411

sediment motion induced by waves alone effectively ceases and where the influence of the sea bed on wave action is small in comparison with the effect of wind. The breaker zone directly seaward of the low tide line.

Stratigraphy   Stratigraphy refers to the study of stratified rocks especially their sequence in time. The character of the rocks and the correlation of beds in different localities.

Event   Event refers to an occurrence meeting specified conditions, e.g. damage, a threshold wave height or a threshold water level.

Plate tectonics   Plate tectonics refer to the theory of geophysical processes that explains the movements of lithospheric plates and the processes that occur at their boundaries.

Observations   Information obtained through one or more of the five senses or through instruments that extend the senses are observations.

Rift valley   Rift valley refers to the fault-bounded valley found along the crest of many ocean ridges; it is created by tensional stresses that accompany the process of sea-floor spreading.

Hydrothermal vents   An opening in the ground from which pour out hot, saline water solutions are called hydrothermal vents.

Geology   Geology refers to the science which treats of the origin, history and structure of the Earth, as recorded in rocks; together with the forces and processes now operating to modify rocks.

Feedback   A kind of system response that occurs when output of the system also serves as input leading to changes in the system is called feedback.

Go to **Cram101.com** for the Practice Tests for this Chapter.
And, **NEVER** highlight a book again!

Mantle	Zone of the earth's interior between its core and its crust. Compare with core, crust is the mantle.
System	A set of components that function and interact in some regular and theoretically predictable manner is called a system.
Plates	Various-sized areas of earth's lithosphere that move slowly around with the mantle's flowing asthenosphere are plates. Most earthquakes and volcanoes occur around the boundaries of these plates.
Lithosphere	Lithosphere refers to outer shell of the earth, composed of the crust and the rigid, outermost part of the mantle outside of the asthenosphere; material found in the earth's plates.
Plate	One of about a dozen rigid segments of Earth's lithosphere that move independently is a plate. The plate consists of continental or oceanic crust and the cool, rigid upper mantle directly below the crust.
Flood basalt	Tremendous outpourings of basaltic lava that form thick, extensive plateaus are called flood basalt.
Basalt	Basalt refers to a dark, fine-grained igneous rock composed of minerals enriched in ferromagnesian silicates.
Convection	Convection refers to the vertical transport of a fluid or the transfer of heat in fluids.
Zone	Division or province of the ocean with homogeneous characteristics is referred to as a zone.
Planet	A smaller, usually nonluminous body orbiting a star is a planet.
Ridge	Ridge refers to the volcanic mountain ranges that lie along the spreading centers on the floors of the oceans.
Mantle plume	The upwelling of hot material into and through the lithosphere, with magma spilling out onto the earth's surface and building thick volcanic piles is called a mantle plume. If the lithosphere is moving relative to the plume or 'hot spot,' a linear trail of volcanoes is produced.
Plume	Plume refers to an arm of magma rising upward from the mantle.
Mud	Mud refers to a mixture of silt and clay sized particles.
Thermal	The energy of the random motion of atoms and molecules is referred to as thermal.
Point	Point refers to the extreme end of a cape, or the outer end of any land area protruding into the water, usually less prominent than a cape. A low profile shoreline promontory of more or less triangular shape, the top of which extends seaward.
Geyser	A hot spring that gushes magmaheated water and steam is called geyser.
Rock	Any material that makes up a large, natural, continuous part of earth's crust is rock.
Reach	Reach refers to an arm of the ocean extending into the land. A straight section of restricted waterway of considerable extent; may be similar to a narrows, except much longer in extent.
Rhyolite	Rhyolite refers to a volcanic rock typical of continents. Typically forms from high viscosity magma.
Ash	The loose debris that is ejected from an erupting volcano is called ash.
Ocean	The great body of salt water which occupies two-thirds of the surface of the Earth, or one of its major subdivisions is called an ocean.
Magma	Molten rock below the earth's surface is referred to as magma.

Heat	Total kinetic energy of all the randomly moving atoms, ions, or molecules within a given substance, excluding the overall motion of the whole object. This form of kinetic energy flows from one body to another when there is a temperature difference betwe is referred to as heat.
Head	A comparatively high promontory with either a CLIFF or steep face. It extends into a large body of water, such as a sea or lake. An unnamed HEAD is usually called a headland. The section of RIP CURRENT which has widened out seaward of the BREAKERS, also called head of
Uplift	The rising of one part of the Earth's crust relative to another part is referred to as uplift.
Continent	Continent refers to lower-density masses of rock, exposed as about 40 percent of the Earth's surface: 29 percent as land and I 1 percent as the floor of shallow seas.
Caldera	Caldera refers to a large, basinshaped volcanic depression. Roughly circular in map view, that forms by a piston-like collapse of a cylinder of overlying rock into an underlying, partially evacuated magma chamber.
Continental crust	The light, buoyant granitic rock that underlies continental masses and averages about 35 kilometers in thickness is called continental crust.
Crust	Solid outer zone of the earth. It consists of oceanic crust and continental crust. Compare core, mantle.
Development	Development refers to change from a society that is largely rural, agricultural, illiterate, and poor, with a rapidly growing population, to one that is mostly urban, industrial, educated, and wealthy, with a slowly growing or stationary population.
Ocean basin	Ocean basin refers to deep-ocean floor made of basaltic crust. Compare with continental margin.
Basin	A large submarine depression of a generally circular, elliptical or oval shape is called a basin.
Climate	Physical properties of the troposphere of an area based on analysis of its weather records over a long period. The two main factors determining an area's climate are temperature, with its seasonal variations, and the amount and distri.
Magnetic field	A region where magnetic forces affect any magnetized bodies or electric currents is the magnetic field. Earth is surrounded by a magnetic field.
Upwelling	The process by which water rises from a deeper to a shallower depths, usually as a result of offshore surface water flow is called upwelling. It is most prominent where persistent wind blows parallel to a coastline so that the resultant Ekman transport moves surface water away from the coast.
Observations	Information obtained through one or more of the five senses or through instruments that extend the senses are observations.
Well	Well refers to a hole, generally cylindrical and usually walled or lined with pipe, that is dug or drilled into the ground to penetrate an aquifer below the zone of saturation.
Rift	Rift refers to the valley created at a pull-apart zone.
Lava	Molten rock that is extruded out of volcanoes is called lava.
Sea	Sea refers to the ocean. A large body of salt water, second in rank to an ocean, more or less landlocked and generally part of, or connected with, an ocean or a larger sea. State of the ocean or lake surface, in regard to waves.
Volcano	Vent or fissure in the earth's surface through which magma, liquid lava, and gases are

released into the environment is a volcano.

Theory	A general explanation of a characteristic of nature consistently supported by observation or experiment is referred to as a theory.
Asthenosphere	Earth materials located at a depth of 100 km to 700 km below the Earth's surface, underneath the tectonic plates is called the asthenosphere.
Map	Map refers to a representation of Earth's surface usually depicting mostly land areas.
Site	A factor considering the summation of all environmental features of a location that influences the placement of a city is a site.
Chalk	A white, soft limestone consisting dominantly of the shells of foraminifera is a chalk.
Rocks	An aggregate of one or more minerals rather large in area are rocks. The three classes of rocks are the following: Igneous rock - crystalline rocks formed from molten material. Sedimentary rock - A rock resulting from the consolidation of loose sediment that has accumulated in layers. Metamorphic rock - Rock that has formed from preexisting rock as a result of heat or pressure.
Recent	A synonym of Holocene is called recent.
Seismic	Referring to vibrations in the Earth produced by earthquakes is referred to as seismic.
Seismic wave	Seismic wave refers to a low-frequency wave generated by the forces that cause earthquakes. Some kinds of seismic waves can pass through Earth.
Wave	An oscillatory movement in a body of water manifested by an alternate rise and fall of the surface is called a wave. Disturbances of the surface of a liquid body, as the ocean, in the form of a ridge, swell or hump. The term wave by itself usually refers to the term surface gravity wave.
Resolution	Resolution in general, refers to a measure of the finest detail distinguishable in an object or phenomenon. In particular, a measure of the finest detail distinguishable in an image.
Volcanic rock	Rock formed by solidification of magma at the Earth's surface is a volcanic rock.
Mass	The amount of material in an object is the mass.
Temperature	Temperature refers to a measure of the average speed of motion of the atoms, ions, or molecules in a substance or combination of substances at a given moment. Compare with heat.
Depth	Depth refers to vertical distance from still-water level to the bottom.
Base	A substance that combines with a hydrogen ion in solution is called the base.
Swell	Waves that have traveled a long distance from their generating area and have been sorted out by travel into long waves of the same approximate period are referred to as a swell.
Atoll	A ring-shaped coral reef that surrounds a lagoon is called an atoll.
Avalanche	A large mass of snow, ice, soil, or rock that moves rapidly downslope under the pull of gravity is referred to as an avalanche.
Core	Core refers to a cylindrical sample extracted from a beach or seabed to investigate the types and DEPTHS of sediment layers. An inner, often much less permeable portion of a BREAKWATER, or BARRIER beach.
Current	Current refers to the flowing of water, or other liquid or gas. That portion of a stream of water which is moving with a velocity much greater than the average or in which the progress of the water is principally concentrated. Ocean currents can be classified in a number of different ways.

Go to **Cram101.com** for the Practice Tests for this Chapter.

Oceanic crust	Oceanic crust refers to the outermost rock shell of the earth, some 5 kilometers thick, that underlies ocean basins; it is composed of basalt and sedimentary layers.
Subduction zone	Elongate region in which the sea floor slides beneath a continent or island arc is referred to as the subduction zone.
Plate tectonics	Plate tectonics refer to the theory of geophysical processes that explains the movements of lithospheric plates and the processes that occur at their boundaries.
Tectonics	The study of the major structural features of the Earth's crust or the broad structure of a region is referred to as tectonics.
Viscosity	Resistance to flow is called viscosity.
Density	Density refers to the ratio of a mass to a unit volume specified as grams per cubic centimeter.
Buoyancy	The resultant upward forces, exerted by the water on a submerged or floating body, equal to the weight of the water displaced by this body is referred to as buoyancy.
Force	Force refers to a push or pull that affects motion. The product of mass and acceleration of a material.
Entrainment	The act of resuspending or remobilizing a sediment grain resting on the sea bottom is an entrainment.
Subsidence	Sinking or down warping of a part of the earth's surface is called subsidence.
Energy	Capacity to do work by performing mechanical, physical, chemical, or electrical tasks or to cause a heat transfer between two objects at different temperatures is an energy.
Decompression melting	The most common process creating magma is by reducing pressure on hot rock, not by adding more heat is a decompression melting.
Granite	A light-colored, coarsegrained, intrusive igneous rock composed mainly of quartz and feldspar and that typifies the continental crust are called the granite.
Crystallization	The growth of minerals in a fluid such as magma is a crystallization.
River	A natural stream of water larger than a brook or creek is a river.
Key	Key refers to a low, insular BANK of sand, coral, etc., as one of the islets off the southern coast of Florida.
Conditions	Conditions refers to physical or chemical attributes of the environment that, while not being consumed, influence biological processes and population growth. Examples are temperature, salinity, and acidity. Compare resources.
Equilibrium	A point of rest. A system that does not tend to undergo any change of its own accord but remains in a single, fixed condition is said to be in equilibrium. Compare with steady state.
Shield volcano	Shield volcano refers to a very wide volcano built of low-viscosity lavas.
Lead	Lead refers to a heavy metal that is an important constituent of automobile batteries and other industrial products. A toxic metal capable of causing environmental disruption and producing a health problem to people and other living organisms.
Sediment	Loose, fragments of rocks, minerals or organic material which are transported from their source for varying distances and deposited by air, wind, ice and water are sediment. Other sediments are precipitated from the overlying water or form chemically, in place. Sediment includes all the unconsolidated materials on the sea floor. The fine-grained material deposited by water or wind.

Mixture	Mixture refers to combination of two or more elements and compounds.
Recycling	Recycling refers to collecting and reprocessing a resource so it can be made into new products. Compare with reuse.
Trace elements	Elements that occur in seawater in tiny quantities, typically measured in concentrations of parts per billion are trace elements.
Island arc	A curved group of volcanic islands associated with a deep oceanic trench and subduction zone is called island arc.
Chemical	One of the millions of different elements and compounds found naturally or synthesized by human is referred to as chemical.
Trace element	A minor constituent of seawater present in amounts less than 1 part per million is called a trace element.
Atoms	Atoms refers to minute units made of subatomic particles that are the basic building blocks of all chemical elements and thus all matter; the smallest unit of an element that can exist and still have the unique characteristics of that element. Compare to ion, molecule.
Atom	The smallest component of an element, comprising neutrons, protons, and electrons is an atom.
Diffraction	The phenomenon occurring when water waves are propagated into a sheltered region formed by a BREAKWATER or similar barrier that interrupts a portion of the otherwise regular train of waves, resulting in the multi-directional spreading of the waves is called diffraction.
Wavelength	The horizontal distance between corresponding points on successive waves, such as from crest to crest or trough to trough is called a wavelength.
Concentration	Amount of a chemical in a particular volume or weight of air, water, soil, or other medium is referred to as concentration.
Parts per million	Parts per million refers to the number of parts of a chemical found in one million parts of a particular gas, liquid, or solid.
Buoy	Buoy refers to a float; especially a floating object moored to the bottom, to mark a CHANNEL, anchor, shoal rock, etc.
Accumulation	Buildup of matter, energy, or information in a system is referred to as accumulation.
Trench	A long narrow submarine depression with relatively steep sides is called a trench.
State	State refers to an expression of the internal form of matter. Water exists in three states: solid, liquid, and gas. A solid has a fixed volume and fixed shape; a liquid has a fixed volume but no fixed shape; and a gas has neither fixed volume nor fixed shape.
Pile	A long substantial pole of wood, concrete or metal, driven into the earth or sea bed to serve as a support or protection is called pile.
Topography	Topography refers to the form of the features of the actual surface of the Earth in a particular region considered collectively.
Range	Land used for grazing is referred to as the range.
Debris	Debris refers to any accumulation of rock fragments; detritus.
Gravity	The attraction between bodies of matter is a gravity.
Tsunami	Tsunami refers to a large, high-velocity wave generated by displacement of the sea floor; also called seismic sea wave. Commonly misnamed tidal wave.
Depression	Depression refers to a general term signifying any depressed or lower area in the ocean floor.

Load	The quantity of sediment transported by a current. It includes the suspended load of small particles in the water, and the bedload of large particles that move along the bottom.
Coral	Coral refers to any of more than 6,000 species of small cnidarians, many of which are capable of generating hard calcareous skeletons.
Sand	Sand refers to an unconsolidated mixture of inorganic soil consisting of small but easily distinguishable grains ranging in size from about .062 mm to 2.0 mm.
Trough	A long and broad submarine depression with gently sloping sides is referred to as a trough.
Seamount	Conical mountain rising 1000 m or more above the sea floor is a seamount.
Oceanography	That science treating of the oceans, their forms, physical features and phenomena is referred to as oceanography.
Cross section	A two-dimensional drawing showing features in the vertical plane as in a canvon wall or road cut is referred to as cross section.
Terrestrial	Terrestrial refers to pertaining to land. Compare with aquatic.
Slumping	The sliding of large, cohering blocks of sediment or rock downslope under the influence of gravity is called slumping.
Earthquake	Shaking of the ground resulting either from the fracturing and displacement of rock, producing a fault, or from subsequent movement along the fault is the earthquake.
Strain	A change in torn or size of a body due to external forces is called strain.
Andesite	Andesite refers to a common volcanic rock found in the volcanic arcs of subduction zones; it is intermediate in composition between the granitic crust of the continents and the basaltic crust of the oceans.
Water vapor	Water vapor refers to the gaseous, invisible form of water.
Silica	A compound with a composition, such as quartz in granite and opal in the shells of radiolaria is referred to as silica.
Slump	In mass wasting, movement along a curved surface in which the upper part moves vertically downward while the lower part moves outward is a slump.
Rips	Rips refer to agitation of water caused by the meeting of currents or by rapid current setting over an irregular bottom.
Fissure	A narrow parting or crack in rock are called the fissure.
Fire	The rapid combination of oxygen with organic material to produce flame, heat, and light is referred to as the fire.
Cinder cone	Steep volcanic hill made of loose pyroclastic debris is referred to as the cinder cone.
Carbon dioxide	Molecule of carbon and oxygen present in the atmosphere at approximately 350 ppm. Emissions of carbon dioxide resulting from burning of fossil fuels are thought to be co:.itributing to potential global warming through an enhanced greenhouse effect.
Smog	Originally a combination of smoke and fog, but now used to describe other mixtures of pollutants in the atmosphere is referred to as smog.
Atmosphere	Atmosphere refers to the whole mass of air surrounding the earth.
Stream	Stream refers to any flow of water; a current. A course of water flowing along a bed in the earth.
Erosion	Erosion refers to wearing away of the land by natural forces. On a beach, the carrying away

of beach material by wave action, tidal currents or by DEFLATION. The wearing away of land by the action of natural forces.

Brittle	Behavior of material where stress causes abrupt fracture are called the brittle.
Relief	The difference in elevation between the highest and lowest points in an area is called relief.
Pahoehoe	Lava flow with a smooth, ropy surface is the pahoehoe.
Crater	Crater refers to an abrupt basin commonly rimmed by ejected material. In volcanoes, craters form by outwarcc explosion, are commonly less than 2 km diameter, and occur at the summit of a volcanic cone. Similar rimmed basins form by impacts with meteorites, asteroids, and.
Littoral	Littoral refers to of, or pertaining to, a shore, especially a seashore. Living on, or occurring on, the SHORE.
Limestone	A sedimentary rock composed dominantly of calcium carbonate, either precipitated from seawater or deposited as shell debris are called the limestone.
Shoreline	Shoreline refers to the intersection of a specified plane of water with the shore. All of the water areas of the state, including reservoirs and their associated uplands, together with the lands underlying them.
Fault	A fracture in rock along which there has been an observable amount of displacement. Faults are rarely single planar units; normally they occur as parallel to sub-parallel sets of planes along which movement has taken place to a greater or lesser extent. Such sets are called fault or fracture-zones.
Coastal processes	Collective term covering the action of natural forces on the shoreline, and the nearshore seabed are called coastal processes.
Unconsolidated	Unconsolidated with regards to sediment grains refers to loose, separate, or unattached to one another.
Shore	That strip of ground bordering any body of water which is alternately exposed, or covered by tides and/or waves is a shore. A shore of unconsolidated material is usually called a beach.
Reef	A ridge of rock or other material lying just below the surface of the sea is called a reef.
Barrier reef	A coral reef growing around the periphery of an island, but separated from it by a lagoon is referred to as a barrier reef.
Continental margin	The drowned edges of continents consisting of the continental shelf, the continental slope, and the continental rise is called continental margin.
Divergent plate boundary	Divergent plate boundary refers to area where earth's lithospheric plates move apart in opposite directions. Compare convergent plate boundary, transform fault.
Fall	Fall refers to a mass moving nearly vertical and downward under the influence of gravity.
Glacier	Large masses of moving ice on land derived by the recrystallization of snow into ice under pressure are referred to as glacier.
Lake	Large natural body of standing fresh water formed when water from precipitation, land runoff, or groundwater flow fills a depression in the earth created by glaciation, earth movement, volcanic activity, or a giant meteorit are called the lake.
Stress	Stress refers to force per unit area. May be compression, tension, or shear.
Failed rift	Site of a spreading center that did not open far enough to create an ocean basin is a failed rift.

Go to **Cram101.com** for the Practice Tests for this Chapter.

427

Aseismic ridge	A submarine ridge with which no earthquakes are associated is called the aseismic ridge.
Creek	Creek refers to a stream, less predominant than a river, and generally tributary to a river. A small tidal CHANNEL through a coastal MARSH.
Wildlife	Wildlife refers to all free, undomesticated species.
Groundwater	Groundwater refers to water that sinks into the soil and is stored in slowly flowing and slowly renewed underground reservoirs called aquifers; underground water in the zone of saturation, below the water table. Compare runoff, surface water.
Dike	Dike refers to sometimes written as dyke; earth structure along a sea or river in order to protect LITTORAL lands from flooding by high water; DIKES along rivers are sometimes called levees.
Hypothesis	In science, an explanation set forth in a manner that can be tested and is capable of being disproved. A tested hypothesis is accepted until and unless it has been disproved.
Magnitude	An assessment of the size of an event is a magnitude. Magnitude scales exist for earthquakes, volcanic eruptions. hurricanes, and tornadoes. For earthquakes, different magnitudes are calculated for the same earthquake when different types of seismic waves are used.
Front	The boundary between two air masses with different temperatures and densitie is referred to as front.
Seafloor spreading	Seafloor spreading refers to the theory that new ocean crust forms at spreading centers, most of which are on the ocean floor, and pushes the continents aside. Power is thought to be provided by convection currents in Earth's upper mantle.
Climate change	Refers to any long-term trend in MEAN SEA LEVEL, wave height, wind speed, drift rate etc are called the climate change.
Species	Group of organisms that resemble one another in appearance, behavior, chemical makeup and processes, and genetic structure is a species. Organisms that reproduce sexually are classified as members of the same species only if they can breed with one another and produce offspring.
Coal	Solid, combustible mixture of organic compounds with 30-98% carbon by weight, mixed with various amounts of water and small amounts of sulfur and nitrogen compounds. It is formed in several stages as the remains of plants are subjected to heat and press is a coal.
Plants	Eukaryotic, mostly multicelled organisms such as algae, mosses, ferns, flowers, cacti, grasses, beans, wheat, rice, and trees are plants. These organisms use photosynthesis to produce organic nutrients for themselves and for other organisms is referred to as plants.
Situation	The relative geographic location of a site that makes it a good location for a city is a situation.
Lower mantle	The rigid portion of Earth's mantle below the asthenosphere are called the lower mantle.
Geophysics	The study of the physical characteristics and properties of the Earth is a geophysics.
Science	Attempts to discover order in nature and use that knowledge to make predictions about what should happen in nature is called science.
Resources	Resources refer to substances that can be consumed by an organism and, as a result, become unavailable to other organisms.
Geology	Geology refers to the science which treats of the origin, history and structure of the Earth, as recorded in rocks; together with the forces and processes now operating to modify rocks.
Feedback	A kind of system response that occurs when output of the system also serves as input leading

to changes in the system is called feedback.

Tectonics	The study of the major structural features of the Earth's crust or the broad structure of a region is referred to as tectonics.
Development	Development refers to change from a society that is largely rural, agricultural, illiterate, and poor, with a rapidly growing population, to one that is mostly urban, industrial, educated, and wealthy, with a slowly growing or stationary population.
Erosion	Erosion refers to wearing away of the land by natural forces. On a beach, the carrying away of beach material by wave action, tidal currents or by DEFLATION. The wearing away of land by the action of natural forces.
Climate	Physical properties of the troposphere of an area based on analysis of its weather records over a long period. The two main factors determining an area's climate are temperature, with its seasonal variations, and the amount and distri.
Valley	An elongated depression, usually with an outlet, between bluffs or between ranges of hills or mountains is a valley.
Point	Point refers to the extreme end of a cape, or the outer end of any land area protruding into the water, usually less prominent than a cape. A low profile shoreline promontory of more or less triangular shape, the top of which extends seaward.
System	A set of components that function and interact in some regular and theoretically predictable manner is called a system.
Sedimentary rocks	Sedimentary rocks refer to rocks that have formed by the compaction and cementation of sediment.
Rocks	An aggregate of one or more minerals rather large in area are rocks. The three classes of rocks are the following: Igneous rock - crystalline rocks formed from molten material. Sedimentary rock - A rock resulting from the consolidation of loose sediment that has accumulated in layers. Metamorphic rock - Rock that has formed from preexisting rock as a result of heat or pressure.
Continent	Continent refers to lower-density masses of rock, exposed as about 40 percent of the Earth's surface: 29 percent as land and I 1 percent as the floor of shallow seas.
Sediment	Loose, fragments of rocks, minerals or organic material which are transported from their source for varying distances and deposited by air, wind, ice and water are sediment. Other sediments are precipitated from the overlying water or form chemically, in place. Sediment includes all the unconsolidated materials on the sea floor. The fine-grained material deposited by water or wind.
Rock	Any material that makes up a large, natural, continuous part of earth's crust is rock.
Sea	Sea refers to the ocean. A large body of salt water, second in rank to an ocean, more or less landlocked and generally part of, or connected with, an ocean or a larger sea. State of the ocean or lake surface, in regard to waves.
Weathering	Physical and chemical processes in which solid rock exposed at earth's surface is changed to separate solid particles and dissolved material, which can then be moved to another place as sediment is referred to as weathering.
Sedimentary rock	Rock that forms from the accumulated products of erosion and in some cases from the compacted shells, skeletons, and other remains of dead organisms is sedimentary rock. Compare with igneous rock, metamorphic roc is a sedimentary rock.
Relief	The difference in elevation between the highest and lowest points in an area is called relief.
Equilibrium	A point of rest. A system that does not tend to undergo any change of its own accord but

Go to **Cram101.com** for the Practice Tests for this Chapter.

remains in a single, fixed condition is said to be in equilibrium. Compare with steady state.

Igneous rocks	Igneous rocks refers to rocks formed from the solidification of magma. They are extrusive if they crystallize on the surface of the Earth and intrusive if they crystallize beneath the surface.
Rift	Rift refers to the valley created at a pull-apart zone.
Flood	Period when tide level is rising; often taken to mean the flood current which occurs during this period. A flow above the CARRYING CAPACITY of a CHANNEL.
Uplift	The rising of one part of the Earth's crust relative to another part is referred to as uplift.
Lava	Molten rock that is extruded out of volcanoes is called lava.
Solar system	The sun together with the planets and other bodies that revolve around it is a solar system.
Meteorite	A stony or metallic body from space that passed through the atmosphere and landed on the surface of the Earth is referred to as a meteorite.
Planet	A smaller, usually nonluminous body orbiting a star is a planet.
Volcano	Vent or fissure in the earth's surface through which magma, liquid lava, and gases are released into the environment is a volcano.
Sand dune	Dune formed of sand is called a sand dune.
Wind	Wind refers to the mass movement of air.
Map	Map refers to a representation of Earth's surface usually depicting mostly land areas.
Subduction zone	Elongate region in which the sea floor slides beneath a continent or island arc is referred to as the subduction zone.
Zone	Division or province of the ocean with homogeneous characteristics is referred to as a zone.
Slope	The degree of inclination to the horizontal is the slope.
Stability	Ability of a living system to withstand or recover from externally imposed changes or stresses is called stability.
River	A natural stream of water larger than a brook or creek is a river.
Groundwater	Groundwater refers to water that sinks into the soil and is stored in slowly flowing and slowly renewed underground reservoirs called aquifers; underground water in the zone of saturation, below the water table. Compare runoff, surface water.
Conditions	Conditions refers to physical or chemical attributes of the environment that, while not being consumed, influence biological processes and population growth. Examples are temperature, salinity, and acidity. Compare resources.
Latitude	Latitude refers to distance from the equator. Compare altitude.
Topography	Topography refers to the form of the features of the actual surface of the Earth in a particular region considered collectively.
Degree	An arbitrary measure of temperature. One degree Celsius _ 1.8 degrees Fahrenheit.
Soil	Soil refers to a layer of weathered, unconsolidated material on top of bedrock; often also defined as containing organic matter and being capable of supporting plant growth.
Peninsula	An elongated portion of land nearly surrounded by water and connected to a larger body of land, usually by a neck or an isthmus is called a peninsula.

435

Limestone	A sedimentary rock composed dominantly of calcium carbonate, either precipitated from seawater or deposited as shell debris are called the limestone.
Eolian	Describes sediment that was deposited by wind are called eolian.
Range	Land used for grazing is referred to as the range.
Well	Well refers to a hole, generally cylindrical and usually walled or lined with pipe, that is dug or drilled into the ground to penetrate an aquifer below the zone of saturation.
Surface water	Precipitation that does not infiltrate the ground or return to the atmosphere by evaporation or transpiration is called surface water.
Cliff	A high steep face of rock is referred to as cliff.
Environment	Environment refers to all external conditions and factors, living and nonliving, that affect an organism or other specified system during its lifetime; the earth's life-support systems for us and for all other forms of life-another term for solar capita.
Weather	Weather refers to short-term changes in the temperature, barometric pressure, humidity, precipitation, sunshine, cloud cover, wind direction and speed, and other conditions in the troposphere at a given place and time. Compare with climate.
Plate	One of about a dozen rigid segments of Earth's lithosphere that move independently is a plate. The plate consists of continental or oceanic crust and the cool, rigid upper mantle directly below the crust.
Reduce	With respect to waste management, reduce refers to practices that will reduce the amount of waste we produce.
Desert	Desert refers to biome in which evaporation exceeds precipitation and the average amount of precipitation is less than 25 centimeters a year. Such areas have little vegetation or have widely spaced, mostly low vegetation. Compare forest, grassland.
State	State refers to an expression of the internal form of matter. Water exists in three states: solid, liquid, and gas. A solid has a fixed volume and fixed shape; a liquid has a fixed volume but no fixed shape; and a gas has neither fixed volume nor fixed shape.
Isostatic equilibrium	Isostatic equilibrium refers to balanced support of lighter material in a heavier, displaced supporting matrix. Analogous to buoyancy in a liquid.
Convergence	Resemblance among species belonging to different taxonomic groups as the result from adaptation to similar environments is the convergence.
Stream	Stream refers to any flow of water; a current. A course of water flowing along a bed in the earth.
Mantle	Zone of the earth's interior between its core and its crust. Compare with core, crust is the mantle.
Depth	Depth refers to vertical distance from still-water level to the bottom.
Brittle	Behavior of material where stress causes abrupt fracture are called the brittle.
Compression	A state of stress that causes a pushing together or contraction is called compression.
Magma	Molten rock below the earth's surface is referred to as magma.
Crust	Solid outer zone of the earth. It consists of oceanic crust and continental crust. Compare core, mantle.
Granite	A light-colored, coarsegrained, intrusive igneous rock composed mainly of quartz and feldspar and that typifies the continental crust are called the granite.

Go to **Cram101.com** for the Practice Tests for this Chapter.

437

Isostasy	Isostasy refers to the tendency of the Earth's crust to maintain a state of near equilibrium, i.e., if anything occurs to modify the existing state, a compensation change will occur to maintain a balance.
Base	A substance that combines with a hydrogen ion in solution is called the base.
Metamorphic rock	Metamorphic rock refers to rock produced when a preexisting rock is subjected to high temperatures, high pressures, chemically active fluids, or a combination of these agents. Compare with igneous rock, sedimentary rock.
Event	Event refers to an occurrence meeting specified conditions, e.g. damage, a threshold wave height or a threshold water level.
Observations	Information obtained through one or more of the five senses or through instruments that extend the senses are observations.
Current	Current refers to the flowing of water, or other liquid or gas. That portion of a stream of water which is moving with a velocity much greater than the average or in which the progress of the water is principally concentrated. Ocean currents can be classified in a number of different ways.
Elevation	Elevation refers to the distance of a point above a specified surface of constant potential; the distance is measured along the direction of gravity between the point and the surface.
Subsidence	Sinking or down warping of a part of the earth's surface is called subsidence.
Mass	The amount of material in an object is the mass.
Altitude	Altitude refers to the height above sea level. Compare to latitude.
Igneous rock	Rock formed when molten rock material wells up from earth's interior, cools, and solidifies into rock masses. Compare metamorphic rock, sedimentary roc is an igneous rock.
Energy	Capacity to do work by performing mechanical, physical, chemical, or electrical tasks or to cause a heat transfer between two objects at different temperatures is an energy.
Resources	Resources refer to substances that can be consumed by an organism and, as a result, become unavailable to other organisms.
Continental crust	The light, buoyant granitic rock that underlies continental masses and averages about 35 kilometers in thickness is called continental crust.
Asthenosphere	Earth materials located at a depth of 100 km to 700 km below the Earth's surface, underneath the tectonic plates is called the asthenosphere.
Ridge	Ridge refers to the volcanic mountain ranges that lie along the spreading centers on the floors of the oceans.
Ocean	The great body of salt water which occupies two-thirds of the surface of the Earth, or one of its major subdivisions is called an ocean.
Plate tectonics	Plate tectonics refer to the theory of geophysical processes that explains the movements of lithospheric plates and the processes that occur at their boundaries.
Bay	Bay refers to a recess or inlet in the shore of a sea or lake between two capes or headlands, not as large as a gulf but larger than a cove.
Lake	Large natural body of standing fresh water formed when water from precipitation, land runoff, or groundwater flow fills a depression in the earth created by glaciation, earth movement, volcanic activity, or a giant meteorit are called the lake.
Topographic map	A map on which elevations are shown by means of contour lines is referred to as a topographic

Fracture	Fracture refers to a general term for any breaks in rock. Fractures include faults, joints, and crack,_.
Chalk	A white, soft limestone consisting dominantly of the shells of foraminifera is a chalk.
Unconformity	A surface that represents a break in the geologic record, with the rock unit immediately above it being considerably younger than the rock beneath is called unconformity.
Dip	The angle of inclination measured in degrees from the horizontal is called dip.
Bedding	The layering of rocks, especially sedimentary rooks is called bedding.
Precipitation	Water in the form of rain, sleet, hail, and snow that falls from the atmosphere onto the land and bodies of water is called precipitation.
Cement	Minerals such as silica and carbonate that are chemically precipitated in the pores of sediments, binding the grain are referred to as cement.
Gulf	A relatively large portion of sea, partly enclosed by land is called gulf.
Moraine	Moraine refers to an accumulation of earth, stones, etc., deposited by a glacier, usually in the form of a mound, ridge or other prominence on the terrain.
Basin	A large submarine depression of a generally circular, elliptical or oval shape is called a basin.
Coast	A strip of land of indefinite length and width that extends from the SEASHORE inland to the first major change in terrain features is referred to as coast.
Impermeable	Impermeable refers to impervious; the condition of rock that does not allow fluids to flow through it.
Spring	A place where groundwater flows out onto the surface is a spring.
Fall	Fall refers to a mass moving nearly vertical and downward under the influence of gravity.
Fault	A fracture in rock along which there has been an observable amount of displacement. Faults are rarely single planar units; normally they occur as parallel to sub-parallel sets of planes along which movement has taken place to a greater or lesser extent. Such sets are called fault or fracture-zones.
Anticline	Anticline refers to a fold where rock layers are compressed into a convex-upward pattern.
Front	The boundary between two air masses with different temperatures and densitie is referred to as front.
Clay	Clay refers to a fine grained sediment with a typical grain size less than 0.004 mm. Possesses electromagnetic properties which bind the grains together to give a bulk strength or cohesion.
Shoreline	Shoreline refers to the intersection of a specified plane of water with the shore. All of the water areas of the state, including reservoirs and their associated uplands, together with the lands underlying them.
Coastal plain	Coastal plain refers to the plain composed of horizontal or gently sloping strata of clastic material fronting the COAST and generally representing a strip of recently emerged sea bottom that has emerged from the sea in recent geologic times. Also formed by AGGRADATION.
Continental margin	The drowned edges of continents consisting of the continental shelf, the continental slope, and the continental rise is called continental margin.
Normal fault	Normal fault refers to high-angle faults with one block dropping down relative to another block; they denote tension and are found at the axial rift valleys of ocean spreading ridges.

Go to **Cram101.com** for the Practice Tests for this Chapter.

Evaporation	Evaporation refers to conversion of a liquid into a gas.
Pleistocene	An epoch of the Quaternary Period characterized by several glacial ages is referred to as the Pleistocene.
Runoff	Fresh water from precipitation and melting ice that flows on the earth's surface into nearby streams, lakes, wetlands, and reservoir is referred to as runoff.
Load	The quantity of sediment transported by a current. It includes the suspended load of small particles in the water, and the bedload of large particles that move along the bottom.
Debris	Debris refers to any accumulation of rock fragments; detritus.
Network	Network refers to a set consisting of stations for which geometric relationships have been determined and which are so related that removal of one station from the set will affect the relationships between the other stations; and lines connecting the stations to show this interdependence.
Bedrock	Solid rock lying beneath loose soil or unconsolidated sediment is called bedrock.
Cross section	A two-dimensional drawing showing features in the vertical plane as in a canvon wall or road cut is referred to as cross section.
Upland	Upland refers to the dry land area above and landward of the ordinary high water mark.
Recent	A synonym of Holocene is called recent.
Basalt	Basalt refers to a dark, fine-grained igneous rock composed of minerals enriched in ferromagnesian silicates.
Hot spot	A zone on a lithospheric plate that overlies unusually hot asthenosphere, and where large volumes of lava commonly are extruded, building a large volcanic pile on the sea floor is called hot spot.
Unconsolidated	Unconsolidated with regards to sediment grains refers to loose, separate, or unattached to one another.
Ash	The loose debris that is ejected from an erupting volcano is called ash.
Neck	Neck refers to the narrow strip of land which connects a peninsula with the mainland, or connects two ridges. The narrow band of water flowing seaward through the surf.
Fissure	A narrow parting or crack in rock are called the fissure.
Mantle plume	The upwelling of hot material into and through the lithosphere, with magma spilling out onto the earth's surface and building thick volcanic piles is called a mantle plume. If the lithosphere is moving relative to the plume or 'hot spot,' a linear trail of volcanoes is produced.
Plume	Plume refers to an arm of magna rising upward from the mantle.
Dike	Dike refers to sometimes written as dyke; earth structure along a sea or river in order to protect LITTORAL lands from flooding by high water; DIKES along rivers are sometimes called levees.
Dam	Dam refers to structure built in rivers or estuaries, basically to separate water at both sides and/or to retain water at one side.
Flooding	The natural process whereby waters emerge from their stream channel to cover part of the floodplain. Natural flooding is not a problem until people choose to build homes and other structures on floodplains.
Ice age	Ice age refers to one of several periods of low temperature during the last million years.

Island arc	A curved group of volcanic islands associated with a deep oceanic trench and subduction zone is called island arc.
Crater	Crater refers to an abrupt basin commonly rimmed by ejected material. In volcanoes, craters form by outwarcc explosion, are commonly less than 2 km diameter, and occur at the summit of a volcanic cone. Similar rimmed basins form by impacts with meteorites, asteroids, and.
Quaternary	The youngest geologic period, includes the present time is a quaternary. The latest period of time in the stratigraphic column, 0 - 2 million years, represented by local accumulations of glacial and post-glacial deposits, which continue, without change of fauna, from the top of the Pliocene. The quaternary appears to be an artificial division of time to separate pre-human from post-human sedimentation. As thus defined, the quaternary is increasing in duration as man's ancestry becomes longer.
Caldera	Caldera refers to a large, basinshaped volcanic depression. Roughly circular in map view, that forms by a piston-like collapse of a cylinder of overlying rock into an underlying, partially evacuated magma chamber.
Fire	The rapid combination of oxygen with organic material to produce flame, heat, and light is referred to as the fire.
Key	Key refers to a low, insular BANK of sand, coral, etc., as one of the islets off the southern coast of Florida.
Geomorphology	Geomorphology refers to that branch of physical geography which deals with the form of the Earth, the general configuration of its surface, the distribution of the land, water, etc. The investigation of the history of geologic changes through the interpretation of topographic forms.
Geology	Geology refers to the science which treats of the origin, history and structure of the Earth, as recorded in rocks; together with the forces and processes now operating to modify rocks.
Feedback	A kind of system response that occurs when output of the system also serves as input leading to changes in the system is called feedback.
Slide	In mass wasting, movement of a descending mass along a plane approximately parallel to the slope of the surface is referred to as slide.
Host	Plant or animal on which a parasite feeds is referred to as host.
Orbit	Orbit in ocean waves refers to the circular pattern of water particle movement at the air-sea interface.

445

Resources	Resources refer to substances that can be consumed by an organism and, as a result, become unavailable to other organisms.
Mineral resources	Elements, chemical compounds, minerals, or rocks concentrated in a form that can be extracted to obtain a usable commodity are referred to as mineral resources.
Natural resources	Natural resources refer to nutrients and minerals in the soil and deeper layers of the earth's crust, water, wild and domesticated plants and animals, air, and other resources produced by the earth's natural processes. Compare with human capital, manufactured capital, and solar capital.
Fall	Fall refers to a mass moving nearly vertical and downward under the influence of gravity.
Resource	Resource refers to anything obtained from the living and nonliving environment to meet human needs and wants. It can also be applied to other species.
Base	A substance that combines with a hydrogen ion in solution is called the base.
Soil	Soil refers to a layer of weathered, unconsolidated material on top of bedrock; often also defined as containing organic matter and being capable of supporting plant growth.
Basalt	Basalt refers to a dark, fine-grained igneous rock composed of minerals enriched in ferromagnesian silicates.
Population	Group of individual organisms of the same species living within a particular area is referred to as a population.
Planet	A smaller, usually nonluminous body orbiting a star is a planet.
Oil	The liquid form of petroleum consisting of a complex mixture of large hydrocarbon molecules is referred to as oil.
Lead	Lead refers to a heavy metal that is an important constituent of automobile batteries and other industrial products. A toxic metal capable of causing environmental disruption and producing a health problem to people and other living organisms.
Ore deposits	Earth materials in which metals are concentrated in high concentrations, sufficient to be mined are called ore deposits.
Conditions	Conditions refers to physical or chemical attributes of the environment that, while not being consumed, influence biological processes and population growth. Examples are temperature, salinity, and acidity. Compare resources.
Weathering	Physical and chemical processes in which solid rock exposed at earth's surface is changed to separate solid particles and dissolved material, which can then be moved to another place as sediment is referred to as weathering.
Nonrenewable energy	Alternative energy sources, including nuclear and geothermal, that are dependent on fuels or a resource that may be used up much faster than it is replenished by natural processes is a nonrenewable energy.
Energy	Capacity to do work by performing mechanical, physical, chemical, or electrical tasks or to cause a heat transfer between two objects at different temperatures is an energy.
Coal	Solid, combustible mixture of organic compounds with 30-98% carbon by weight, mixed with various amounts of water and small amounts of sulfur and nitrogen compounds. It is formed in several stages as the remains of plants are subjected to heat and press is a coal.
Natural gas	Underground deposits of gases consisting of 50-90% by weight methane gas and small amounts of heavier gaseous hydrocarbon compounds such as propane and butane are called natural gas.
Power	Power refers to the time rate of doing work.

Renewable energy	Alternative energy sources, such as solar, water, wind, and biomass, that are more or less continuously made available in a time framework useful to people is renewable energy.
Solar energy	Collecting and using energy from the sun directly is called solar energy.
Wind power	Wind power refers to an alternative energy source that has been used by people for centuries. More recently, thousands of windmills have been installed to produce electric energy.
Tidal power	Tidal power refers to a form of water utilizing ocean tides in places where favorable topography allows for construction of a power plant.
Geothermal energy	Geothermal energy refers to heat transferred from the earth's underground concentrations of dry steam, wet steam, or hot water trapped in fractured or porous rock.
Plate tectonics	Plate tectonics refer to the theory of geophysical processes that explains the movements of lithospheric plates and the processes that occur at their boundaries.
Tectonics	The study of the major structural features of the Earth's crust or the broad structure of a region is referred to as tectonics.
Range	Land used for grazing is referred to as the range.
Reserves	Reserves refer to resources that have been identified from which a usable mineral can be extracted profitably at present prices with current mining.
Fossil	The remains or traces of organisms preserved in rocks or ancient sediment are referred to as fossil.
Forest	Forest refers to biome with enough average annual precipitation to support growth of various species of trees and smaller forms of vegetation. Compare desert, grassland.
Sedimentary rocks	Sedimentary rocks refer to rocks that have formed by the compaction and cementation of sediment.
Rocks	An aggregate of one or more minerals rather large in area are rocks. The three classes of rocks are the following: Igneous rock - crystalline rocks formed from molten material. Sedimentary rock - A rock resulting from the consolidation of loose sediment that has accumulated in layers. Metamorphic rock - Rock that has formed from preexisting rock as a result of heat or pressure.
Mineral resource	Mineral resource refers to the concentration of naturally occurring solid, liquid, or gaseous material in or on the earth's crust, in a form and amount such that extracting and converting it into useful materials or items is currently or potentially profitable.
Hydrothermal deposits	Minerals precipitated from or altered by very hot water circulating through rocks are referred to as hydrothermal deposits.
Clastic rocks	Rocks built up of fragments which have been produced by the processes of weathering and EROSION, and in general transported to a point of deposition are referred to as clastic rocks.
Stream	Stream refers to any flow of water; a current. A course of water flowing along a bed in the earth.
Sand	Sand refers to an unconsolidated mixture of inorganic soil consisting of small but easily distinguishable grains ranging in size from about.062 mm to 2.0 mm.
Gravel	Gravel refers to loose, rounded fragments of rock, larger than sand, but smaller than cobbles. Small stones and pebbles, or a mixture of these with sand.
Placer deposits	Mineral deposits consisting of dense, resistant and often economically valuable minerals which have been weathered from terrigenous rocks, transported to the sea and concentrated in

marine sediments by wave or current action are placer deposits.

Loess	Loess refers to extensive deposits of wind-blown fine sediment commonly winnowed from glacially dumped debris.
Chemical	One of the millions of different elements and compounds found naturally or synthesized by human is referred to as chemical.
Evaporite	A type of sediment precipitated from an aqueous solution, usually by the evaporation of water from a basin with restricted circulation is the evaporite.
Halite	A mineral composed of sodium chloride and commonly referred to as rock salt are called the halite.
Gypsum	An evaporite deposit composed of hydrous calcium sulfat is a gypsum.
Sediment	Loose, fragments of rocks, minerals or organic material which are transported from their source for varying distances and deposited by air, wind, ice and water are sediment. Other sediments are precipitated from the overlying water or form chemically, in place. Sediment includes all the unconsolidated materials on the sea floor. The fine-grained material deposited by water or wind.
Phosphate	An essential plant macronutrient is phosphate.
Limestone	A sedimentary rock composed dominantly of calcium carbonate, either precipitated from seawater or deposited as shell debris are called the limestone.
Hydrocarbon	Hydrocarbon refers to organic compounds composed of hydrogen, oxygen, and carbon that are the main components of petroleum.
Metamorphism	The changes in minerals and rock textures that occur with the elevated temperatures and pressures below the Earth's surface are referred to as metamorphism.
Asbestos	Asbestos refers to a term for several minerals that have the form of small-elongated particles. Some types of particles are believed to be carcinogenic or to carry with them carcinogenic materials.
Groundwater	Groundwater refers to water that sinks into the soil and is stored in slowly flowing and slowly renewed underground reservoirs called aquifers; underground water in the zone of saturation, below the water table. Compare runoff, surface water.
Residual	The components of water level not attributable to astronomical effects are called residual.
Brines	With respect to mineral resources, refers to waters with a high salinity that contain useful materials such as bromine, iodine, calcium chloride, and magnesium, we have brines.
Clay	Clay refers to a fine grained sediment with a typical grain size less than 0.004 mm. Possesses electromagnetic properties which bind the grains together to give a bulk strength or cohesion.
Crust	Solid outer zone of the earth. It consists of oceanic crust and continental crust. Compare core, mantle.
Parts per million	Parts per million refers to the number of parts of a chemical found in one million parts of a particular gas, liquid, or solid.
Concentration	Amount of a chemical in a particular volume or weight of air, water, soil, or other medium is referred to as concentration.
Mantle	Zone of the earth's interior between its core and its crust. Compare with core, crust is the mantle.
Reach	Reach refers to an arm of the ocean extending into the land. A straight section of restricted

waterway of considerable extent; may be similar to a narrows, except much longer in extent.

Igneous rock	Rock formed when molten rock material wells up from earth's interior, cools, and solidifies into rock masses. Compare metamorphic rock, sedimentary roc is an igneous rock.
Rock	Any material that makes up a large, natural, continuous part of earth's crust is rock.
Host	Plant or animal on which a parasite feeds is referred to as host.
Magma	Molten rock below the earth's surface is referred to as magma.
State	State refers to an expression of the internal form of matter. Water exists in three states: solid, liquid, and gas. A solid has a fixed volume and fixed shape; a liquid has a fixed volume but no fixed shape; and a gas has neither fixed volume nor fixed shape.
Crystallization	The growth of minerals in a fluid such as magma is a crystallization.
Density	Density refers to the ratio of a mass to a unit volume specified as grams per cubic centimeter.
Meteorite	A stony or metallic body from space that passed through the atmosphere and landed on the surface of the Earth is referred to as a meteorite.
Volcano	Vent or fissure in the earth's surface through which magma, liquid lava, and gases are released into the environment is a volcano.
Cross section	A two-dimensional drawing showing features in the vertical plane as in a canvon wall or road cut is referred to as cross section.
Nuclear reactors	Nuclear reactors refer to devices that produce controlled nuclear fission, generally for the production of electric energy.
Silicate minerals	The most important group of rock-forming minerals is called silicate minerals.
Heat	Total kinetic energy of all the randomly moving atoms, ions, or molecules within a given substance, excluding the overall motion of the whole object. This form of kinetic energy flows from one body to another when there is a temperature difference betwe is referred to as heat.
Rift	Rift refers to the valley created at a pull-apart zone.
Stone	Stone refers to quarried or artificially broken rock for use in construction.
Igneous rocks	Igneous rocks refers to rocks formed from the solidification of magma. They are extrusive if they crystallize on the surface of the Earth and intrusive if they crystallize beneath the surface.
Marble	Marble refers to metamorphosed limestone.
Continental crust	The light, buoyant granitic rock that underlies continental masses and averages about 35 kilometers in thickness is called continental crust.
Fault	A fracture in rock along which there has been an observable amount of displacement. Faults are rarely single planar units; normally they occur as parallel to sub-parallel sets of planes along which movement has taken place to a greater or lesser extent. Such sets are called fault or fracture-zones.
Granite	A light-colored, coarsegrained, intrusive igneous rock composed mainly of quartz and feldspar and that typifies the continental crust are called the granite.
Oceanic crust	Oceanic crust refers to the outermost rock shell of the earth, some 5 kilometers thick, that underlies ocean basins; it is composed of basalt and sedimentary layers.

Ocean	The great body of salt water which occupies two-thirds of the surface of the Earth, or one of its major subdivisions is called an ocean.
Erosion	Erosion refers to wearing away of the land by natural forces. On a beach, the carrying away of beach material by wave action, tidal currents or by DEFLATION. The wearing away of land by the action of natural forces.
Bed load	Bed load refers to heavy or large sediment particles that travel near or on the bed.
Load	The quantity of sediment transported by a current. It includes the suspended load of small particles in the water, and the bedload of large particles that move along the bottom.
Sediment transport	Sediment transport refers to the main agencies by which sedimentary materials are moved. Running water and wind are the most widespread transporting agents. In both cases, three mechanisms operate, although the particle size of the transported material involved is very different, owing to the differences in density and viscosity of air and water.
Clastic sediments	Clastic sediments refers to deposits of fragments of preexisting rocks that have been transported from their point of origin.
River	A natural stream of water larger than a brook or creek is a river.
Fuel	Air,/ substar ce that produces heat by combustion is a fuel.
Current	Current refers to the flowing of water, or other liquid or gas. That portion of a stream of water which is moving with a velocity much greater than the average or in which the progress of the water is principally concentrated. Ocean currents can be classified in a number of different ways.
Meander	Loop-like bends or curves in the flow path of a current are a meander.
Drainage basin	Total area drained by a stream and its tributaries is referred to as drainage basin.
Basin	A large submarine depression of a generally circular, elliptical or oval shape is called a basin.
Point	Point refers to the extreme end of a cape, or the outer end of any land area protruding into the water, usually less prominent than a cape. A low profile shoreline promontory of more or less triangular shape, the top of which extends seaward.
Chert	Chert refers to a hard siliceous rock composed of opaline silica derived from the hard parts of microscopic plants and animals.
Atmosphere	Atmosphere refers to the whole mass of air surrounding the earth.
Evaporation	Evaporation refers to conversion of a liquid into a gas.
Lake	Large natural body of standing fresh water formed when water from precipitation, land runoff, or groundwater flow fills a depression in the earth created by glaciation, earth movement, volcanic activity, or a giant meteorit are called the lake.
Well	Well refers to a hole, generally cylindrical and usually walled or lined with pipe, that is dug or drilled into the ground to penetrate an aquifer below the zone of saturation.
Key	Key refers to a low, insular BANK of sand, coral, etc., as one of the islets off the southern coast of Florida.
Smelting	Process in which a desired metal is separated from the other elements in an ore mineral is referred to as smelting.
Upwelling	The process by which water rises from a deeper to a shallower depths, usually as a result of offshore surface water flow is called upwelling. It is most prominent where persistent wind blows parallel to a coastline so that the resultant Ekman transport moves surface water away

Go to **Cram101.com** for the Practice Tests for this Chapter.

from the coast.

Temperature	Temperature refers to a measure of the average speed of motion of the atoms, ions, or molecules in a substance or combination of substances at a given moment. Compare with heat.
Valley	An elongated depression, usually with an outlet, between bluffs or between ranges of hills or mountains is a valley.
Chemical weathering	The decomposition of rocks under attack of base- or acidladen waters is called chemical weathering.
Relief	The difference in elevation between the highest and lowest points in an area is called relief.
Clay minerals	Clay minerals refers to layered, platy minerals composed of hydrous aluminum silicates. Examples include kaolinite and chlorite.
Nutrients	Nutrients refer to chemicals such as phosphorus and nitrogen that, when released into water sources, may cause pollution events such as eutrophication.
Plants	Eukaryotic, mostly multicelled organisms such as algae, mosses, ferns, flowers, cacti, grasses, beans, wheat, rice, and trees are plants. These organisms use photosynthesis to produce organic nutrients for themselves and for other organisms is referred to as plants.
Mixture	Mixture refers to combination of two or more elements and compounds.
Climate	Physical properties of the troposphere of an area based on analysis of its weather records over a long period. The two main factors determining an area's climate are temperature, with its seasonal variations, and the amount and distri.
Nonrenewable resource	Resource that exists in a fixed amount in various places in the earth's crust and has the potential for renewal only by geological, physical, and chemical processes taking place over hundreds of millions to billions of years is a nonrenewable resource.
Recharge	The addition of new water to an aquifer or to the zone of saturation is called a recharge.
Water table	The upper surface of a zone of saturation, where the body of groundwater is not confined by an overlying impermeable formation is a water table. Where an overlying confining formation exists, the aquifer in question has no water table.
Nuclear energy	Energy released when atomic nuclei undergo a nuclear reaction such as the spontaneous emission of radioactivity or nuclear fusion is nuclear energy.
Situation	The relative geographic location of a site that makes it a good location for a city is a situation.
Recent	A synonym of Holocene is called recent.
Radiation	Fast-moving particles or waves of energy are called radiation.
Benefits	Benefits refers to the economic value of a scheme, usually measured in terms of the cost of damages avoided by the scheme, or the valuation of perceived amenity or environmental improvements.
Technology	Technology refers to the creation of new products and processes intended to improve our efficiency, chances for survival, comfort level, and quality of life. Compare with science.
System	A set of components that function and interact in some regular and theoretically predictable manner is called a system.
Theory	A general explanation of a characteristic of nature consistently supported by observation or experiment is referred to as a theory.

Combustion	Act of burning is referred to as combustion.
Carbon dioxide	Molecule of carbon and oxygen present in the atmosphere at approximately 350 ppm. Emissions of carbon dioxide resulting from burning of fossil fuels are thought to be co:.itributing to potential global warming through an enhanced greenhouse effect.
Potential energy	Potential energy refers to energy stored in an object because of its position or the position of its parts. Compare with kinetic energy.
Water vapor	Water vapor refers to the gaseous, invisible form of water.
Kinetic energy	Kinetic energy refers to energy that matter has because of its mass and speed or velocity. Compare potential energy.
Development	Development refers to change from a society that is largely rural, agricultural, illiterate, and poor, with a rapidly growing population, to one that is mostly urban, industrial, educated, and wealthy, with a slowly growing or stationary population.
Wilderness	Area where the earth and its community of life have not been seriously disturbed by humans and where humans are only temporary visitors is referred to as wilderness.
Dam	Dam refers to structure built in rivers or estuaries, basically to separate water at both sides and/or to retain water at one side.
Life expectancy	Average number of years a newborn infant can be expected to live is a life expectancy.
Greenhouse gases	Greenhouse gases refers to gases in the earth's lower atmosphere that cause the greenhouse effect. Examples are carbon dioxide, chloro.
Noise pollution	Noise pollution refers to any unwanted, disturbing, or harmful sound that impairs or interferes with hearing, causes stress, hampers concentration and work efficiency, or causes accidents.
Pollution	Pollution refers to an undesirable change in the physical, chemical, or biological characteristics of air, water, soil, or food that can adversely affect the health, survival, or activities of humans or other living organisms.
Depth	Depth refers to vertical distance from still-water level to the bottom.
Thermal	The energy of the random motion of atoms and molecules is referred to as thermal.
Subduction	Subduction refers to a process in which one lithospheric plate descends beneath another.
Pore	Pore refers to an opening or void space in; oil or rock.
Zone	Division or province of the ocean with homogeneous characteristics is referred to as a zone.
Permeability	Permeability refers to the property of bulk material which permits movement of water through its pores.
Convection	Convection refers to the vertical transport of a fluid or the transfer of heat in fluids.
Geyser	A hot spring that gushes magmaheated water and steam is called geyser.
Caldera	Caldera refers to a large, basinshaped volcanic depression. Roughly circular in map view, that forms by a piston-like collapse of a cylinder of overlying rock into an underlying, partially evacuated magma chamber.
Air pollution	One or more chemicals in high enough concentrations in the air to harm humans, other animals, vegetation, or materials are called air pollution. Excess heat and noise can also be considered forms of air pollution.
Output	Output refers to matter, energy, or information leaving a system. Compare with input, throughput.

Go to **Cram101.com** for the Practice Tests for this Chapter.

Discharge	The volume of water flowing in a stream per unit of time is the discharge.
Subsidence	Sinking or down warping of a part of the earth's surface is called subsidence.
Tides	The periodic rise and fall of the Earth's water surface as a consequence of the gravitational attraction of the Moon and the Sun, which are called tides.
Bay	Bay refers to a recess or inlet in the shore of a sea or lake between two capes or headlands, not as large as a gulf but larger than a cove.
Entrance	The entrance to a navigable BAY, HARBOR or CHANNEL, INLET or mouth separating the ocean from an inland water body.
Tidal current	The alternating horizontal movement of water associated with the rise and fall of the tide caused by astronomical tide-producing forces is called tidal current.
Water pollution	Any physical or chemical change in surface water or groundwater that can harm living organisms or make water unfit for certain uses is referred to as water pollution.
Matter	Matter refers to anything that has mass and takes up space. On earth, where gravity is present, we weigh an object to determine its mass.
Barrier islands	Long, thin, low offshore islands of sediment that generally run parallel to the shore along some coasts are referred to as barrier islands.
Coast	A strip of land of indefinite length and width that extends from the SEASHORE inland to the first major change in terrain features is referred to as coast.
Peat	Peat refers to an organic deposit consisting predominantly of partly decayed plant matter.
Environment	Environment refers to all external conditions and factors, living and nonliving, that affect an organism or other specified system during its lifetime; the earth's life-support systems for us and for all other forms of life-another term for solar capita.
Mud	Mud refers to a mixture of silt and clay sized particles.
Lagoon	A shallow body of water, as a pond or lake, which usually has a shallow restricted INLET from the se is referred to as lagoon.
Sea	Sea refers to the ocean. A large body of salt water, second in rank to an ocean, more or less landlocked and generally part of, or connected with, an ocean or a larger sea. State of the ocean or lake surface, in regard to waves.
Debris	Debris refers to any accumulation of rock fragments; detritus.
Rings	Rings refers to large whirl-like eddies created by meander cutoffs of strong geostropic currents such as the Gulf Stream. They either have warm-water centers or cold-water centers.
Fossils	Skeletons, bones, shells, body parts, leaves, seeds, or impressions of such items that provide recognizable evidence of organisms that lived long ago are called fossils.
Nearshore	Nearshore in beach terminology refers to an indefinite zone extending seaward from the shoreline well beyond the breaker zone. The zone which extends from the swash zone to the position marking the start of the offshore zone, typically at water depths of the order of 20
Bar	An offshore ridge or mound of sand, gravel, or other unconsolidated material which is submerged, especially at the mouth of a river or estuary, or lying parallel to, and a short distance from, the beach is referred to as a bar.
Offshore	Offshore in beach terminology refers to the comparatively flat zone of variable width, extending from the shoreface to the edge of the continental shelf. It is continually submerged. The direction seaward from the shore. The zone beyond the nearshore zone where sediment motion induced by waves alone effectively ceases and where the influence of the sea

Go to Cram101.com for the Practice Tests for this Chapter.

bed on wave action is small in comparison with the effect of wind. The breaker zone directly seaward of the low tide line.

Barrier island	A detached portion of a barrier beach between two inlets is a barrier island.
Shoreline	Shoreline refers to the intersection of a specified plane of water with the shore. All of the water areas of the state, including reservoirs and their associated uplands, together with the lands underlying them.
Compaction	The decrease in volume and porosity of a sediment via burial is referred to as compaction.
Compression	A state of stress that causes a pushing together or contraction is called compression.
Strip mining	Strip mining refers to a form of surface mining in which bulldozers, power shovels, or stripping wheels remove large chunks of the earth's surface in strips.
Acid	A substance that releases a hydrogen ion in solution is an acid.
Precipitation	Water in the form of rain, sleet, hail, and snow that falls from the atmosphere onto the land and bodies of water is called precipitation.
Extinction	Extinction refers to complete disappearance of a species from the earth. This happens when a species cannot adapt and successfully reproduce under new environmental conditions or when it evolves into one or more new species. Compare speciation.
Ash	The loose debris that is ejected from an erupting volcano is called ash.
Sludge	Gooey mixture of toxic chemicals, infectious agents, and settled solids, removed from wastewater at a sewage treatment plant is called sludge.
Hydrocarbons	Compounds containing only carbon and hydrogen are a large group of organic compounds, including petroleum products, such as crude oil and natural gas are hydrocarbons.
Molecule	Molecule refers to a combination of two or more atoms of the same chemical element or different chemical elements held together by chemical bonds. Compare with atom, ion.
Crude oil	Gooey liquid consisting mostly of hydrocarbon compounds and small amounts of compounds containing oxygen, sulfur, and nitrogen. Extracted from underground accumulations, it is sent to oil refineries, where it is converted to heating oil, diesel fuel, ga is a crude oil.
Algae	Algae refers to simple marine and freshwater plants, unicellular and multicellular, that lack roots, stems, and leaves.
Migration	A term that refers to the habit of some animals is a migration.
Anticline	Anticline refers to a fold where rock layers are compressed into a convex-upward pattern.
Dip	The angle of inclination measured in degrees from the horizontal is called dip.
Impermeable	Impermeable refers to impervious; the condition of rock that does not allow fluids to flow through it.
Salt dome	Salt dome refers to an anticlinal fold with a columnar salt plug at its core.
Stratum	A layer or bed of sedimentary rock is a stratum.
Accumulation	Buildup of matter, energy, or information in a system is referred to as accumulation.
Map	Map refers to a representation of Earth's surface usually depicting mostly land areas.
Seismic	Referring to vibrations in the Earth produced by earthquakes is referred to as seismic.
Seismic waves	A long-period wave caused by an underwater seismic disturbance or volcanic eruption is a seismic waves.

Go to **Cram101.com** for the Practice Tests for this Chapter.

Go to **Cram101.com** for the Practice Tests for this Chapter.
And, **NEVER** highlight a book again!

Geology	Geology refers to the science which treats of the origin, history and structure of the Earth, as recorded in rocks; together with the forces and processes now operating to modify rocks.
Sedimentary rock	Rock that forms from the accumulated products of erosion and in some cases from the compacted shells, skeletons, and other remains of dead organisms is sedimentary rock. Compare with igneous rock, metamorphic roc is a sedimentary rock.
Horizon	Horizon refers to the line or circle which forms the apparent boundary between Earth and sky. A plane in rock strata characterized by particular features, as occurrence of distinctive fossil species. One of the series of distinctive layers found in a vertical cross-section of any well-developed soil.
Oil shale	Oil shale refers to a fine-grained rock containing various amounts of kerogen, a solid, waxy mixture of hydrocarbon compounds. Heating the rock to high temperatures converts the kerogen into a vapor that can be condensed to form slow-flowing heavy oil called shale oil.
Annual	Annual refers to plant that grows, sets seed, and dies in one growing season. Compare to perennial.
Politics	Process through which individuals and groups try to influence or control government policies and actions that affect the local, state, national, and international communities is called politics.
Conservation	The protection of an area, or particular element within an area, accepting the dynamic nature of the environment and therefore allowing change is a conservation.
Bacteria	Bacteria refer to prokaryotic, one-celled organisms. Some transmit diseases. Most act as decomposers and get the nutrients they need by breaking down complex organic compounds in the tissues of living or dead organisms into simpler inorganic nutrient compounds.
Volatile	Substances that readily become gases when pressure is decreased, or temperature increased are referred to as volatile.
Shore	That strip of ground bordering any body of water which is alternately exposed, or covered by tides and/or waves is a shore. A shore of unconsolidated material is usually called a beach.
Global warming	Warming of the earth's atmosphere as a result of increases in the concentrations of one or more greenhouse gase is called global warming.
Climate change	Refers to any long-term trend in MEAN SEA LEVEL, wave height, wind speed, drift rate etc are called the climate change.
Ice age	Ice age refers to one of several periods of low temperature during the last million years.
Continental slope	Continental slope refers to the declivity from the offshore border of the CONTINENTAL SHELF to oceanic depths. It is characterized by a marked increase in slope.
Slope	The degree of inclination to the horizontal is the slope.
Tsunami	Tsunami refers to a large, high-velocity wave generated by displacement of the sea floor; also called seismic sea wave. Commonly misnamed tidal wave.
Feedback	A kind of system response that occurs when output of the system also serves as input leading to changes in the system is called feedback.
Nucleus	Extremely tiny center of an atom, making up most of the atom's mass is the nucleus. It contains one or more positively charged protons and one or more neutrons with no electrical charge.
Isotopes	Two or more forms of a chemical element that have the same number of protons but different mass numbers because of different numbers of neutrons in their nuclei is referred to as isotopes.

Mass	The amount of material in an object is the mass.
Heat energy	Heat energy refers to energy of the random motion of atoms and molecules.
Thermal pollution	Thermal pollution refers to an increase in water temperature that has harmful effects on aquatic life.
Chain reaction	Multiple nuclear fissions, taking place within a certain mass of a fissionable isotope, that release an enormous amount of energy in a short time is a chain reaction.
Meltdown	Meltdown refers to the melting of the core of a nuclear reactor.
Rhyolite	Rhyolite refers to a volcanic rock typical of continents. Typically forms from high viscosity magma.
Surface water	Precipitation that does not infiltrate the ground or return to the atmosphere by evaporation or transpiration is called surface water.
Continent	Continent refers to lower-density masses of rock, exposed as about 40 percent of the Earth's surface: 29 percent as land and I 1 percent as the floor of shallow seas.
Brine	Brine refers to water having a much higher.
Fertilizer	Fertilizer refers to substance that adds inorganic or organic plant nutrients to soil and improves its ability to grow crops, trees, or other vegetation.
Longshore drift	Movement of sediments approximately parallel to the COASTLINE is called longshore drift.
Ridge	Ridge refers to the volcanic mountain ranges that lie along the spreading centers on the floors of the oceans.
Hot spot	A zone on a lithospheric plate that overlies unusually hot asthenosphere, and where large volumes of lava commonly are extruded, building a large volcanic pile on the sea floor is called hot spot.
Volcanic rock	Rock formed by solidification of magma at the Earth's surface is a volcanic rock.
Shear	The failure of a body where the mass on one side slides past the portion on the other side is a shear.
Fold	Fold refers to wavy geologic structures formed by the compression and bending of sedimentary layers.
Transform fault	Transform fault refers to area where earth's lithospheric plates move in opposite but parallel directions along a fracture in the lithosphere. Compare with convergent plate boundary, divergent plate boundary.
Plume	Plume refers to an arm of magna rising upward from the mantle.
Flood	Period when tide level is rising; often taken to mean the flood current which occurs during this period. A flow above the CARRYING CAPACITY of a CHANNEL.
Population density	Number of organisms in a particular population found in a specified area is the population density. Compare with ecological population density.
Growth rate	The net increase in some factor per unit time. In ecology, the growth rate of a population is sometimes measured as the increase in numbers of individuals or biomass per unit time and sometimes as a percentage increase in numbers or biomass per unit time.
Food	General term for organic molecules capable of providing energy to heterotrophs when combined with oxygen during biochemical respiration is called food.
Species	Group of organisms that resemble one another in appearance, behavior, chemical makeup and processes, and genetic structure is a species. Organisms that reproduce sexually are

467

classified as members of the same species only if they can breed with one another and produce offspring.

Poverty	Inability to meet basic needs for food, clothing, and shelter is poverty.
Weather	Weather refers to short-term changes in the temperature, barometric pressure, humidity, precipitation, sunshine, cloud cover, wind direction and speed, and other conditions in the troposphere at a given place and time. Compare with climate.
Ecosystem	Ecosystem refers to the living organisms and the nonliving environment interacting in a given area.
Deforestation	Removal of trees from a forested area without adequate replanting is a deforestation.
Habitat	The place where an organism lives is called habitat.
Soil erosion	Soil erosion refers to movement of soil components, especially topsoil, from one place to another, usually by exposure to wind, flowing water, or both. This natural process can be greatly accelerated by human activities that remove vegetation from soil.
Environmental degradation	Depletion or destruction of a potentially renewable resource such as soil, grassland, forest, or wildlife by using it faster than it is naturally replenished. If such use continues, the resource can become nonrenewable or nonexis is referred to as environmental degradation.
Degradation	The geologic process by means of which various parts of the surface of the earth are worn away and their general level lowered, by the action of wind and water is referred to as degradation.
Chaos	Chaos refers to behavior that never repeats itself exactly. Examples of chaotic behavior are the waves of an ocean, the movement of leaves in the wind, and day-to-day variations in weather.
Crater	Crater refers to an abrupt basin commonly rimmed by ejected material. In volcanoes, craters form by outwarcc explosion, are commonly less than 2 km diameter, and occur at the summit of a volcanic cone. Similar rimmed basins form by impacts with meteorites, asteroids, and.
Observations	Information obtained through one or more of the five senses or through instruments that extend the senses are observations.
Delta	ALLUVIAL DEPOSIT, usually triangular, at the mouth of a river of other stream. It is normally built up only where there is no tidal or CURRENT action capable of removing the sediment as fast as it is deposited, and hence the DELTA builds forward from the COASTLINE. A TIDAL DELTA is a similar deposit at the mouth of a tidal INLET, put there by TIDAL CURRENTS. A WAVE DELTA is a deposit made by large waves which run over the top of a SPIT or BAR beach and down the landward side.
Continental margin	The drowned edges of continents consisting of the continental shelf, the continental slope, and the continental rise is called continental margin.
Uplift	The rising of one part of the Earth's crust relative to another part is referred to as uplift.
Biomass	Biomass refers to organic matter produced by plants and other photosynthetic producers; total dry weight of all living organisms that can be supported at each trophic level in a food chain or web; dry weight of all organic matter in plants and animals in an ecosystem; pla.
Fossil fuel	Products of partial or complete decomposition of plants and animals that occur as crude oil, coal, natural gas, or heavy oils as a result of exposure to heat and pressure in earth's crust over millions of year are fossil fuel.
Photovoltaic	Photovoltaic cell refers to a device in which radiant energy is converted directly into

Go to **Cram101.com** for the Practice Tests for this Chapter.

Go to **Cram101.com** for the Practice Tests for this Chapter.
And, **NEVER** highlight a book again!

cell	electrical energy.
Cell	Smallest living unit of an organism. Each cell is encased in an outer membrane or wall and contains genetic material and other parts to perform its life function. Organisms such as bacteria consist of only one cell, but most of the organisms we are.
Recycle	Recycle is integral part of waste management that attempts to identify resources in the waste stream that may be collected and reused.
Syncline	Syncline refers to a fold where rock layers are compressed into a concave-upward position.
Slide	In mass wasting, movement of a descending mass along a plane approximately parallel to the slope of the surface is referred to as slide.

470

Go to **Cram101.com** for the Practice Tests for this Chapter.

Solar system	The sun together with the planets and other bodies that revolve around it is a solar system.
System	A set of components that function and interact in some regular and theoretically predictable manner is called a system.
Geology	Geology refers to the science which treats of the origin, history and structure of the Earth, as recorded in rocks; together with the forces and processes now operating to modify rocks.
River	A natural stream of water larger than a brook or creek is a river.
Forest	Forest refers to biome with enough average annual precipitation to support growth of various species of trees and smaller forms of vegetation. Compare desert, grassland.
Crater	Crater refers to an abrupt basin commonly rimmed by ejected material. In volcanoes, craters form by outwarcc explosion, are commonly less than 2 km diameter, and occur at the summit of a volcanic cone. Similar rimmed basins form by impacts with meteorites, asteroids, and.
Lava	Molten rock that is extruded out of volcanoes is called lava.
Planet	A smaller, usually nonluminous body orbiting a star is a planet.
Density	Density refers to the ratio of a mass to a unit volume specified as grams per cubic centimeter.
Radar	An instrument for determining the distance and direction to an object by measuring the time needed for radio signals to travel from the instrument to the object and back, and by measuring the angle through which the instrument's antenna has traveled is referred to as radar.
Map	Map refers to a representation of Earth's surface usually depicting mostly land areas.
Ocean	The great body of salt water which occupies two-thirds of the surface of the Earth, or one of its major subdivisions is called an ocean.
Flood	Period when tide level is rising; often taken to mean the flood current which occurs during this period. A flow above the CARRYING CAPACITY of a CHANNEL.
Conditions	Conditions refers to physical or chemical attributes of the environment that, while not being consumed, influence biological processes and population growth. Examples are temperature, salinity, and acidity. Compare resources.
Temperature	Temperature refers to a measure of the average speed of motion of the atoms, ions, or molecules in a substance or combination of substances at a given moment. Compare with heat.
Rocks	An aggregate of one or more minerals rather large in area are rocks. The three classes of rocks are the following: Igneous rock - crystalline rocks formed from molten material. Sedimentary rock - A rock resulting from the consolidation of loose sediment that has accumulated in layers. Metamorphic rock - Rock that has formed from preexisting rock as a result of heat or pressure.
Atmosphere	Atmosphere refers to the whole mass of air surrounding the earth.
Heat	Total kinetic energy of all the randomly moving atoms, ions, or molecules within a given substance, excluding the overall motion of the whole object. This form of kinetic energy flows from one body to another when there is a temperature difference betwe is referred to as heat.
Uplift	The rising of one part of the Earth's crust relative to another part is referred to as uplift.
Stream	Stream refers to any flow of water; a current. A course of water flowing along a bed in the earth.

Go to **Cram101.com** for the Practice Tests for this Chapter.

Erosion	Erosion refers to wearing away of the land by natural forces. On a beach, the carrying away of beach material by wave action, tidal currents or by DEFLATION. The wearing away of land by the action of natural forces.
Eolian	Describes sediment that was deposited by wind are called eolian.
Crust	Solid outer zone of the earth. It consists of oceanic crust and continental crust. Compare core, mantle.
Plates	Various-sized areas of earth's lithosphere that move slowly around with the mantle's flowing asthenosphere are plates. Most earthquakes and volcanoes occur around the boundaries of these plates.
Development	Development refers to change from a society that is largely rural, agricultural, illiterate, and poor, with a rapidly growing population, to one that is mostly urban, industrial, educated, and wealthy, with a slowly growing or stationary population.
Mantle	Zone of the earth's interior between its core and its crust. Compare with core, crust is the mantle.
Asteroids	Stony or metallic masses that orbit around the Sun are referred to as asteroids.
Comets	Comets refers to icy bodies moving through outer space.
Thermal	The energy of the random motion of atoms and molecules is referred to as thermal.
Gradient	A measure of slope in meters of rise or fall per meter of horizontal distance. More general, a change of a value per unit of distance, e.g. the GRADIENT in longshore transport causes EROSION or ACCRETION. With reference to winds or currents, the rate of increase or decrease in speed, usually in the vertical; or the curve that represents this rate.
Heat energy	Heat energy refers to energy of the random motion of atoms and molecules.
Energy	Capacity to do work by performing mechanical, physical, chemical, or electrical tasks or to cause a heat transfer between two objects at different temperatures is an energy.
Orbit	Orbit in ocean waves refers to the circular pattern of water particle movement at the air-sea interface.
Volatile	Substances that readily become gases when pressure is decreased, or temperature increased are referred to as volatile.
Carbon dioxide	Molecule of carbon and oxygen present in the atmosphere at approximately 350 ppm. Emissions of carbon dioxide resulting from burning of fossil fuels are thought to be co:.itributing to potential global warming through an enhanced greenhouse effect.
Silicate	Silicate refers to any compound that comprises the most abundant mineral group in the earth's crust.
Silicate minerals	The most important group of rock-forming minerals is called silicate minerals.
Event	Event refers to an occurrence meeting specified conditions, e.g. damage, a threshold wave height or a threshold water level.
Rock	Any material that makes up a large, natural, continuous part of earth's crust is rock.
Core	Core refers to a cylindrical sample extracted from a beach or seabed to investigate the types and DEPTHS of sediment layers. An inner, often much less permeable portion of a BREAKWATER, or BARRIER beach.
Range	Land used for grazing is referred to as the range.

Wave	An oscillatory movement in a body of water manifested by an alternate rise and fall of the surface is called a wave. Disturbances of the surface of a liquid body, as the ocean, in the form of a ridge, swell or hump. The term wave by itself usually refers to the term surface gravity wave.
Meteorite	A stony or metallic body from space that passed through the atmosphere and landed on the surface of the Earth is referred to as a meteorite.
Kinetic energy	Kinetic energy refers to energy that matter has because of its mass and speed or velocity. Compare potential energy.
Point	Point refers to the extreme end of a cape, or the outer end of any land area protruding into the water, usually less prominent than a cape. A low profile shoreline promontory of more or less triangular shape, the top of which extends seaward.
Compression	A state of stress that causes a pushing together or contraction is called compression.
Bedrock	Solid rock lying beneath loose soil or unconsolidated sediment is called bedrock.
Fall	Fall refers to a mass moving nearly vertical and downward under the influence of gravity.
Debris	Debris refers to any accumulation of rock fragments; detritus.
Metamorphism	The changes in minerals and rock textures that occur with the elevated temperatures and pressures below the Earth's surface are referred to as metamorphism.
Slump	In mass wasting, movement along a curved surface in which the upper part moves vertically downward while the lower part moves outward is a slump.
Mass	The amount of material in an object is the mass.
Maria	Dark, low-lying areas of the Moon is called Maria.
Wind	Wind refers to the mass movement of air.
Accretion	The accumulation of sediment, deposited by natural fluid flow processes is accretion.
Basin	A large submarine depression of a generally circular, elliptical or oval shape is called a basin.
Basalt	Basalt refers to a dark, fine-grained igneous rock composed of minerals enriched in ferromagnesian silicates.
Well	Well refers to a hole, generally cylindrical and usually walled or lined with pipe, that is dug or drilled into the ground to penetrate an aquifer below the zone of saturation.
Depth	Depth refers to vertical distance from still-water level to the bottom.
Magma	Molten rock below the earth's surface is referred to as magma.
Flood basalt	Tremendous outpourings of basaltic lava that form thick, extensive plateaus are called flood basalt.
Harbor	A water area nearly surrounded by land, sea walls, BREAKWATERS or artificial dikes, forming a safe anchorage for ships is referred to as harbor.
Seismic	Referring to vibrations in the Earth produced by earthquakes is referred to as seismic.
Soil	Soil refers to a layer of weathered, unconsolidated material on top of bedrock; often also defined as containing organic matter and being capable of supporting plant growth.
Chemical	One of the millions of different elements and compounds found naturally or synthesized by human is referred to as chemical.
Hypothesis	In science, an explanation set forth in a manner that can be tested and is capable of being

disproved. A tested hypothesis is accepted until and unless it has been disproved.

Flooding	The natural process whereby waters emerge from their stream channel to cover part of the floodplain. Natural flooding is not a problem until people choose to build homes and other structures on floodplains.
Host	Plant or animal on which a parasite feeds is referred to as host.
Observations	Information obtained through one or more of the five senses or through instruments that extend the senses are observations.
Spring	A place where groundwater flows out onto the surface is a spring.
State	State refers to an expression of the internal form of matter. Water exists in three states: solid, liquid, and gas. A solid has a fixed volume and fixed shape; a liquid has a fixed volume but no fixed shape; and a gas has neither fixed volume nor fixed shape.
Rift	Rift refers to the valley created at a pull-apart zone.
Recent	A synonym of Holocene is called recent.
Lithosphere	Lithosphere refers to outer shell of the earth, composed of the crust and the rigid, outermost part of the mantle outside of the asthenosphere; material found in the earth's plates.
Climate	Physical properties of the troposphere of an area based on analysis of its weather records over a long period. The two main factors determining an area's climate are temperature, with its seasonal variations, and the amount and distri.
Gravel	Gravel refers to loose, rounded fragments of rock, larger than sand, but smaller than cobbles. Small stones and pebbles, or a mixture of these with sand.
Deflation	Deflation refers to the removal of loose material from a beach or other land surface by wind action.
Volcano	Vent or fissure in the earth's surface through which magma, liquid lava, and gases are released into the environment is a volcano.
Base	A substance that combines with a hydrogen ion in solution is called the base.
Caldera	Caldera refers to a large, basinshaped volcanic depression. Roughly circular in map view, that forms by a piston-like collapse of a cylinder of overlying rock into an underlying, partially evacuated magma chamber.
Convection	Convection refers to the vertical transport of a fluid or the transfer of heat in fluids.
Plate tectonics	Plate tectonics refer to the theory of geophysical processes that explains the movements of lithospheric plates and the processes that occur at their boundaries.
Tectonics	The study of the major structural features of the Earth's crust or the broad structure of a region is referred to as tectonics.
Pore	Pore refers to an opening or void space in; oil or rock.
Electron	Tiny particle moving around outside the nucleus of an atom. Each electron has one unit of negative charge and almost no mass. Compare neutron, proton.
Meander	Loop-like bends or curves in the flow path of a current are a meander.
Photomosaic	An assemblage of photographs, each of which shows part of a region, put together in such a way that each point in the region appears once and only once in the assemblage, and scale variation is minimized is a photomosaic.
Terrestrial	Terrestrial refers to pertaining to land. Compare with aquatic.

Go to **Cram101.com** for the Practice Tests for this Chapter.

Bacteria	Bacteria refer to prokaryotic, one-celled organisms. Some transmit diseases. Most act as decomposers and get the nutrients they need by breaking down complex organic compounds in the tissues of living or dead organisms into simpler inorganic nutrient compounds.
Mineral	A naturally occurring, inorganic, crystalline solid that has a definite chemical composition and possesses characteristic physical properties is a mineral.
Hydrocarbons	Compounds containing only carbon and hydrogen are a large group of organic compounds, including petroleum products, such as crude oil and natural gas are hydrocarbons.
Fossils	Skeletons, bones, shells, body parts, leaves, seeds, or impressions of such items that provide recognizable evidence of organisms that lived long ago are called fossils.
Mixture	Mixture refers to combination of two or more elements and compounds.
Fossil	The remains or traces of organisms preserved in rocks or ancient sediment are referred to as fossil.
Molecule	Molecule refers to a combination of two or more atoms of the same chemical element or different chemical elements held together by chemical bonds. Compare with atom, ion.
Acid	A substance that releases a hydrogen ion in solution is an acid.
Topographic map	A map on which elevations are shown by means of contour lines is referred to as a topographic
Elevation	Elevation refers to the distance of a point above a specified surface of constant potential; the distance is measured along the direction of gravity between the point and the surface.
Oceanography	That science treating of the oceans, their forms, physical features and phenomena is referred to as oceanography.
Recycle	Recycle is integral part of waste management that attempts to identify resources in the waste stream that may be collected and reused.
Hot spot	A zone on a lithospheric plate that overlies unusually hot asthenosphere, and where large volumes of lava commonly are extruded, building a large volcanic pile on the sea floor is called hot spot.
Crystallization	The growth of minerals in a fluid such as magma is a crystallization.
Morphology	Morphology refers to river/estuary/lake/seabed form and its change with time.
Viscous	Ease of flow is viscous. The more viscous a substance, the less readily it flows.
Fault	A fracture in rock along which there has been an observable amount of displacement. Faults are rarely single planar units; normally they occur as parallel to sub-parallel sets of planes along which movement has taken place to a greater or lesser extent. Such sets are called fault or fracture-zones.
Fissure	A narrow parting or crack in rock are called the fissure.
Forcing	With respect to global change, processes capable of changing global temperature, such as changes in solar energy emitted from the sun, or volcanic activity, we have forcing.
Friction	The resistance to motion of two bodies in contact is a friction.
Rings	Rings refers to large whirl-like eddies created by meander cutoffs of strong geostropic currents such as the Gulf Stream. They either have warm-water centers or cold-water centers.
Resolution	Resolution in general, refers to a measure of the finest detail distinguishable in an object or phenomenon. In particular, a measure of the finest detail distinguishable in an image.
Ash	The loose debris that is ejected from an erupting volcano is called ash.

Go to **Cram101.com** for the Practice Tests for this Chapter.

Hydrothermal vents	An opening in the ground from which pour out hot, saline water solutions are called hydrothermal vents.
Fracture	Fracture refers to a general term for any breaks in rock. Fractures include faults, joints, and crack,_.
Haze	Fine dust, smoke, water and salt particles that reduce the clarity of the atmosphere is referred to as the haze.
Smog	Originally a combination of smoke and fog, but now used to describe other mixtures of pollutants in the atmosphere is referred to as smog.
Hydrocarbon	Hydrocarbon refers to organic compounds composed of hydrogen, oxygen, and carbon that are the main components of petroleum.
Storm	Local or regional atmospheric disturbance characterized by strong winds often accompanied by precipitation is referred to as a storm.
Geyser	A hot spring that gushes magmaheated water and steam is called geyser.
Force	Force refers to a push or pull that affects motion. The product of mass and acceleration of a material.
Carbon monoxide	Colorless, odorless gas that at very low concentrations is extremely toxic to humans and animals are carbon monoxide.
Nucleus	Extremely tiny center of an atom, making up most of the atom's mass is the nucleus. It contains one or more positively charged protons and one or more neutrons with no electrical charge.
Head	A comparatively high promontory with either a CLIFF or steep face. It extends into a large body of water, such as a sea or lake. An unnamed HEAD is usually called a headland. The section of RIP CURRENT which has widened out seaward of the BREAKERS, also called head of
Star	A massive sphere of incandescent gases powered by the conversion of hydrogen to helium and other heavier elements is a star.
Big bang	Big bang refers to the hypothetical event that started the expansion of the universe from a geometric point. The beginning of time.
Matter	Matter refers to anything that has mass and takes up space. On earth, where gravity is present, we weigh an object to determine its mass.
Milky Way	Milky Way refers to the name of our galaxy. Sometimes applied to the field of stars in our home spiral arm, which is correctly called the Orion arm.
Galaxy	A large rotating aggregation of stars, dust, gas, and other debris held together by gravity. There are perhaps 50 billion galaxies in the universe and 50 billion stars in each galaxy.
Concentration	Amount of a chemical in a particular volume or weight of air, water, soil, or other medium is referred to as concentration.
Condensation	Conversion of a gas to a liquid is referred to as condensation.
Gravity	The attraction between bodies of matter is a gravity.
Degree	An arbitrary measure of temperature. One degree Celsius _ 1.8 degrees Fahrenheit.
Shooting star	Tiny space particles that burn up with a flash of friction-generated light in the Earth's atmosphere are referred to as a shooting star.
Plume	Plume refers to an arm of magna rising upward from the mantle.
Hydrologic cycle	Biogeochemical cycle that collects, purifies, and distributes the earth's fixed supply of

Go to **Cram101.com** for the Practice Tests for this Chapter.

water from the environment to living organisms, and then back to the environment is the hydrologic cycle.

Extinction

Extinction refers to complete disappearance of a species from the earth. This happens when a species cannot adapt and successfully reproduce under new environmental conditions or when it evolves into one or more new species. Compare speciation.

Sea

Sea refers to the ocean. A large body of salt water, second in rank to an ocean, more or less landlocked and generally part of, or connected with, an ocean or a larger sea. State of the ocean or lake surface, in regard to waves.

Peninsula

An elongated portion of land nearly surrounded by water and connected to a larger body of land, usually by a neck or an isthmus is called a peninsula.

Sediment

Loose, fragments of rocks, minerals or organic material which are transported from their source for varying distances and deposited by air, wind, ice and water are sediment. Other sediments are precipitated from the overlying water or form chemically, in place. Sediment includes all the unconsolidated materials on the sea floor. The fine-grained material deposited by water or wind.

Theory

A general explanation of a characteristic of nature consistently supported by observation or experiment is referred to as a theory.

Depression

Depression refers to a general term signifying any depressed or lower area in the ocean floor.

Mass extinction

A catastrophic, widespread, often global event in which major groups of species are wiped out over a short time compared to normal extinctions is a mass extinction. Compare with background extinction.

Plankton

Plankton refers to small plant organisms and animal organisms that float in aquatic ecosystems. Compare with benthos, nekton.

Key

Key refers to a low, insular BANK of sand, coral, etc., as one of the islets off the southern coast of Florida.

Resources

Resources refer to substances that can be consumed by an organism and, as a result, become unavailable to other organisms.

Feedback

A kind of system response that occurs when output of the system also serves as input leading to changes in the system is called feedback.

Science

Attempts to discover order in nature and use that knowledge to make predictions about what should happen in nature is called science.

CPSIA information can be obtained at www.ICGtesting.com
Printed in the USA
LVOW021548021111

253223LV00001B/2/A